ECONOMICS OF PRECISION FARMING

MJP PUBLISHERS

ECONOMICS OF PRECISION FARMING

C. Sekhar

M.Sc.(Ag)., M.B.A. (PM)., M.B.A. (HR)., Ph.D.(FE)., F.AB.M. (USA)

Professor in Agricultural Economics

Department of Agricultural Economics

Centre for Agricultural and Rural Development Studies

Tamil Nadu Agricultural University, Coimbatore - 641 003.

A. Vidhyavathi

Associate Professor in Agricultural Economics

Department of Agricultural Economics

Centre for Agricultural and Rural Development Studies

Tamil Nadu Agricultural University, Coimbatore - 641 003.

MJP PUBLISHERS

Chennai Trichy Tirunelveli New Delhi

MJP PUBLISHERS

ISBN 978-81-8094-359-1 **MJP Publishers**

Printed and bound in India

No. 44, Nallathambi Street,
Triplicane, Chennai 600 005

MJP 600

Publisher **C. Janarthanan**

Project Editor **C. Ambica**

FOREWORD

Foreword is a short introduction to a book, typically by a person other than the author. That when the person other than the author(s) being a teacher to the authors, the responsibility becomes more because the foreword must not be biased. Given this caution, I cannot shy away from appreciating the authors for having written the book very comprehensively. Being a book with chapters contributed by many authors, ensuring proper sequence and cohesion of subjects becomes a challenge. I think that the authors have been successful in this aspect. May be that the senior author who is a lead author or co-author in all the chapters has helped to achieve this purpose.

The authors being social scientists have succeeded in giving a book which has successfully covered both agronomic and socio-economic dimensions of precision farming. In the past decades, precision farming has emerged as the most efficient package for narrowing yield gaps in crops and is being adopted incrementally across India; and disseminating precision farming is critical to double farmers' income as envisaged by Government of India. Given closing cultivation frontier and rapidly growing demand for water, precision farming is the most appropriate crop production package.

The book combines both production and marketing aspects with a focus on horticultural crops. In the case of field crops, Systems Rice Intensification (SRI), is presented as a case study. In addition, the book attempts to give institutional dimension covering organic certification, trading, farmer producer companies, investments in horticulture and policy framework for precision farming, which is a pre-requisite for up scaling precision farming. The senior author's specialization in fisheries and forestry is, obviously reflected through the inclusion of precision aquaculture and precision forestry which are all allied sectors. This is a welcome inclusion because aquaculture addresses the issue of under nutrition and forestry adds to carbon neutrality.

I strongly feel that the book will greatly benefit the researchers, students, practitioners of precision farming and policy makers.

Prof. Dr. C. Ramasamy

Agricultural Economist and World Bank Consultant

September 01, 1998

Coimbatore – 641 003

PREFACE

Precision Farming is one of the techniques which is widely used for enhancing the production and productivity of farm grown fruits and vegetables after assessing the impact of this technology as a Turnkey Project in the Dharmapuri District of Tamil Nadu. After assessing the efficacy, the technology was spread to different parts of Tamil Nadu for achieving higher productivity in a unit area. One such effort was taken up through World Bank assisted IAM WARM Project and the same was implemented in basin wise. This book has addressed the impact of precision farming specifically in Theni District of Tamil Nadu and hence a Case Approach is adopted to elucidate the merits of Precision Farming and the same was compared to the traditional ways of cultivation of fruits and vegetables. Besides these, the Precision Aquaculture, Precision Agriculture, Precision Silviculture, Precision Approaches for Carbon Neutrality, Organic Certification, Farmer Producer Organizations and their role and impact on the production and marketing has also been discussed by inviting the paper from experts of related field.

Moreover, the methodological approaches to Economics of Precision Farming is highlighted and discussed. Besides, it brings out the cost involvement in respect of adoption of Precision Farming Techniques in the farm lands and this book will definitely serve as an experience sharing tool. It encompasses present scenario of area production and productivity of fruits and vegetables in India and Tamil Nadu. It addresses on the various concepts and its related past studies relevant to production, productivity, marketing and export of vegetables and its issues. Besides, the export of fruits and vegetables both fresh and processed and the countries to which the fruits and vegetables are exported and importing of the same and the potentials of revenue generation are also discussed elaborately. Above all, the book has also addressed the measures to overcome the constraints faced by the adopters and executors of Precision Farming Technique and appropriate strategies also suggested to alleviate the same in the Indian Context.

ACKNOWLEDGEMENT

I praise the almighty for his grace, guidance and blessings showered on me to put forth this novel Project Work on Precision Farming.

Words would not suffice to express the gratitude I feel towards my mentor and my beloved Teacher honorable personality Dr. C. Ramasamy, Former Vice Chancellor, Tamil Nadu Agricultural University, Coimbatore for his illuminating and transcendent suggestions, cordial treatment and manifold help bestowed throughout the progress of all my initiatives and work accomplishment.

I wish to express my profound, sincere and fervent thanks to Dr. R. Sankaranarayanan, Former Dean, Horticultural College and Research Institute, Periyakulam and Dr. A. Rathinasamy, Former Professor and Head, Department of Social Sciences, for their keen interest and constant encouragement during the course of this investigation.

I owe my gratitude to Dr. M. Selvarajan, Professor in Horticulture, Department of Fruit Crops, Horticultural College and Research Institute, Periyakulam for his guidance and counseling during the course of this Project work .

My sincere appreciation to the authors who had responded on time in providing few chapters to make the book on "Economics of Precision Farming" most relevant and useful to the student community in Agriculture and Horticulture.

I wish to thank Dr. K. Mani, Professor and Head, Department of Agricultural Economics for the motivation and help rendered in accomplishing the task on time.

My sincere and heartfelt thanks to Mr. S. Sivakumar, R.Loganathan, C. Arunprakash and S. Dhanalakshmi, Post Graduate Scholars in Department of Physical Sciences and Information Technology, Tamil Nadu Agricultural University, Coimbatore for updating and revising the manuscript and delivering the same in the form of a book.

Special recognition and heartfelt thanks is due to my student friends

M.T. Andrew Peter Leon, R. Palani, R. Siranjeevi, M. Satham Hussain and R. Nagaraj for their assistance in data collection, compilation and extensive help and keen interest shown throughout the period of this study.

Last but not least, I extend my sincere and heartfelt thanks to my beloved Parents, Thiru.K.S.A. Chinniah – Tmt. C. Amaravathi Ammal for their blessings and sacrifice and My Children, J.S. Paarry and J.S. Thenndrall for their moral support in all of my professional development and accomplishments.

(Dr. C.SEKHAR)

CONTENTS

LIST OF TABLES

LIST OF FIGURES

CHAPTER I

An Overview of Precision Farming Technologies

Dr. C. Sekhar and Dr. I. Muthuvel

Objectives

- To introduce the students on the importance and practice of Cropping Systems in different regions
- To highlight the needs of Proven Technologies like Precision Farming and its impact to the farming community and
- To delineate the role of vegetables in promoting the Indian Economy

Organization of the Chapter

- Introduction to Vegetables
- Classification of Vegetable Crops
- Cropping System for Vegetable Crops
- The Concept of Precision Farming
- The History of Precision Farming
- Documentation and Record Maintenance
- Summary

1.1 INTRODUCTION TO VEGETABLES

Vegetable cultivation has a checkered history in India. Mention of different kinds of vegetables, like Bottle gourd (*Suraikkai, alabu*), Watermelon (*Dharbusani, kakinga*), Onion (*Ulli, palandu*), Brinjal (*Katharikkai, vartaku*), Hyacinth bean and Radish are available in ancient Sanskrit literature and in the epic like Ramayana. Vegetable cultivation took momentum during the Aryan invasion (1800 BC) and later developed by the Mughal rulers. It was flourished in India by the colonist traders from Portugal, France and Britain, who introduced some new vegetables like Cabbage, Cauliflower, Carrot, Potato, Knool-khol, Chilli, Tomato, Asparagus, etc., At present, India is the second largest producer of vegetables in the world after China because of the efforts took by the colonists.

1.1.1 Etymology

The word vegetable was first recorded in English in the early 15th century from Old French which was taken from Medieval Latin '*vegetabilis*' means "growing , flourishing". In 1767, the meaning of the term "vegetable" was specified to mean "plant cultivated for food, edible herb or root".

1.1.2 Vegetables in the Economy

India is the world's second largest producer of vegetables next only to China Vegetable crops in India occupy only 2.80 per cent of the total cultivated land producing 87.50 million tonnes of vegetables annually including potato from an area of 5.80 million ha.

Vegetable crops give higher yield per unit area as compared to other crops. Most of the vegetables, if properly grown, can give six to ten times more yield than any cereal crop. There exists lot of scope for increasing yield in most of the vegetables by growing high yielding varieties/hybrids and adopting improved production technologies like precision farming. Cultivation of vegetable crops being more labour intensive than cereal crops will be helpful for seasonally under-employed farm workers in increasing their income. Most vegetable crops are of short-duration. With proper planning, three or more crops can be grown in the same plot in one year.

1.1.3 Export

Today, horticultural crops contribute about 25 per cent of the total agricultural exports of the country. Of all the fresh vegetables exported, the share of onion is around 75 per cent. Other traditional vegetables which are presently exported are potato, gherkin, okra, bitter gourd, chilli, cauliflower and some root vegetables.

In addition to these traditional vegetables, non-traditional vegetables which have been identified by the Agricultural and Processed Food Products Export Development Authority (APEDA) to have export potential are asparagus, celery, sweet pepper, green pea, French bean and Cherry tomato. Among the processed vegetables, major demand is for dehydrated onion, garlic and preserved gherkin.

Countries which import traditional vegetables like onion, potato, okra, bitter gourd and chilli are Gulf countries, Singapore, Malaysia, Sri Lanka, Bangladesh and Nepal. Non-traditional vegetables are exported to European countries, Australia, Gulf countries and South-East Asian nations. Cole crops and root vegetables are must in demand in temperate countries during winter. In India, these vegetables are grown in the open during winter and thus the cost of production is less as compared to those grown under protective cover. There is lot of opportunity for exporting these vegetables to the European and North American countries from India.

1.1.4 Role of Vegetables in Nutrition

Vegetables provide all the nutrient components, like carbohydrates, protein, fat, vitamins, minerals and water along with roughages which are the essential constituents of a balanced diet. The balanced diet keeps the human being healthy and provides high energy value for performing heavy tasks. Hence it is always advisable to consume higher percentage of fruits and vegetables. Preferably one can take the vegetables in the form of salads while they take juices or the cool drinks.

Vegetables contain a great variety of other phytochemicals, some of which have been claimed to have antioxidant, antibacterial, antifungal, antiviral and anticarcinogenic properties. Diets containing recommended amounts of fruits and vegetables may help lower the risk of heart diseases and type 2 diabetes. These diets may also protect against some cancers and decrease bone loss. The potassium provided by both fruits and vegetables may help prevent the formation of kidney stones.

Most of the patients who visit ortho hospitals due to breakage of bones are deficient of calcium and phosphorous in their diet. That's why the breakage of bones are visible in performing simpler tasks either at home or the office. Hence it is advisable to take higher share of Fruits and Vegetables in their diet for healthy life.

1.1.5 Toxic and Anti-Nutrient Compounds in Vegetables

Toxins are produced by plants as a defence mechanism to protect from man and animals. During the course of evolution and continuous cultivation, these undesirable qualities might have been eliminated by selection. Still a few harmful chemicals like trypsin inhibitors, phytates, oxalates and nitrates exist in cultivated vegetables.

Trypsin inhibitors are widely distributed in legumes, especially in raw soybeans. They inhibit activity of trypsin in the gut and interfere with digestibility of dietary proteins and reduce their utilization by affecting hydrolysis of proteins to amino acids.

Phytic acid present in mature seeds of peas and beans binds iron, zinc, calcium and magnesium and reduces bioavailability of iron. On germination of grains, phytate content reduces due to enzymatic break down which improves iron availability.

Green leaf vegetables and legumes ae rich source of oxalates and oxalic acid. Maximum oxalate concentrationis noticed in amaranth (772 mg per 100 gram). Oxalic acid converts calcium from the food consumed to insoluble calcium oxalates making it unavailable for absorption. Calcium oxalate crystals present in leaves and tubers of Amorphophallus and colocasia cause itching of skin, tongue and throat. Dietary oxalates contribute to increased excretion of oxalate in urine leading to urinary stones.

Vegetables are the single largest source of nitrate in human diet. Leaf vegetables have high concentration of nitrate. Nitrate accumulation results in serious deleterioius effects. Within gastrointestinal tract, nitrate is reduced to nitrite which is absorbed into blood stream where it binds with haemoglobin, oxidizing ferrous ion to form methaemoglobin. Methaemoglobin is incapable of oxygen transport and results in anoxia specifically referred to as "Methaemoglobinaemia". When nitrate ions combine with secondary or tertiary amines, N-nitroso compounds are formed, a few of which are potent carcinogens. On cooking, nitrate content gets reduced drastically (Peter, 2007).

However, vegetables often also contain toxins and antinutrients such as α-chaconine, enzyme inhibitors (of cholinesterase, protease, amylase etc.,), cyanide and cyanide precursors, oxalic acid and more. Depending on the concentration, such compounds may reduce the edibility, nutritional value and health benefits of dietary vegetables. Cooking and/or other processing may be necessary to eliminate or reduce them.

1.2 CLASSIFICATION OF VEGETABLE CROPS

Vegetable crops consist of 1200 species from 78 families. Among them, more than 860 species under 59 families belong to dicotyledoneae and the remaining species under the rest 19 families to monocotyledoneae. In the tropical and subtropical parts of the world, about 90 species of vegetable crops are cultivated, but hardly 15 are commercially important. Vegetable crops are not only botanically different, their cultural and climatic adaptation are also varied. It is, therefore, necessary to classify the vegetable crops for their easy recognition. Vegetable crops can be classified according to the following ways.

Table 1.1 Nutritive Value of Different Vegetables per Hundred Gram of Edible Portion on Fresh Weight Basis

Sl. No	Name of the Vegetables	Moisture (g)	Carbohydrates (g)	Protein (g)	Fat (g)	Calorie (Energy)	Vit.A (IU)	Thiamine (mg)	Riboflavin (mg)	Ascorbic Acid (mg)	Calcium (mg)	Iron (mg)	Phosphorus (mg)
01	Potato	74.70	22.60	1.60	0.10	97	040	0.40	0.04	17.00	10.00	0.70	35.00
02	Tomato	94.00	3.60	1.90	0.10	23	307	0.07	0.01	31.00	20.00	1.80	00.00
03	Chilli	85.70	3.00	2.90	0.60	29	292	0.19	0.39	111.00	30.00	1.20	80.00
04	Brinjal	92.70	4.00	1.40	0.30	24	118	0.04	0.11	12.00	18.00	0.90	47.00
05	Cabbage	92.40	5.30	1.40	0.20	29	080	0.06	0.05	100.00	46.00	0.80	38.00
06	Bitter Gourd	92.40	4.20	1.60	0.20	25	210	0.07	0.09	88.00	20.00	1.80	70.00
07	Pumpkin	86.00	4.60	1.40	0.10	25	2180	0.06	0.04	2.00	10.00	0.70	30.00
08	Carrot	82.20	10.60	0.90	0.20	48	12000	0.04	0.02	3.00	48.00	0.60	30.00
09	Okra	89.60	6.40	1.90	0.20	35	88	0.07	0.10	13.00	66.00	1.50	56.00
10	Drumstick Leaves	75.90	12.50	6.70	1.70	92	10848	0.06	0.05	220.00	440.00	7.00	70.00

(Source: Hazra and Som, 2015)

Table 1.2 Details of Classification of Vegetables

S.No.	Type of Classification	Basis of Classification
01	Botanical	Based on Botanical Relationships
02	Hardiness	Endurance against Frost
03	Growing Season	Suitable Crops during different Climate and Season
04	Tolerance to Soil Reaction	Tolerance to Soil Acidity
05	Tolerance to Salt	Tolerance to Soil Salinity
06	Plant Parts used	According to their Edible Plant parts
07	Methods of Culture	According to the Cultural Practices of Vegetables.

1.2.1 Classification Based on Methods of Culture

This is best for the growers because the vegetable crops requiring the same cultural practices (but not same season for cultivation) are grouped together though they may be divergent botanically (Thompson, 1959).

Group 1 : **Potato**

Group 2 : **Solanaceous Fruits**

Tomato, Brinjal, Chilli, Sweet Pepper

Group 3 : **Cucurbits**

Pumpkin, Summer Squash, Winter Squash, Musk Melon, Water Melon, Cucumber, Bottle Gourd, Ridge Gourd, Snake Gourd, Bitter Gourd

Group 4 : **Cole Crops**

Cabbage, Cauliflower, Sprouting Broccoli, Brussels sprouts, Knool Khol, Chinese Cabbage

Group 5 : **Root Crops**

Radish, Carrot, Beet, Turnip, Horse Radish, Rutabaga, Salsify, Turnip-rooted Chervil, Spanish Salsify, Black Salsify

Group 6 : Bulb Crops

Onion, Garlic, Leek, Shallot, Welsh Onion, Chive

Group 7 : Salad Crops

Lettuce, Celery, Endive, Chicory, Parsley, Sorrel, Water Cress, Garden Cress, Upland Cress

Group 8 : Greens and Pot Herbs

Amaranthus, Palak, Spinach, New Zealand Spinach, Coriander, Indian Spinach, French Spinach, Dandelion, Pigweed, Terragon, Water Spinach, Parsley

Group 9 : Peas and Beans

Pea, French Bean, Cowpea, Hyacinth Bean, Asparagus Bean, Cluster Bean, Winged Bean, Lima Bean

Group 10: Tuber Crops other than Potato

Taro, Giant Taro, Yam, Cassava, Elephant foot-yam, Chinese Potato

Group 11 : Sweet Potato

Group 12 : Okra

Group 13 : Pointed Gourd

Group 14 : Chayote (Chow Chow)

Group 15 : Tropical Perennial Vegetable Crops

Drumstick, Agathi, Chekurmanis, Curry Leaf

Group 16 : Temperate Perennial Vegetable Crops

Asparagus, Rhubarb, Seakale, Globe Artichoke

1.3 CROPPING SYSTEM FOR VEGETABLE CROPS

Cropping system is defined as the cropping pattern followed on a farm and its interactions with farm resources, other farm enterprises and production technology. The yearly sequence and spatial arrangement of crops and fallow on a given area is termed as cropping pattern.

Realising the low production of vegetables against huge requirement it has been targeted to increase area under vegetable cultivation. In this situation, total production per unit area can be increased by increasing the productivity per unit area through,

a. growing improved high yielding varieties,

b. adequate and scientific crop management practices,

c. increasing cropping intensities following different cropping patterns

1.3.1 Multiple Cropping

It is a cropping system representing the philosophy of maximum crop production per unit area of land within a calendar year.

Multiple cropping in vegetable crops has several merits, like (i) generation of more income from the farm through increased production per unit area in the year, (ii) generation of more employment through intensive cropping throughout the year, (iii) maintenance of soil fertility and health through balanced nutrition and scientifically based crop alterations, (iv) complete and economical use of natural resources, (v) rational use of capital and time and, (vi) management of pest and diseases and weeds.

1.3.2 Crop Rotation

It is a cropping system of growing different crops in a regular sequence on the same land repeatedly for the period of two, three or more years. Here the cycle of cropping sequence takes more than one year to complete.

1.3.3 Succession Cropping

The system of growing two or more crops in succession on the same land within a year with the aim to keep the high-priced land occupied with the cash crop for most of the year. E.g. Potato (Oct to Jan) – Onion (Jan to May) – Okra (May to Oct).

1.3.4 Relay Cropping

It is also a system of growing different crops on the same land within a year but in this system, the succeeding crop is sown/planted before the preceding crop is ready for harvest. So the growing span of two crops overlaps for a short period.

eg. Brinjal – Ridge Gourd/Bitter Gourd/Watermelon (brinjal is grown during September to February. Seedlings of the cucurbits grown in polyethylene packets are to be planted in December and continued up to April)

1.3.5 Inter Cropping

It is a cropping system where two or more crops are grown simultaneously in alternate rows or otherwise in the same land showing significant amount of intercrop competition.

eg. Tomato + Palak (Palak can be grown successfully in between tomato for 75 to 80 days duration. Harvesting of palak by uprooting the whole plant starts 35 days after sowing.)

1.4 PRECISION FARMING

Precision Farming or Precision Agriculture is a concept of using the new technologies and collected field information, doing the right thing, in the right place, at the right time. Collected information may be used to more precisely evaluate optimum sowing density, estimate fertilizers and other inputs need, and to more accurately predict crop yields.

It helps in avoiding unwanted practices to a crop, regardless of local soil/climate conditions, i.e., it reduces labour, water, inputs such as fertilizers, pesticides etc. and assures quality produce.

1.4.1 History of Precision Farming

The concept of precision farming first emerged in the United States in the early 1980's. In 1985, researchers at the University of Minnesota varied lime inputs in crop fields. It was also at this time that the practice of grid sampling appeared. Towards the end of the 1980's this technique was used to derive the first input recommendation maps for fertilizers and pH corrections. The use of yield sensors developed from new technologies, combined with the advent of GPS receivers, has been gaining ground ever since.

Precision agriculture has already been fine tuned to the protection of microlevel environment as well as mentoring of ecological inputs. It conserves water, soil, bio-resources and enhances proficiency of inputs even at the molecular level.

Precision farming involves the application of technologies and principles to manage spatial and temporal variability associated with all aspects of agricultural production for improving crop performance and environment quality. Precision farming calls for an efficient management of resources through location-specific hi-tech interventions.

The advance crop production encompasses a variety of interventions such as micro irrigation, fertigation, protected greenhouse cultivation, soil and leaf nutrient-based fertilizer management, mulching moisture conservation, micro propagation, genetically-modified crops, use of bio fertilizers, vermin culture, high-density planting, hi-tech mechanization, green food, soil-less culture and biological control. Some of the other terminologies used for precision farming are

- Precision Agriculture (PA),
- Site-Specific Farming (SSF),
- Site-Specific Management (SSM),
- Farming-by-the-Foot,
- Variable-Rate Technology (VRT).

Utilization of these interventions orchestrated together having the aim of achieving higher output in a given time period leads to precision farming, which is largely a knowledge driven.

1.4.2 Importance of Precision Farming

- One can improve the crop yield
- Precision farming provides information to make better management decisions
- Reduce Inorganic Fertilizer application and permits efficient use of such resources in the production process
- Motivate Maintenance of Farm Accounting and Records
- Reduce Pollution in the Soil and in the Environment
- Enhance the Profit Margin of the Producer-Farmers

1.4.3 Application of Precision Farming

While discussing the application aspects of Precision Farming, the following aspects are much important. They are

- Drip Irrigation and Fertigation Arrangements
- Mulching Practices
- Protected Cultivation Method

1.4.3.1 Drip Irrigation and Fertigation Arrangements

Drip irrigation is nothing but the regulated form of irrigation or slow application of irrigation water through defined pores or emitters at frequent defined interval near the root zone of the plant over a longer period of time. It saves water by 50 to 70 per cent, reduces labor and energy cost and weed infestation is almost negligible.

Drip Fertigation includes both water and fertilizers which are delivered to the crop through drip irrigation. Fertigation provide essential element directly to the active root zone, thereby increasing the fertilizer use efficiency which ultimately helps in improving the productivity and quality of the farm produce.

Pond Lining is an impermeable geomembrane used for water retention, including the lining of lakes, garden ponds and artificial streams in parks and gardens. This is normally carried out to protect the water from percolation and conserving the water so as to grow the crops without any water loss.

1.4.3.2 Mulching Practices

It is used to reduce or increase the temperature, suppress weed growth and conserve soil moisture. Mulching is a process or practice of covering the soil or ground to make more favorable conditions for plant growth, development and efficient crop production. Mulch is a technical term means "Covering of Soil". While natural mulches such as leaf, straw, dead leaves and compost have been used for centuries, during the last 60 years the advent of synthetic materials has altered the methods and benefits of mulching. When compared to other mulches plastic mulches are completely impermeable to water; it therefore prevents direct evaporation of moisture from the soil and thus limits the water losses and soil erosion over the surface. In this manner it plays a positive role in water conservation. The suppression of evaporation also has a supplementary effect; it prevents the rise of water containing salt, which is important in countries with high salt content in their water resources.

Table 1.3 Selection of Film for Mulching Practices in Various Crops

Sl. No	Thickness of the Film in Microns	Life Expectancy	Crop Duration
01	15 - 25	03 – 04 Months	Short Duration Crops (Vegetables)
02	50	11 – 12 Months	Medium Duration (Fruit Crops, Coffee, Papaya and Sugarcane)
03	100	12 – 20 Months	Long Durataion (Mango, Citrus Fruits and Medium Grown Trees)

Table 1.4 Cost of Mulching

Thickness in Microns	Area (Sq. Meter) / Kg	Required Kgs Per Ha	Cost of Mulching in Rs per Ha
25	43.50	138.00	16500.00
50	21.75	275.00	33000.00
100	10.90	550.00	66000.00

(Source: S.V. Kottiswaran, 2018)

Table 1.3 and 1.4 revealed the details of plastic mulching requirements and their cost involvement per ha for effective implementation. The less micron sheets particularly 25 microns, are costing around Rs 16500 per ha. The small farmers can adopt the 25 micron thickness sheets for the mulching process as other micron sheets are bit costlier.

Impact of Plastic Mulching on Different Crops

Ashrafuzzaman et.al (2011) has reported the effect of coloured plastic mulch on growth and yield of chilli during 2006. The plastic mulches were transparent, blue, and black and bare soil was the control. Different mulches generated higher soil temperature and soil moisture under mulch over the control. Transparent and blue plastic mulches encouraged weed population which were suppressed under black plastic. Plant height, number of primary branches, stem base diameter, number of leaves and yield were better for the plants on plastic. At the mature green stage, fruits had the highest vitamin-C content on the black plastic. Mulching produced the fruits with the highest chlorophyll-a, chlorophyll-b and total chlorophyll contents and also increased the number of fruits per plant and yield. However, mulching did not affect the length and diameter of the fruits and number of seeds per fruit. Plants on black plastic mulch had the maximum number of fruits and highest yield. Thus, mulching appears to be a viable tool to increase the chilli production under tropical conditions.

Research was conducted by Flarios-Larios during 1994 to evaluate the use of plastic mulches in the production of cucumber under micro-irrigation in tropical conditions. Clear (C) plastic mulch installed on beds increased the fruit number and the yield of cucumber (*Cucumis sativus* L. cv. Fanci Pack) in comparison to unmulched (U) soil, during 1992–1993. White (W) and black (B) plastic mulches showed a moderated effect. Marketable yields of 63.3, 46.2, 44.8 and 21.6 tonnes per ha were achieved under C, W, B and U treatments respectively ($P<0.05$). The plastic mulches reduced the number of days to flowering and to first harvest and increased the fruit number. It is concluded that clear plastic mulch is more adequate for tropical conditions than black or white plastic mulches.

1.4.3.3 Protected Cultivation

Green House Technology is the technique of providing favorable environment condition to the plants or crops. It is rather used to protect the plants from the adverse climatic conditions such as wind, cold, precipitation, excessive radiation, excessive temperature, insects and diseases. It is also of vital importance to create an ideal micro climate around the plants. Environmental conditions are so modified that one can grow any plant in any place at any time by providing suitable environmental conditions with minimum labor force. There are 3 types of green houses. They are

- Hi-tech Green House
- Medium Cost Green House
- Low Cost Green House

1.4.3.3.1 Hi-Tech Green House

Many commercial glass greenhouses or hothouses are high tech production facilities for vegetables or flowers. The glass greenhouses are filled with equipment including screening installations, heating, cooling, lighting, and may be controlled by a computer to optimize conditions for plant growth. Different techniques are then used to evaluate optimality-degrees and comfort ratio of greenhouse micro-climate (i.e., air temperature, relative humidity and vapor pressure deficit) in order to reduce production risk prior to cultivation of a specific crop. The Features of the HI-Tech Green House is furnished as follows.

- Generation and Distribution of Carbon-Di Oxide
- Evaporative Cooling
- Heating System
- Humidification System
- Lighting System
- Shading System either partially or fully
- Nutrient Application System

The cost estimate for the 500 Sq. Meter size is arrived at Rs 8.25 Lakhs; 1000 Sq. Meter area cost about 14.65 Lakhs; 2000 Sq. Meter size costing around Rs 28.40 lakhs and 3000 square meter size is costing around Rs. 42 lakhs. The costs are on the upper side considering the transportation and erection of the unit.

1.4.3.3.2 Medium Cost Green House

Greenhouse users prefers to have manually or semi-automatic control arrangement owing to minimum investment. This type of greenhouse is constructed using galvanized iron (G.I) pipes. The canopy cover is attached with structure with the help of screws. Whole structure is firmly fixed with the ground to withstand the disturbance against the wind. Exhaust fans with thermostat are provided to control the temperature. Evaporative cooling pads and misting arrangements are also made to maintain a favourable humidity inside the greenhouse. As these system are semi-automatic, hence, require a lot of attention and care, and it is very difficult and cumbersome to maintain uniform environment throughout the cropping period. These greenhouses are suitable for dry and composite climatic zones. The features of the green houses are naturally ventilated and in possession of cooling systems.

The Cost estimate for different size of the medium cost green house is

- 500 square meter costs around Rs 5.30 lakhs. When we compare the hi-tech system, it is less of Rs 3 lakhs;
- 1000 Squre meter costs around Rs 9.35 lakhs. When we compare the hi-tech system, it is less of Rs 5.30 lakhs;

- 2000 Square meter costs around Rs 17.80 lakhs. When one could compare the hi-tech system, it is less of Rs 10 lakhs;
- 3000 Square meter costs around Rs 25.30 lakhs. When one could compare the hi-tech system, it is less of Rs 17 lakhs. However, one should pay much attention in regulating or maintaining the green house as it is semi-automatic system.

1.4.3.3.3 Low Cost Green House

Low cost or Low-tech greenhouse is a simple structure constructed with locally available materials such as bamboo, timber etc. The ultra violet (UV) film is used as cladding materials. Unlike conventional or hi-tech greenhouses, no specific control device for regulating environmental parameters inside the greenhouse are provided. Simple techniques are, however, adopted for increasing or decreasing the temperature and humidity. Even light intensity can be reduced by incorporating shading materials like nets. The temperature can be reduced during summer by opening the side walls. Such structure is used as rain shelter for crop cultivation. Otherwise, inside temperature is increased when all sidewalls are covered with plastic film. This type of greenhouse is mainly suitable for cold climatic zone. Normally it is made up of wooden or bamboo sticks for its full construction. Normally 500 square meter area is preferred for this type of green houses. It can be erected with either wooden frames or with bamboo poles or frames at a cost of Rs 2.50 lakhs to 3.00 lakhs.

Precision Agriculture has created scope of transforming the traditional agriculture, through proper resource utilization and management, to an environmentally friendly sustainable agriculture. The basic goal of precision agriculture is to optimize the yield with minimum resource use and reduced environmental pollution is highly required for developing countries to face the challenge of sustainable production. Rapid socio-economic changes in some developing countries are creating new scopes for the application of precision agriculture. Application of the balanced soft and hard precision agriculture technologies based on the need of specific socio-economic condition of the country will make in the overall perspective, with the introduction and adoption of modern technologies, agriculture sector is expected to achieve a vertical growth to face the changing needs of ever increasing population.

1.4.4 TAMIL NADU PRECISION FARMING PROJECT

It is a Tamil Nadu State sponsored turnkey project being implemented at Dharmapuri and Krishnagiri districts in 400 ha with a total budget of 7.20 Million for period of three years. It is a State Mega demo to train the farmers and empower them as agri-entrepreneurs. The project has adequately prepared the farmer for market driven production at a time when MNCs like Wall Mart, Carry Four, Tesco, Wools Worth and Reliance, Bharti-Telecom are all set for invasion into retail vegetable market in India.

The Precision Farming Project was first started in Tamil Nadu in Dharmapuri and Krishnagiri during 2004-05. The Tamil Nadu Agricultural University was the nodal agency that implemented this project for a period of three years. An amount of Rs. 75,000 for the installation of drip irrigation and Rs. 40,000 for crop production expenses was given to the farmers. The first crop was taken up under the total guidance of scientists from the university, while the subsequent five crops were taken up by the farmers in three years. In the first year, the farmers were unwilling to undertake this project because of their frustration due to the continuing drought in that area for four years since 2002. But after seeing the success of the first 100 farmers and the high market rate for the produce obtained from this scheme, farmer started registering in the large numbers for the second year (with 90 per cent of subsidy) and the third year (with 80 per cent of the subsidy).

1.4.4.1 Objectives of Tamil Nadu Precision Farming Project

The farmers are to be sensitized on the market forces and reorient the production system as a whole to respond to the market. The participatory farming through registered growers associations are to be provided with precision technologies in production and make them understand the food safety issues, which in turn helps to generate, produce acceptable to international consumers there by getting linked to the market.

1.4.4.2 Paradigm Shift in the Horticultural Sector

Between 1906-1994, the State Agricultural Universities, Development Department and Policy makers made best of their efforts to maximize the productivity *vis a vis* production. The whole exercise was production of what was possible in the given land, water resource and the environmental conditions. It was predominantly a production led horticultural activity. WTO has necessitated a shift in the focus far beyond the farm into the market sector, taking into account the market forces in globalized trade system and demanded a new market driven production systems.

1.4.4.3 Production for Domestic Market

a. Production system for local market characterized by small area (1-2 ha), too many varieties and intensive nature of cultivation.

b. Production system for interstate market (internal export) characterized by larger horticultural estates, one or two specific varieties and extensive nature of cultivation.

1.4.4.4 Production for Export

a. It involves Good Agricultural Practices, organically certified production, Precision farming etc with total concern for consumer's preference of the importing countries.

b. For export of value added products to developed countries, duly following the prescribed sanitary and phytosanitary standards, HACCP and traceability ensuring 100 per cent food safety. The precision farming project is an effort to strengthen the production systems for export in general and raw produces in particular.

1.4.4.5 Key Technologies

1.4.4.5.1 Remote Sensing Technology

Making use of the GIS, the physiographic map, soil map and land use map for the location were developed. The remote sensing programme enabled the scientists to exactly locate the actual field to scale map its boundary. The data on N, P, K, Ca, Mg, Mn, Fe and Cu were collected and this enabled to precisely apply the required nutrient in accurate quantities to the root zone during critical phases of crop growth.

1.4.4.5.2 Chisel Plough

The chisel plough technology ensured better aeration to root zone and effective drainage during rainy days. Further it helps the plants to develop root system with characteristic uniformity in pattern, architecture and in adequate mass. The chisel plough needs to be operated once in two years.

1.4.4.5.3 Hi-Tech Community Nursery

The seedlings were raised in protrays under net houses with insect proof netting making use of EC and pH adjusted coco peat media treated with Pseudomonas. The seedlings produced were strikingly uniform with similar physical and physiological age thus ensuring 100 per cent field stand and sustained 100 per cent productive plants.

1.4.4.5.4 Drip and Fertigation System

Drip and fertigation system of the project ensured water economy, precise application of water-soluble fertilizers to root zone and kept an ideal moisture regime of 60 per cent. In fertigated fields, the moisture regime was 60 per cent and air gradient was 40 per cent. Thus the growth became unchecked from seedling to harvest.

1.4.4.5.5 Growing Crops and Growing with the Crops

The field scientists stayed at villages and extended the technical support to the farmers through out the crop period. The growing with the crop has made all the difference.

1.4.4.5.6 Market Support

The marketing scientists have introduced markets in Cochin, Chennai, Bangalore, Safal and Coimbatore and sensitized them on the importance of minimal grading and sorting and timely delivery at the market.

The farmers were provided with plastic crates and trained sufficiently on grading and sorting to different markets. The marketing scientists also studied the market situations and flow of vegetables and guided the farmers to select crop and variety for cultivation during particular season, thus making the production as **MARKET LED PRODUCTION** in real sense.

1.4.4.5.7 Empowerment of Farmers and Farmers' Forum

The beneficiary farmers' were organized under various commodity Forums. The organisations were registered under societies act and they were imparted skills to develop awareness on latest state of art technologies. The forum has helped them to buy the inputs directly from the manufacturers cutting the cost down to minimum and strengthened their bargaining power while selling their produce.

1.4.4.5.8 Insurance Cover

The Farmers' Forum and the United India Assurance officials meeting was organized and eventually, the insurance cover was extended to the crop with a low premium of 4 per cent.

1.4.4.5.9 Market Tie-Ups

Market tie ups for the crops raised are being negotiated between the Farmers' Forum and the buyers. The buyers are now frequenting the production site for tie up for purchase of vegetables since, the quality of the produce is always highest. Safal market has rated the banana from the precision farming site as super quality, a new grade status.

1.4.4.5.10 Reduction in Pesticide Usage

The practice of going for 16 rounds of combination sprays have been avoided and judicious combination has been put into practice. The drip and fertigation technology always maintain a dry soil for a depth of 3 cm and this helps to reduce the weed growth and spore multiplication.

1.4.4.5.11 Promotion of Brand Value

The Precision Farmer's Association has developed a 'logo' and brand of their own and in each crate, the logo was pasted. The logo is now gaining popularity with the buyers offering higher rate for crates with such logo.

1.4.4.5.12 Farmers Corporate

The corporate have four virtues with which they are successful viz.,

1. They have the capability to honour any order of high volume (Quantity)
2. They do the cultivation professionally there by the physical and nutritive qualities are ensured (Quality)
3. The first lot and last lot are strikingly uniform (Uniformity)
4. They are able to deliver in time (Timely delivery)

These virtues viz., quantity, quality, uniformity and timely delivery are inculcated in the minds of the farmers and Farmer's Forum is now gaining the corporate qualities and thus becoming a Farmer's Corporate, a new model to emulate in countries like India where holdings are too fragmented and empowerment of Farmers' Forum is the only way.

1.5 FARM DOCUMENTATION AND RECORD MAINTENANCE

Each farmer has been maintaining a Daily Maintenance Record to record all the activities carried out in a week, including inputs used and the harvests made. This record also helps to work out the economics of production and arrive at the Benefit Cost Ratio or the Output Input Ratio. Besides the Theoretical and Technical aspects of Precision Farming, a Micro Level study is also carried out in Theni District of Tamil Nadu which is a green belt for raising horticultural crops. The study was carried out to fulfil the following specific objectives to highlight how efficient the Precision Farming activities are. So that, the young minds who are interested in establishing agriculture and horticulture in an enterprising way, the book on precision farming will be a helping tool to them.

1.6 PROBLEM FOCUS

Precision Technology in Agriculture and allied sciences are emerging for adoption in a big way among the farmers and agripreneurs. The documented evidence of success stories for the student community and the other stakeholders is not available in the form of a book and hence special effort has been taken to convert the student project experiences and by conducting some case studies, the Book on Economics of Precision Farming was developed and presented here with the following specific objectives.

1.7 OBJECTIVES

- To assess the area, production and productivity of vegetables in Tamil Nadu and India
- To assess the area, production and productivity of select vegetables exclusively under the precision farming techniques in the sample farms
- To document the economics of Precision Farming Practices
- To assess the marketing practices prevalent in respect of vegetables produced by the precision farming practitioners
- To document the status of export of precision vegetables to different countries from India
- To document the status of investment made in horticulture under five year plan and to suggest suitable policy measures

All these objectives stated above has been discussed with the outcome of field research and are stated in different chapters for better understanding and follow up by the readers and the student community who are interested in Hi-Tech Horticulture. The Economic aspects will be giving an idea to the users on the extent of reaping the returns from precision farming.

1.8. SUMMARY

This chapater has addressed the method of classification of vegetables, cropping system prevalent in respect of vegetables, precision farming and the components involved in precision farming of vegetables, History of precision farming, Tamil Nadu Precision Farming Project and its implementation with key components are discussed. Besides the need of farm documentation and its importance is also highlighted.

References

1. Ashrafuzzaman, M., M. Abdul Halim,M.R. Ismail, S.M. Shahidullah and M.A. Hossain (2011) "Effect of Plastic Mulch on Growth and Yield of Chilli" Brazilian Archives of Biology and Technology, 54(2) 2011.
2. Farios-Larios, J., S.Guzman, A.C. Michel (1994) "Effect of Plastic Mulch on the Growth and Yield of Cucumber", Journal of Biological Agriculture and Horticulture, 10 (4): 1994.
3. Hazra and Som, (2015). "Vegetable Science", (Kalyani Publishers: New Delhi) pp. 14-34.
4. Kottiswaran, S.V., Balaji Kannan, M. Rajasekar and M. Shanthini (2018) "A Guide on Protected Cultivation", (Tamil Nadu Agricultural University, Coimbatore) pp. 01-07
5. Peter, K.V. (2007). "Vegetable Crops", (New India Publishing Agency: New Delhi), pp. 4-14.
6. Thompson, H.C., Kelly, C.W. (1959). "Vegetable Crops", (Tata McGraw Hill Publishing Company: New Delhi), pp. 17-23

Web References

- http://agritech.tnau.ac.in/horticulture/horti_Greenhouse%20cultivation.html

CHAPTER II

Production of Vegetables in India

Dr. C. Sekhar

Objectives

- To expose the students on the status of Area Production and Productivity of different horticultural crops due to the adoption of Precision Farming in India
- To provide the information on the status of Area Production and Productivity of precision output of Fruits and Vegetables in India

Organization of the Chapter

- Introduction
- Productivity of Vegetables in the World
- Area Production and Productivity of Vegetables
- Area Production and Productivity of Tomato in India
- Area Production and Productivity of Brinjal in India
- Area Production and Productivity of Onion in India
- Details of Vegetable crops / varieties Ruling Indian Farms
- Area Production and Productivity of Fruits in India
- Summary

2.1 INTRODUCTION

The state wise area production and productivity is analyzed with respect to Vegetables at national and regional level and the same is analyzed and discussed under the following heads for providing clear picture for further promotion of area under vegetables. Production is defined as realization of output of Vegetables with the support of labor, capital and other resources by adopting indigenous and hi-tech techniques suitable and accessible to the Farmer producer and productivity is conceptualized as the production of vegetables in tonnes per ha.

2.1.1 Productivity of Vegetables in the World

The area under different vegetable crops and its production details are unable to be accessed and hence the productivity details are furnished here for comparative analysis with India and Tamil Nadu and the details are presented in Table 2.1.

Table 2.1 Average Productivity of Important Vegetables in India and the World

Sl. No	Name of the Vegetable Crop	Productivity in Tonnes per Ha		Productivity of Leading Country in Tonnes per Ha
		India	World	
01	Cauliflower	18.10	17.90	China (21.60)
02	Cabbage	22.00	19.60	Korea (55.30)
03	Potato	19.40	16.20	The Netherlands (44.80)
04	Peas	06.70	08.70	France (16.60)
05	Brinjal	16.60	16.10	India (16.60)
06	Bhendi (Okra)	09.60	06.70	NA

(Source: K.V. Peter, 2007)

Table 2.1 revealed the details of productivity of principal vegetables in the world. Among the vegetables considered, the Cabbage yield was highest realized by Korea to the tune of 55 tonnes per ha. Whereas, India could realize not even 50 per cent of the Productivity obtained by Korea. But the global average productivity in respect of Cabbage was arrived at only 19.60 tonnes per ha. Hence India has to make sincere effort in transfer of technology and related capacity building activities to the extension functionaries as well as the producers of Cabbage. Preference of Cabbage by the consumers are from organic origin and hence appropriate technology development and transfer is warranted immediately.

The Netherlands is the country capable of realizing the Potato productivity to the tune of 45 tonnes per ha followed by India could realize only 19.40 tonnes per ha which is also lesser than 50 per cent of the Netherlands productivity and the global average productivity of potato is arrived at only 16 tonnes. The other countries has to follow the methods of producing the Potato by Netherlands need to be studied by performing an exposure visit to this nation so as to impart such level of productivity.

In respect of Cauliflower, China is the leader had achieved around 22 tonnes per ha. The Indian Productivity in respect of Cauliflower and the Global productivity is almost equal which stands at 18 tonnes per ha. There is no much difference between the leader and the followers in terms of Productivity of Cauliflower.

The Productivity realized in respect of Brinjal is in favor of India. India is the leader in the world in achieving the highest productivity in Brinjal which is arrived at 16.60 tonnes per ha. The global average productivity is arrived at only 16 tonnes per ha. India is only marginally high. But when one could compare the productivity realized from precision farming technologies, the productivity of Brinjal in India is several fold increase over the global average. What is needed here is intensifying the practice of adopting the Precision Farming Technologies by the farmer and the extension functionaries should be assigned the task of enhancing the productivity of vegetables higher than the average productivity obtained by China. This is possible if we take concerted efforts and follow up in adoption and monitoring the implementation of the technologies at field.

To sum up, one could compare the productivity of vegetables realized by India and the global average productivity, India has to take larger effort in achieving the productivity obtained by the leaders in the world. Those practices could be made realizable if the budgetary allocation to Agriculture and Horticulture in a big way which is delineated in Chapter – XV.

2.1.2 Area Production and Productivity of Vegetables in India

The details on area production and productivity of Vegetables in India over the years are analyzed and the results are discussed in Table 2.2. It provides an overall understanding on whether the area is increasing or decreasing at macro level and the magnitude of production over years provide the information to the policy makers on the extent of adoption of technology and its efficacy and the need based efforts to enhance the productivity of Vegetables to meet the requirements of both domestic and international markets.

Table 2.2 State – Wise Area, Production and Productivity of Vegetables in India

Sl. No	State/Union Territories	Area (in 000'ha)			Production in '000 Tonnes			Productivity in Tonnes / Ha			Percentage Change in Productivity (2001-02 to 2014-15)
		2001-02	2010-11	2014-15	2001-02	2010-11	2014-15	2001-02	2010-11	2014-15	
1	West Bengal	1139.00	1349.70	1387.20	18075.30	26725.50	26354.60	15.90	19.80	18.99	19.43
2	Uttar Pradesh	0777.90	0829.40	1134.00	15044.80	17679.40	23575.60	19.30	21.30	20.78	07.67
3	Bihar	0578.90	0845.00	0842.00	8022.90	14630.20	14467.20	13.90	17.30	17.18	23.60
4	Andhra Pradesh	0222.50	0651.20	0354.90	2586.70	11847.60	6445.60	11.60	18.20	18.16	56.55
5	Gujarat	0232.20	0515.90	0578.40	3278.20	9379.50	11543.30	14.10	18.20	19.95	41.49
6	Karnataka	0358.10	0466.30	0462.90	4173.20	9056.40	8564.00	11.70	19.40	18.50	58.12
7	Tamil Nadu	0213.80	0277.30	0238.20	5444.60	8279.90	5682.80	25.50	29.90	23.80	(-) 06.67
8	Orissa	0643.40	0553.80	0668.80	7447.40	7790.10	9425.10	11.60	14.10	14.09	21.47
9	Maharashtra	0402.40	0611.00	0580.50	5128.30	7504.00	8136.30	12.70	12.30	14.01	10.31
10	Haryana	0150.40	0346.40	0368.40	2151.90	4649.30	5291.50	14.30	13.40	14.36	00.42
11	Chhattisgarh	0104.10	0345.80	0418.60	1355.30	4248.80	5739.50	13.00	12.30	13.71	05.46
12	Jharkhand	0158.50	0259.50	0316.70	1736.30	4112.40	4279.30	11.00	15.85	13.51	22.82
13	Madhya Pradesh	0136.40	0283.70	0677.60	0117.50	3698.60	14315.40	13.30	13.00	21.12	58.80
14	Punjab	0135.00	0174.10	0197.10	2275.60	3585.80	4066.20	16.90	20.60	20.63	22.07
15	Kerala	0114.30	0149.50	0115.40	2541.90	3392.70	2882.50	22.20	22.70	24.97	12.48
16	Assam	0237.40	0260.10	0296.20	2935.20	2925.50	4752.10	12.40	11.20	16.04	29.35
17	Jammu&Kashmir	0050.80	0069.70	0063.10	0728.90	1559.10	1395.50	14.40	22.40	22.11	53.54
18	Himachal Pradesh	0034.60	0080.40	0083.70	0639.10	1474.90	1585.40	18.50	18.30	18.94	02.38
19	Uttarakhand	0093.80	0085.80	0091.00	0737.30	1030.90	1085.50	07.90	12.00	11.92	50.89

Sl. No	State/Union Territories	Area (in 000'ha)			Production in '000 Tonnes			Productivity in Tonnes / Ha			Percentage Change in Productivity (2001-02 to 2014-15)
		2001-02	2010-11	2014-15	2001-02	2010-11	2014-15	2001-02	2010-11	2014-15	
20	Rajasthan	0099.30	0140.30	0163.80	0432.50	0885.00	1270.10	04.40	06.30	07.75	76.14
21	Tripura	0031.30	0036.00	0048.10	0353.20	0532.30	0806.40	11.30	14.80	16.76	48.32
22	Delhi	0111.00	0029.80	0022.80	0741.40	0496.80	0391.90	06.70	16.70	17.18	156.42
23	Meghalaya	0035.70	0041.80	0044.60	0265.90	0356.50	0534.00	07.40	08.50	11.97	61.76
24	Manipur	0010.60	0022.20	0027.80	0066.10	0236.50	0297.70	06.20	10.70	10.70	72.58
25	Sikkim	0014.20	0023.90	0026.10	0060.00	0120.90	0134.40	04.20	05.10	05.14	24.76
26	Mizoram	0006.80	0017.50	0044.00	0044.10	0115.60	0273.80	06.50	06.60	06.22	(-) 04.31
27	Nagaland	0026.30	0010.70	0038.60	0286.00	0079.40	0492.40	10.90	07.40	12.75	16.97
28	Goa	0007.60	0005.70	0007.20	0076.00	0057.80	0082.00	10.00	10.10	11.38	13.80
29	Arunachal Pradesh	0020.80	0004.20	0001.70	0083.90	0038.50	0041.00	04.00	09.20	24.11	502.75
30	Andaman &Nicobar	0003.10	0005.70	0006.70	0015.80	0034.50	0050.9	05.10	06.10	07.59	48.82
31	Lakshadweep	0000.20	0000.40	0000.30	0000.20	0014.10	0000.7	01.00	35.30	02.33	133.00
32	Pondicherry	0003.70	0000.60	0001.50	0054.20	0008.80	0043.7	14.60	14.70	29.13	99.52
33	Dadra&Nagar Haveli	0001.50	0001.10	0001.10	0013.50	0005.50	0005.5	09.00	05.00	05.00	(-) 44.44
34	Chandigarh	0000.10	0000.10	0000.00	0001.70	0001.70	0000.00	17.00	17.00	00.00	00.00
35	Daman&Diu	0000.10	0000.00	0000.10	1.10.00	0000.00	0003.60	11.00	00.00	36.00	227.27
	Total	5016.70	8494.60	9309.10	70546.70	146554.50	164015.5	14.06	17.25	17.62	25.32

(Source: Agricultural Statistics at a Glance 2016)

Table 2.2 revealed the details of state wise area under vegetables, production and productivity of vegetables in a decade of time in India so as to assess the changes in proudctivity. Total area under vegetables in India is arrived at 93 lakh ha. Of which, Tamil Nadu had an area of only 2.38 lakh ha which is accounted for 2.56 per cent to the total area under Vegetables in India. Among the States, West Bengal found to have the highest area (14.90 Per cent) and holding the place of front runner followed by Utter Pradesh. The area under vegetables in India rose from 50 lakh ha to 93 lakh ha in a decade of time and the increase in area is accounted to be 86 per cent over the year 2001-02. The second highest area under vegetables in India is with the Utter Pradesh state (11.34 lakh ha) which had 12.18 per cent to the total area. Third place goes to Bihar in respect of area under Vegetables which had 8.42 lakh ha. Among the four constituent states of South India, the Andhra Pradesh, Karnataka, Kerala and Tamil Nadu put together had the share of 12 per cent to the total area under Vegetables in India revealed that the south India is trailing behind in respect of area under vegetables. When one could compare the figures for 2010-11 and 2014-15, the southern states have slipped heavily. The decrease in area is accounted to be around 6 per cent between 2010-11 and 2014-15. The slippage in the area was much pronounced in Andhra Pradesh. It might be due to continuous monsoon failure in Andhra Pradesh and hence concerted effort should be taken to bring higher area under vegetables following the example of West Bengal which is the leader in area in India. The West Bengal state is taking continuous effort in bringing higher area under vegetables. It is visible through the absolute increase in area presented in Table 2.2.

In respect of production of Vegetables in India, the South Indian States had the poor share in vegetable production which is accounted for only 14.37 per cent to the total production of Vegetables in India during the year 2014-15. When one could compare the production of vegetables in South India, it has also slipped from 22 per cent to 14 per cent between the period of 2010-11 and 2014-15. Front runner in production of vegetables in India is West Bengal and Utter Pradesh states respectively had 16.07 per cent and 14.37 per cent to the total vegetables produced in India. Tamil Nadu is the state which had the production of only 3.46 per cent of total vegetables produced in India. The decadal change in the production of vegetables in India is arrived at 108 per cent over the year 2001-02. But the same increase was found to be 132.50 per cent between 2001 and 2015. The increase in the production of vegetables is attributed mainly due to the implementation of National Horticulture Mission (NHM) through its area expansion and yield enhancement efforts. However one has to assess why the Southern India particularly Andhra Pradesh had failed to enhance area and production between 2010-11 and 2014-15 need to be thoroughly analyzed and remedial measures to be taken for augmenting or enhancing the production and productivity of vegetables.

In respect of productivity of Vegetables, the scenario is totally different. Daman and Diu is the Island state in India capable of realizing around 36 tonnes per ha followed by Pondicherry, Kerala and Tamil Nadu achieved the productivity of 29, 25 and 23 tonnes of vegetables respectively. When one could compare the productivity of Tamil Nadu with respect to the national productivity of vegetables, the percentage increase in Tamil Nadu is accounted for 26 per cent. It is considerably high due to the implementation of precision farming technology across the state of Tamil Nadu following the experiences of Dharmapuri District in the yester years. The percentage change in the productivity of vegetables in India during the year 2014-15 is arrived at 25.32 per cent over the base year of 2001-02 due to the implementation of farm technologies at field for enhancement of productivity due to the NHM.

To sum up, in respect of area and production of Vegetables, West Bengal, Utter Pradesh and Bihar are the leaders particularly the West Bengal state is the leader. When one could compare the productivity of Vegetables, Daman and Diu, an Island state is capable of achieving the highest productivity followed by Pondicherry, Kerala and Tamil Nadu.

2.1.3 Area Production and Productivity of Horticultural Crops in Tamil Nadu

The horticultural crops are contributing higher returns to the farmer and hence increased production and productivity will definitely contribute to the upliftment of rural economy. In this context, the area production and productivity of horticultural crops in Tamil Nadu is analyzed and the results are presented in Table 2.3.

Table 2.3 Area Production and Productivity of Horticultural Crops in Tamil Nadu (2015-16)

Sl.No	Name of the Crops	Area in Lakh Ha	Production in Lakh Tonnes	Productivity in Tonnes / Ha	Percentage to Total
01	Fruits	03.11	68.29	21.96	35.52
02	Vegetables	02.91	77.72	26.71	40.42
03	Flowers	06.36	43.32	06.81	22.53
04	Spices	03.44	00.13	00.037	00.07
05	Plantation Crops	01.13	02.12	01.88	01.10
06	Honey	00.00	00.15	00.00	00.08
07	Medicinal and Aromatic Plants	02.57	00.55	00.21	00.28
Total		19.52	192.28	09.85	100.00

(Source: Agricultural Statistics at a Glance, 2016)

Table 2.3 revealed that the area under horticultural crops in Tamil Nadu is arrived at 19.52 lakh ha which is capable of producing the horticultural output to the tune of 192.28 lakh tonnes leaving the average productivity of around 10 tonnes per ha. Among the horticultural outputs, the share of Vegetables is found to be only marginal which is arrived at 2.91 lakh tonnes with a productivity of 26.71 tonnes per ha which is accounted for only around 40 per cent to the total horticultural output in Tamil Nadu and hence a detailed analysis was made to highlight the area production and productivity of Vegetables in Tamil Nadu.

2.2 AREA PRODUCTION AND PRODUCTIVITY OF SELECT VEGETABLES IN INDIA

In the preceding section, we have seen the area under horticultural crops. The ensuing section deals with area production and productivity of select Vegetables like Tomato, Brinjal and Onion in India. The details are analyzed and the results are presented in subsequent Tables.

2.2.1 Area Production and Productivity of Tomato in India

Tomato is an important vegetable crop variety which is demanded heavily during one season and the same is priceless during the off season. However, it is important to analyze the trend in area production and productivity of Tomato across the country and hence the same is analyzed and the results are presented in Table 2.4.

Table 2.4 Trend in Area Production and Productivity of Tomato in India

Sl.No	Year	Area in Thousand Ha	Percentage of Total Vegetable Area	Production in Thousand Tonnes	Percentage of Total Vegetable Production	Pro ductivity in Tonnes per Ha
1	1991-92	289.10	05.20	4243.40	07.20	14.70
2	2001-02	458.10	07.40	7462.30	08.00	16.30
3	2002-03	478.80	07.90	7616.70	09.00	15.90
4	2003-04	502.80	08.00	8125.60	08.70	16.20
5	2004-05	505.40	07.50	8825.40	08.70	17.50
6	2005-06	546.10	07.60	9820.40	08.90	18.00
7	2006-07	596.00	07.90	10055.00	08.70	16.90
8	2007-08	566.00	07.20	10303.00	08.00	18.20
9	2008-09	599.00	07.50	11149.00	08.60	18.60
10	2009-10	634.40	07.90	12433.20	09.30	19.60
11	2010-11	865.00	10.20	16826.00	11.50	19.50
12	2011-12	907.1	10.09	18653.3	11.93	20.56
13	2012-13	879.63	9.55	18226.64	11.23	20.72
14	2013-14	882.03	9.38	18735.91	11.50	21.24
15	2014-15	893.72	9.41	19166.69	11.47	21.44
Percentage Change During 2014-15 over 2001-02		95.09		156.84		31.15

Table 2.4 revealed that the area under Tomato in India showed an increase of only 95 per cent during the year 2014-15 over the year 2001-02. In respect of production

of Tomato, the percentage increase during 2014-15 over the year 2001-02 is arrived at 157 per cent. The decadal (1.5 Decade) change in respect of area under Tomato and its production is found to be considerably increased. Concerted efforts had been taken to bring higher area under Tomato as its price shoots up during off season and the same is not taken care off by the consumers and hence value addition opportunities in Tomato need to be brought in just like Tomato paste and the same may be marketed on the lines of Tamarind Pulp which is demanded heavily in Arab countries.

In respect of productivity of Tomato in India, the productivity rose from 16 tonnes per ha to 21 tonnes between the year 2001-02 and 2014-15 registering a percentage change of 31 per cent. The increase in the productivity could be realized gradually because of adoption of improved varieties and adoption of precision technologies at farms.

To sum up, the trend in area production and productivity of Tomato has shown that the area is increased to the tune of 95 per cent while the production increased to the level of 157 per cent during the year 2014-15 with the increase in productivity level of 21 tonnes per ha which is accounted for 31 per cent increase over the base year 2001-02.

2.2.2. State Wise Area Production and Productivity of Tomato in India

Decadal analysis revealed that the area production and productivity in respect of Tomato is increasing due to technological interventions made by the NHM. However it is important to analyze the state wise area production and productivity of Tomato and hence the details are analyzed and the results are presented in Table 2.5.

Table 2.5 State Wise Area Production and Productivity of Tomato in India

(Area in Thousand Ha; Production in Thousand Tonnes; Productivity in Tonnes per Ha)

Sl. No	Name of the State / Union Territories	2010 – 2011			2014 – 2015			
		Area	Pro duction	Produc tivity	Area	Pro duction	Pro ducti vity	% Change
1	Andhra Pradesh	296.30	5926.20	20.00	167.70	3354.50	20.00	00.00
2	Karnataka	051.20	1756.70	34.30	64.30	2031.20	31.58	(-) 07.93
3	Orissa	096.60	1367.20	14.10	96.60	1374.90	14.23	00.92
4	Maharashtra	052.00	0738.00	14.20	47.70	1122.00	23.52	65.63

Sl. No	Name of the State / Union Territories	2010 - 2011 Area	Pro duction	Produc tivity	2014 - 2015 Area	Pro duction	Pro ducti vity	% Change
5	West Bengal	054.10	1063.70	19.60	56.90	1149.60	20.20	03.06
6	Bihar	046.80	1056.20	22.60	47.70	1046.40	21.93	(-) 02.96
7	Gujarat	038.80	0978.40	25.20	44.60	1259.00	28.22	11.98
8	Chhattisgarh	042.90	0627.90	14.60	53.10	886.10	16.68	14.25
9	Tamil Nadu	27.20	580.60	21.40	28.20	349.90	12.40	(-) 42.06
10	Jharkhand	022.30	0401.60	18.00	26.90	259.90	9.66	(-) 46.33
11	Others	136.60	2330.00	17.10	260.00	6333.20	24.35	42.40
	Total	**864.90**	**16826.40**	**19.50**	893.70	19166.70	21.45	09.98

Table 2.5 revealed that among many states in India, Andhra Pradesh alone had higher area under Tomato cultivation. During the year 2010-11, the area under Tomato in India was found to be 8.65 lakh ha. But the same was increased to 8.94 lakh ha under Tomato during the year 2014-15 registering a percentage change of 3.30 per cent over the year 2010-11. All other states had faced either stagnant in area under Tomato or a decrease in the area under Tomato. In general the country had a 44.37 per cent increase alone in respect of area under Tomato during the year 2014-15. All other states should follow the formula adopted to enhance the area under Tomato so that the higher production of Tomato could be achieved. However, Andhra Pradesh had failed to increase the area during 2014-15. The reduction in area is arrived at 43.40 per cent during the year 2014-15 over the base year 2010-11.

In respect of Proudction of Tomato, India had only 14 per cent increase during the year 2014-15 over the year 2010-11. Andhra Pradesh state is a leader in production of Tomato contributed 35 per cent of Tomato to the nation followed by Karnataka contributing to the level of 10 per cent. Andhra Pradesh and Karnataka are the two states contributing around 45 per cent of total production of Tomato. All other states are trailing behind with stagnant production. But the statistical data proved that it is not so between the block years 2010-11 and 2014-15. The production of Tomato found to decline during the period. It might be due to the continued drought and another could be unassured price per unit of Tomato.

In respect of Productivity of Tomato, Karnataka state has increased its productivity from the level of 29.50 tonnes during the year 2008-09 to the level of 34.30 tonnes per ha during the year 2010-11. During the year 2014-15, the Productivity was found to degenerate from the level of 34 tonnes to 31.50 tonnes per ha. The increase might be due to the adoption of hybrids and the precision farming technologies to enhance the productivity. Other states should listen to the efforts taken to enhance the productivity of Tomato by Karnataka state and the area expansion effort taken by the Andhra Pradesh state so that the demand during the off season could be overcome in the country. The decrease in the productivity might be due to sudden influx of rain during the peak season and scored less productivity.

2.3 AREA PRODUCTION AND PRODUCTIVITY OF BRINJAL IN INDIA

The experiences of Tomato cultivation in India revealed that Andhra Pradesh is the leader in Area Expansion Program and the Karnataka State is the leader in adoption of hi-tech practices to enhance the productivity of Tomato. In this respect, let us discuss how the Brinjal cultivation is practiced in the country by analyzing the trend in cultivation of Brinjal and the area production and productivity of Brinjal state wise to understand the contribution of individual state in the country.

2.3.1 Trend in Area Production and Productivity of Brinjal in India

In respect of Brinjal, the trend in area production and productivity is analyzed to mark the decadal developments in Brinjal cultivation and hence the results are analyzed and the details are presented in Table 2.6.

Table 2.6 Trend in Area, Production and Productivity of Brinjal in India

Sl.No	Year	Area (in 000" Ha)	Percentage of Total Vegetable Area	Production in Thousand Tonnes	Percentage of Total Vegetable Production	Productivity in Tonnes per Ha
1	2001-02	502.40	08.20	08347.7	09.40	16.60
2	2002-03	507.30	08.30	08001.2	09.40	15.80
3	2003-04	516.40	08.20	08477.3	09.10	16.40
4	2004-05	526.50	07.80	08600.8	08.50	16.30
5	2005-06	559.70	08.30	09364.6	09.20	16.70

Sl.No	Year	Area (in 000" Ha)	Percentage of Total Vegetable Area	Production in Thousand Tonnes	Percentage of Total Vegetable Production	Productivity in Tonnes per Ha
6	2006-07	568.00	07.50	09453.0	08.20	16.60
7	2007-08	561.00	07.10	09678.0	07.50	17.30
8	2008-09	600.00	07.50	10378.0	08.00	17.30
9	2009-10	612.00	07.70	10563.0	07.90	17.30
10	2010-11	680.00	08.00	11896.0	08.10	17.50
11	2011-12	691.6	7.69	12634.1	8.08	18.26
12	2012-13	722.07	7.84	13443.59	8.28	18.61
13	2013-14	711.31	7.57	13557.82	8.32	19.06
14	2014-15	679.40	7.15	12438.65	7.44	18.30
Percentage Change in 2014-15 Over 2001-02		35.23		49.01		10.24

Table 2.6 revealed that the area under Brinjal in a decade of time is found to be only marginally increased from 5.02 lakh ha to 6.8 lakh ha. The increase in respect of area during the year 2010-11 is found to be only 35 per cent over the year 2001-02. In respect of Production of Brinjal in India, the same is arrived at only 49 per cent during the year 2014-15 over the year 2001-02. In respect of productivity of Brinjal in India, the overall productivity of Brinjal is stagnating between 16.60 tonnes and 17.50 tonnes per ha. This is very low when comparing the success stories of Brinjal established under Precision Farming Technologies. Tamil Nadu is the pioneering state which had entered into the practice of adopting the Precision Farming Technologies way back during the middle of the decade and achieved the productivity of Brinjal to the level of 102 tonnes per ha. Still the technology is not spread to or not adopted by the Department of Horticulture and Plantation Crops of different states in the country. In this circumstance, let us discuss the area production and productivity of Brinjal state wise in the country to learn about the adopters and non adopters of precision farming technology or hi-tech cultivation of Brinjal in their respective states.

2.3.2 State Wise Area Production and Productivity of Brinjal in India

State wise area production and productivity of Brinjal is analyzed and the results are presented in Table 2.7. It revealed that 12 states in the country are actively participated in production of Brinjal in India. Among these 12 States, West Bengal and Orissa found to be the top players in bringing higher area under Brinjal cultivation. Andhra Pradesh and Gujarat are states found to hold the third and fourth position in area under Brinjal cultivation. West Bengal and Orissa are the two states had 43 per cent of Country's brinjal area being their preference for brinjal.

Table 2.7 Statewise Area, Production and Productivity of Brinjal in India

(Area in Thousand Ha; Production in Thousand Tonnes; Productivity in Tonnes per Ha)

Sl. No	State	2008-2009			2010-2011		
		Area	Production	Pro ductivity	Area	Production	Pro ductivity
1	West Bengal	155.30	2758.60	17.80	158.40	2870.60	18.10
2	Orissa	130.00	1965.20	15.10	133.20	2204.80	16.60
3	Andhra Pradesh	026.60	0531.30	20.00	075.10	1502.40	20.00
4	Gujarat	062.60	1046.30	16.70	072.00	1236.30	17.20
5	Bihar	055.10	1186.10	21.50	055.70	1215.60	21.80
6	Maharashtra	028.90	0471.10	16.30	035.00	0490.00	14.00
7	Chhattisgarh	024.20	0355.50	14.70	026.70	0439.50	16.50
8	Karnataka	015.30	0372.90	24.30	014.20	0354.50	25.00
9	Madhya Pradesh	021.40	0321.30	15.00	023.40	0281.00	12.00
10	Haryana	Included in Others			014.60	0255.20	17.40
11	Jharkhand	022.70	0453.90	20.00	Jharkhand & Assam Included In Others		
12	Assam	015.50	0275.90	17.80			
13	Others	042.70	0639.90	15.00	071.70	1046.00	14.60
	Total	600.30	10377.60	17.30	680.00	11895.80	17.50
	Percentage Change During 2010-11 over 2008-09				13.33	14.63	01.16

Table 2.8 Percentage Change in the Area, Production and Productivity of Brinjal in India

Sl. No	State	2013-2014			2014-2015		
		Area	**Production**	**Pro ductivity**	**Area**	**Pro duction**	**Pro ductivity**
1	West Bengal	161.5	2977.0	18.43	161.9	2985	18.43
2	Orissa	125.2	2158.3	17.23	122.7	2103.3	17.14
3	Andhra Pradesh	58	1160.5	20.00	17.2	381.0	22.15
4	Gujarat	76.0	1477.0	19.43	76.0	1477.0	19.43
5	Bihar	57.5	1240.5	21.57	576.0	1141.6	1.98
6	Maharashtra	30.0	690.0	23.00	27.1	546.8	20.17
7	Chhattisgarh	33.1	586.0	17.70	33.4	606.7	18.16
8	Karnataka	15.8	402.5	25.47	16.5	427.2	25.89
9	Madhya Pradesh	42.7	1066.7	24.98	46.1	1154.0	25.03
10	Haryana	19.0	400.8	21.09	18.5	296.0	16.00
11	Jharkhand	21.2	272.5	12.85	21.4	275.2	12.85
12	Assam	17.0	270.7	15.92	17.3	286.4	16.55
13	Others	54.3	855.3	15.75	454.7	758.5	1.66
	Total	711.3	13557.8	253.4723	1588.8	12438.7	215.4945
	Percentage Change During 2014-15 over 2013-14				12.33	(-) 8.25	(-) 0.15

In respect of Production of Brinjal, West Bengal, Orissa and Andhra Pradesh states found to hold the first, second and third position respectively during 2010-11. Whereas, during the year 2014-15, West Bengal and Odisha alone were in the first two ranks in respect of production of Brinjal. Andhra Pradesh is replaced from third place by Gujarat, Madhya Pradesh and Bihar. All other states contribution is dismal towards Brinjal production. The percentage increase in the production of Brinjal during the year 2010-11 is found to be only around 15 per cent while the area registered an increase of only 13 per cent over the year 2008-09. However, the percentage change in the area under Brinjal during 2014-15 is 12 per cent increase over the year 2013-14. But the production and productivity of Brinjal during 2014-15 was found to be negative. It was mainly due to continuous drought after 2010-11.

The productivity of Brinjal is almost stagnating and the percentage increase in the productivity of Brinjal during the year 2010-11 is found to be 1.16 per cent only Among the States, Karnataka is the topper in achieving the productivity of Brinjal (25 Tonnes per Ha) followed by Bihar has achieved only 21.80 tonnes per ha. Andhra Pradesh State had gained the productivity of Brinjal only to the level of 20 tonnes per ha. Concerted efforts are needed to enhance the productivity by adopting the precision farming technologies to enhance the productivity of Brinjal in India. The Department of Horticulture has to conduct some special exercise to examine the actual reasons for the failure in production and productivity of Brinjal during the year 2014-15.

2.4 AREA PRODUCTION AND PRODUCTIVITY OF ONION IN INDIA

Onion is an important vegetables produced in India. Some times the demand and supply situation of Onion is erratic and hence it played significant role in changing the Government too because of the turbulence among the common public. In this situation, the trend in area production and productivity of onion and the state wise production of onion is important and hence these are discussed as follows.

2.4.1 Trend in Area Production and Productivity of Onion in India

Based on the policy guidelines of Government of India Onion are exported to different countries creates demand in the local markets. To meet the demand, the people are increasing the area under Onion. In this context, the data on Area production and productivity of onion from 2001-02 to 2010-11 are analyzed and the results are presented in Table 2.9.

Table 2.9 Trend in Area Production and Productivity of Onion in India

Sl. No	Year	Area in Thousand Ha	Percentage of Total Vegetable Area	Production in Thousand Tonnes	Percentage of Total Vegetable Production	Productivity in Tonnes per Ha
1	1991-1992	331.80	05.90	04705.80	08.00	14.20
2	2001-02	495.80	08.10	05252.10	05.60	10.60
3	2002-03	424.70	07.00	04209.50	05.00	09.90
4	2003-04	553.80	08.80	06267.60	06.70	11.30
5	2004-05	613.80	09.10	07760.60	07.70	12.60

Sl. No	Year	Area in Thousand Ha	Percentage of Total Vegetable Area	Production in Thousand Tonnes	Percentage of Total Vegetable Production	Productivity in Tonnes per Ha
6	2005-06	703.60	09.80	09432.50	08.50	13.40
7	2006-07	768.00	10.10	10847.00	09.40	14.10
8	2007-08	821.00	10.50	13900.00	10.80	16.90
9	2008-09	834.00	10.40	13565.00	10.50	16.30
10	2009-10	756.20	09.50	12158.80	09.10	16.10
11	2010-11	1064.00	12.50	15118.00	10.30	14.20
12	2011-12	1087.3	12.09	17511.1	11.20	16.10
13	2012-13	1051.53	11.42	16813.01	10.36	15.98
14	2013-14	1203.57	12.80	19401.68	11.91	16.12
15	2014-15	1149.841	12.11	18736.46	11.21	16.29
Percentage Change During 2014-15 over 2001-02		131.91		256.74		53.67

Table 2.9 revealed that the area under onion between 2001-02 and 2010-11 is found to be steadily increasing year by year. The absolute area under Onion is found to be increased to 10.64 lakh ha during the year 2010-11 from 4.95 lakh ha during the year 2001-02. The percentage increase in area in a decade of time is found to be at 115 per cent. Whereas, the percentage increase accounted for between the period 2001-02 and 2014-15 was found to be 132 per cent, an increase of 17 per cent in a five year period. In respect of production of Onion the increase in production was accounted to be 188 per cent during the year 2010-11 and 257 per cent between 2001-02 and 2014-15..

The absolute increase in the production was from 52 lakh tonnes during 2001-02 to 151 lakh tonnes during the year 2010-11 and 187 lakh tonnes during 2014-15. This information revealed that the farmers are establishing onion in their farm lands based on the demand and supply situation but the effort taken from the technical departments to enhance the productivity is found to be dismal. It is reflected in the overall productivity of onion in India. The productivity of Onion in India is almost stagnating around 16 tonnes per ha. Even the productivity of onion found to be reduced during the later part of the decade called for technical intervention of Government Departments to introduce the hybrids and the assistance to the farming community should be extended through some special schemes to enhance the productivity of onion which is the need of the hour so as to meet both the domestic and export markets.

2.4.2 State Wise Area Production and Productivity of Onion in India

The data on state wise area production and productivity of Onion in India is analyzed and the details are presented in Table 2.10

Table 2.10 State Wise Area Production and Productivity of Onion in India

(Area in Thousand Ha; Production in Thousand Tonnes; Productivity in Tonnes per Ha)

Sl.No	Name of the State / Union Territories	2008-09			2010-11		
		Area	Production	Pro ductivity	Area	Production	Pro ductivity
1	Maharashtra	250.00	3932.50	15.70	415.00	4905.00	11.80
2	Karnataka	165.10	3031.80	18.40	190.50	2592.20	13.60
3	Gujarat	057.60	1409.60	24.50	062.00	1514.10	24.40
4	Bihar	051.60	0946.60	18.30	053.30	1082.00	20.30
5	Madhya Prdesh	053.00	0881.80	16.60	058.30	1021.50	17.50
6	Andhra Pradesh	039.00	0662.60	17.00	047.80	0812.60	17.00
7	Rajasthan	041.00	0369.10	09.00	049.00	0494.20	10.10
8	Haryana	018.80	0347.90	18.50	022.20	0453.90	20.50
9	Tamil Nadu	035.00	0305.50	08.70	Included In Others		
10	Orissa	Included Others			034.80	0385.90	11.10
11	Uttar Pradesh	022.30	0308.00	13.80	023.20	0368.60	15.90
12	Others	100.70	1369.10	13.60	107.80	1487.70	13.80
	Total	834.20	13564.50	16.30	1063.80	15117.70	14.20
	Percentage Increase in 2010-11 over 2008-09				27.58	11.46	(-) 12.88

Sl.No	Name of the State / Union Territories	2013-14			2014-15		
		Area	Pro duction	Pro ductivity	Area	Production	Pro ductivity
1	Maharashtra	468.0	5864.0	12.52991	427.9	5362.0	12.53097
2	Karnataka	136.6	2065.2	15.11859	164.7	2985.8	18.12872
3	Gujarat	72.8	1851.2	25.42857	44.3	1126.5	25.42889
4	Bihar	54.3	1304.2	24.01842	54.3	1247.3	22.97053
5	Madhya Prdesh	117.3	2826.0	24.09207	123.2	2967.4	24.08604
6	Andhra Pradesh	55.8	1004.6	18.00358	38.4	575.6	14.98958
7	Rajasthan	57.5	705.0	12.26087	62.6	800.1	12.78115
8	Haryana	30.2	672.2	22.25828	28.7	667.1	23.2439
9	Tamilnadu	40.0	472.2	11.805	29.8	289.5	9.714765
10	Orissa	35.8	432.1	12.06983	33.2	396.0	11.92771
11	Uttar Pradesh	24.3	410.0	16.87243	24.5	413.4	16.87347
12	Others	111.0	1,795.0	16.17117	156.6	1905.8	12.16986
	Total	1203.6	19401.7	210.6287	1188.2	18736.5	204.8456
	Percentage Increase in 2014-15 over 2013-14				(-)1.28	(-)3.42	(-) 2.74

Table 2.10 revealed that the total area under Onion during the year 2008-09 was found to be 8.34 lakh ha which was expanded to 10.64 lakh ha during the year 2010-11 and the same was further increased to 11.88 lakh ha during 2014-15. State wise analysis revealed that Maharashtra and Karnataka are the two states capable of raising onion in higher area and reap higher production. But their productivity is hovering around 11 tonnes to 13 tonnes per ha which is very low. But the State of Gujarat has taken higher productivity in Onion which is arrived at 24 tonnes per ha. But the general trend in productivity is found to be facing on the negative side which is sliding from 17 tonnes to 14 tonnes. The overall decrease between these block years is found to be 12.88 per cent and 2.74 per cent between 2014-15 and 2013-14 called for urgent regulatory measures to infuse the technologies at field to enhance the yield of onion. This level of productivity was reached during the year 1991-92 itself. Again going back to the year 1991-92 may not add colors to the officials of the Department of Horticulture and Plantation crops across the country. The Ministry of Agriculture should take concerted efforts to enhance the productivity.

2.5 DETAILS OF VEGETABLE CROP VARIETIES RULING INDIAN FARMS

The production and productivity of select vegetables and its performance at field depends on the potentials associated with the varieties chosen by the farmers for cultivation. In this respect, what are the ruling varieties in respect of the select vegetables discussed in the above sections are of much importance and hence the details of crop varieties ruling the Indian farm lands are analyzed and the results are presented in Table 2.11.

Table 2.11 Details of Vegetable Crop Varieties Ruling Indian Farms

SI. No	Details of Varieties	Source / Origin	Important Varieties	Number of Varieties
1	Tomato	Peruvian and Mexican Region	Arka Saurabh, Arka Vikash, ARTH3, ARTH4, Avinash2, BSS 90, CO3, HS 101, HS 102, HS 110, Hisar Anmol, Hisar Arun, Hisar Lalima, Hisar Lalit, Krishna, KS 2, Matri, MTH6, NA601, Naveen, Pusa 120, Punjab Chuhara, Pant Bahar, Pusa Divya, Pusa Early Dwarf, Pusa Hybrid 2, Pusa Hybrid 4, Pusa Ruby, Pusa Sheetal, Pusa Uphar, Rajini, Rashmi, Ratna and Rupali, PKM-1, PKM -2, PKM-3, PKM-4 and PKM-5.	34
2.1	Brinjal (Early Maturity)	India (Asia)	ABH 1, ABH 2, Arka Navneet, Arka Neelkantha, Arka Nidhi, Arka Kranti, Mysore Green, NDBH 1, Pant Samrat, Punjab Chamkila, Punjab Sadabahar, Pusa Bhairav, Pusa Bindu, Pusa Hybrid 5, Pusa Hybrid 6, Pusa Hybrid 9, Pusa Upkar, Pusa Uttam and Pusa Purple Long.	19 (44.19)
2.2	Brinjal (Medium Maturity)	India (Asia)	Annamalai, Arka Keshav, Arka Kusumakar, Arka Sheel, Arka Shirish, Azad B1, BR 112, CO1, Gujarat 6, Hisar Jamuni, Hisar Shyamal, Jamuni Gole, Junagarh Long, MDU 1, Pant Rituraj, PH 4, Punjap Barsati, Punjap Neelum, Pusa Anupam,Pusa Kranti, Pusa Purple Cluster and Vaishali.	22 (51.16)

Sl. No	Details of Varieties	Source / Origin	Important Varieties	Number of Varieties
2.3	Brinjal (Late Maturity)	India (Asia)	Manjari Gota, Punjab Bahar.	02 (4.65)
3	Onion	Asia	Agrifound Dark Red, Agrifound Light Red, Agrifoud Red, Agrifound Rose, Arka Bindu, Arka Kalian, Arka Niketan, Arka Pragati, Baswant 780, Brown Spanish, Co-1, Co-2, Co-3, Co-4, Early Grano, Hisar 2, Kalyanpur Red Round, MDU-1, N-2-4-1, N-53, N-257-9-1, Punjab-48 (S48), Pusa Madhavi, Punjab Red Round, Pusa Ratner, Punjab Selection, Pusa Red, Pusa White Flat ,Pusa White Round, Udaipur 101,102 &103	32
4	Potato	South America	Kufri Sindhuri, Kufri Chandramukhi, Kufri Jyoti, Kufri Muthu, Kufri Lauvkar, Kufri Dewa, Kufri Badshan, Kufri Bahar, Kufri Lalima, Kufri Swarna, Kufri Ashoka, Kufri Jawahar and Kufri Sutlej.	13

(Figures in Parentheses indicate percentage to total Brinjal Varieties)

Table 2.11 revealed that the varieties ruling the Indian farms at National level were 34 in the case of Tomato; 43 in the case of Brinjal; 32 in the case of Onion and 13 in the case of Potato. In respect of Brinjal, it is again classified based on the time of maturity viz., early maturity category had 19 varieties; medium maturity type varieties are 22 and the late maturity type varieties are only two. In all these varieties, the regional varieties which are contributing in terms of higher productivity is not reflected. But at regional level, the regional varieties in respect of Tomato and Brinjal played a dominant role as these varieties are inseparable from the farm families as it related to taste and preferences among the common public. Besides, the Regional Research Institutions of State Agricultural Universities in India also play a vital and dominant role in identifying good varieties by way of selection and hybrid development and release the varieties every year and these varieties are also contributing to the productivity of vegetables. In this respect, analyzing the regional varieties which are playing a dominant role in the case of Tomato, Brinjal and Onion are important.

2.6 AREA PRODUCTION AND PRODUCTIVITY OF FRUITS IN INDIA

Table 2.12 Area Production and Productivitgy of Fruits in India

Year	Area under Fruits in Lakh Ha	Production of Fruits in Lakh Tonnes	Productivity of Fruits in Tonnes per Ha
2001-02	41.27	431.15	10.45
2002-03	39.05	453.17	11.60
2003-04	47.67	460.63	09.66
2004-05	51.55	509.88	09.89
2005-06	54.54	555.05	10.18
2006-07	56.87	597.14	10.50
2007-08	59.89	657.64	10.98
2008-09	62.38	686.38	11.00
2009-10	64.71	717.09	11.08
2010-11	63.82	748.78	11.73
2011-12	67.05	764.24	11.40
2012-13	68.77	805.57	11.71
2013-14	72.16	889.77	12.33
2014-15	63.58	888.19	13.97
CGR (%)	05.06	06.28	

Table 2.12 revealed that the growth rate of area under fruits is arrived at 5 per cent and the growth rate in respect of production of fruits is only 6 per cent revealed that there is no much appreciable increase in productivity. The productivity is hovering around only 11 tonnes per ha. But during the year 2011-12 to 2014-15, the productivity is found to increase from 11 tonnes to 14 tonnes per ha.

Table 2.13 Area Production and Productivity of Guava in India

Year	Area under Guava in Lakh Ha	Production of Guava in Lakh Tonnes	Productivity of Guava in Tonnes per Ha
2001-02	01.55	17.16	11.07
2002-03	01.55	17.93	11.57
2003-04	01.66	18.31	11.03
2004-05	01.61	16.83	10.45
2005-06	01.67	17.37	10.40
2006-07	01.76	18.31	10.40
2007-08	01.79	19.81	11.07
2008-09	02.04	22.70	11.13
2009-10	02.20	25.72	11.69
2010-11	02.05	24.62	12.00
2011-12	02.20	25.70	11.68
2012-13	02.29	26.13	11.41
2013-14	02.68	36.68	13.68
2014-15	02.51	40.83	16.26
CGR (%)	03.96	04.63	

(Jennifer, 2014)

Table 2.13 revealed that the growth rate in area production of Guava is arrived at 4 per cent and 4.6 per cent respectively. When compared to the growth rate of general fruits, it is marginally low. The productivity of Guava is also found to be hovering around 11 tonnes per ha between the year 2001-02 to 2009-10. Beyond that marginal increase in productivity is visible and it rose to 16 tonnes per ha from 11 tonnes. It might be due to readily available market for the fresh as well as value added guava which are distributed through Farmer Producer Company established in the Ayakkudi Village of Palani Taluk of Dindigul District of Tamil Nadu.

According to Malik et.al (2013) the Guava producers are facing lot of problems. They are high cost of transportation, crude system of packaging and lack of technical guidance related to scientific methods of packing. Maximum post harvest loss was observed at the producers' level. This loss can be minimized by adoption of integrated approach of pre and post harvest management by selecting fruit cultivars having longer shelf life, adopting scientific cultural practices and educating guava growers about the post harvest technology. They also concluded that processing of guava was economically feasible and hence needs to be encouraged. Wide scope existed for processing industry as production was higher than the demand particularly during the peak period. Though a variety of guava products could be prepared, but a very little portion of the surplus production of Guava was utilized for processing and value addition. To avoid the glut during the peak period, processing industry can prove very successful in better marketing and boosting the interest of the guava growers towards this popular fruit for their overall prosperity.

2.7. SUMMARY

In this chapter, the area production productivity of Fruits and vegetable crops have been discussed in respect of India and the same details on Statewise. The selected crops for discussion was Brinjal, Onion and Tomato. In all the crop varieties discussed, the regional varieties which are contributing in terms of higher productivity is not reflected. But at regional level, the regional varieties in respect of Tomato and Brinjal played a dominant role as these varieties are inseparable from the part of farm families as it closely related to taste and preferences among the common public.

References

1. Government of India (2016). "Agricultural Statistics at a Glance – 2016", (Ministry of Agriculture and Farmers' Welfare, Department of Agriculture, Cooperation and Farmers' Welfare: New Delhi) pp. 74-230.

2. Jennier, F.M. (2014). "Value Chain Analysis of Guava in Dindigul District of Tamil Nadu", (Unpublished M.B.A. Thesis Submitted to the Department of Agricultural Rural Management, Tamil Nadu Agricultural University, Coimbatore).

3. Malik, S.H. and S.A.Saraf (2013) "Economic Analysis of Processing of Guava in Utter Pradesh State of India, *Journal of Agricultural Sciences*, 5 (6): pp. 44-57

4. Peter, K.V. (2007). "Vegetable Crops", (New India Publishing Agency: New Delhi), pp. 4-14.

CHAPTER III

Techniques of Precision Farming

Dr. I. Muthuvel and Dr. C. Sekhar

Objectives

- To give an orientation to the students of Agriculture and Horticulture on the Precision Farming Technologies
- To highlight to what are all the crops, the Precision Farming Technologies are being practiced across the country
- To Spell out the experiences of Precision Farming Practitioners to inculcate the knowledge to the students

Organization of the Chapter

- Introduction
- Precision Farming Techniques for Tomato
- Precision Farming Techniques for Brinjal
- Precision Farming Techniques for Banana
- Precision Farming Techniques for Bittergourd
- Precision Farming Techniques for Bottlegourd
- Precision Farming Techniques for Cabbage
- Precision Farming Techniques for Cauliflower
- Precision Farming Techniques for Turmeric
- Summary

3.0 INTRODUCTION

3.1 PRECISION FARMING (PF) TECHNOLOGIES

PF is a management philosophy or approach to the farm. It is essentially more precise farm management made possible by modern technology with using various technological tools such as GPS, GIS, Remote Sensing, VRT and Drip Fertigation. The variations occurring in crop or soil properties within a field are recorded, mapped and then management actions are planned and initiated as a consequence of continued assessment of the spatial and temporal variability within that field (Mandal Debashis and Ghosh K.S., 2000).

Global Positioning Systems (GPS)

Global Positioning Systems (GPS) provide farmers with the capabilities to manage field variability on a site-specific basis. GPS is a constellation of 24 satellites orbiting 12,660 miles above the Earth that is used for determining latitude and longitude (position). When installed on agriculture equipment, a GPS receiver provides position information for soil sampling, crop scouting, applying inputs, planting, harvesting and other applications. A GPS receiver allows farmers to continually return to the same location in a field

Geographical Information Systems (GIS)

A GIS retrieves, stores, and analyzes spatial data and can be used to generate field and prescription maps for application of crop inputs. GIS data and maps are used for management decisions such as management zone creation, variable-rate application, planting decisions, and targeted soil sampling. Delineated management zones in GIS are used to site-specifically manage cropland according to measured field and crop variation. Farmers are able to use GIS to produce detailed harvest reports, determine trends from harvest to harvest, and compare production capabilities of different varieties and crop inputs. Agriculture GIS software can also serve as valuable accounting, record keeping, and decision support tools for farmers.

Remote Sensors (RS)

This tool are being used to determine crop stress, soil properties, pest incidence, etc. as the tractor or any other mobile device passes over the field, as a scout goes over the field on foot. Measurement of plant and soil properties as the tractor or combine travels over the field is a developing area in precision farming.

Variable Rate Technology (VRT)

Variable-rate technology (VRT) describes any technology which enables producers to vary the rate of crop inputs. VRT combines a variable-rate (VR) control system with application equipment to apply inputs at a precise time and location to achieve site-specific application rates of inputs. A site-specific approach allows growers to apply products only where they are needed in a field. Varying the application of inputs can reduce input and labor costs, maximize productivity, and reduces the impact over-application may have on the environment. Examples of VR applications for agriculture include fertilizer, lime, seeding and pesticides (Srinivasan, 2001).

Drip Fertigation

Traditionally, irrigation water is applied to the entire field, resulting is loss of water. Drip irrigation was a modern irrigation method in which water is delivered directly into the root zone of the plant. Drip systems are also easily integrated in fertigation systems. It has many advantages over other irrigation methods, including elimination of surface runoff, high uniformity of water distribution, high water usage efficiency, flexibility in fertilization, prevention of weed growth and plant disease.

Yield Monitoring (YM) and Yield Mapping

This is the most direct method to assess the productivity of land and how it should be better managed. A yield monitor (YM) measures the crop as it is harvested. YM usually measure crop weight, impact forces by passing an array of light beams that are broken. Other methods are also being investigated that show promise for future yield monitors. As the yield is measured, data are stored on a computer along with the GPS coordinates at the point where the yield was measured. Mapping software can then create a yield map. The yield map can immediately provide two important pieces of information, yield variability, and yield production. Yield variability is illustrated on a map by a change in color, where each color represents a range in yield. As a convention red is suggested to represent low yields and green high yields. The map legend will tell you how to read what each color indicates. In all mapping situations, it is important to check the legend first. Lack of yield variability would mean that the map shows mostly one color. Yield data can be obtained by calculating the yield for the entire field. Yield monitors are in the development stages. As these systems move from testing to application, hopefully commercial systems will not be far behind (Sreenivasan, 2001).

Keeping in view of the above techniques, the selection of field, Mapping and other formalities will be fulfilled for implementation. Under the head "Precision Farming Technologies", how these technologies are being used in the production process of the following crops and its suggested technologies given by the technocrats are furnished from seed or seedling development to harvest and yield of crops and the same are compared with traditional farming practices wherever needed. If one could follow these technologies scrupulously, one can get the highest yield of the vegetable crops. How exactly the yield increase was obtained are discussed in respect of individual crops elsewhere. The crops discussed are;

- Tomato
- Brinjal
- Banana
- Bitter Gourd
- Bottle Gourd
- Cabbage
- Cauliflower
- Turmeric

3.1.1 Precision Farming Techniques For Tomato

1. Quality Seedling Production

Protected Nursery

The nursery area is covered with 50 per cent shade net and seedlings are raised in Protrays. The sterilized cocopeat is used for raising seedlings at 300 kg/ha is mixed with 5 kg of neem cake along with *Azospirillum* and *Phosphobacteria* each at 1 kg.

Sowing

About 100g of hybrid seeds are required to produce seedlings for 1 ha. The individual seeds are sown in cells of protrays and maintained with sufficient watering and nutrients and Panchakavya spray.

2. Planting

The well ploughed field is applied with farm yard manure, Super Phosphate, bio-fertilizers and neem cake. Raised beds are prepared and watered with drip systems for 8 to 12 hours. Planting is done at a spacing of 90 × 60 × 60 cm in the paired row system. One row of African marigold seedlings of 40 days old are planted simultaneously for every 16 row of tomato seedlings.

3. Fertigation Techniques

For tomato F_1 hybrid, a dose of 200:250:250 kg of NPK / ha is applied throughout the cropping period through split application. The split dose is applied once in every three days through fertigation.

1. Tomato

Quantity of fertilizers Required per hectare : 200-250-250 kg of NPK

Quantity of fertilizers through fertigation : 200-62.5-250 kg of NPK

S.No.	Name of the Fertilizers	Quantity (kg/hectare)
01.	Super phosphate	1172
02.	19-19-19	132
03.	13-00-45	500
04.	12-61-00	62
05.	Urea	223

Application of Basal dose of 1172 kgs of super phosphate during the last plough.

Fertigation Schedule for Tomato

Stage of the crop	Ferilizer application interval	Fertilizer to be applied	No. of times to be applied	Quantity (kg/time)
First stage	From the day of planting to 10th day after planting (3,6 and 9th day)	19-19-19	3 times	21.93
		13-00-45	3 times	9.26
		Urea	3 times	2.82

Stage of the crop	Ferilizer application interval	Fertilizer to be applied	No. of times to be applied	Quantity (kg/time)
		12-61-00	10 times	3.07
Second stage	11th day to 40th day	13-00-45	10 times	16.66
		Urea	10 times	7.53
		19-19-19	10 times	6.60
Third stage	41st day to 70th day	13-00-45	10 times	13.90
		Urea	10 times	6.40
		12-61-00	27 times	1.14
Fourth stage	71st day to 150th day	13-00-45	27 times	6.17
		Urea	27 times	2.80

1. Intercultivation

Staking is done by using either bamboo or eucalyptus poles on 30th day after planting. Weeding is done on 30 and 60th days after planting depending upon the weed growth. Foliar spraying with *Pseudomonas fluorescens* 0.5% at 15 days interval for 6 times is given to the crops.

2. Harvesting and Post Harvest Management

Harvesting of tomato fruits is carried out either at breaker stage or at half red depending upon the market preference. First harvest starts at 60 days after planting.

Grading is done based on the size of the fruits. Three grades viz., uniform big sized fruits, small size damaged and mis-shaped fruits and are packed separately in plastic crates.

3.1.2 Precision Farming Techniques For Brinjal

1. Quality Seedling Production

Protected Nursery

The seedlings are raised in protrays using cocopeat (300 kg/ha) as the medium which is mixed with 5 kg of neem cake with bio fertilizers each at 1 kg.

Sowing

About 200g of brinjal seeds are taken and mixed with *Azospirillum* at 200g/kg and planted in the protrays.

2. Planting

The well ploughed field is mixed with farm yard manure, bio fertilizers, and Super Phosphate. Raised beds of 120 cm width are formed at an interval of 30 cm and the laterals are placed at the centre of each bed. 35 days old seedlings are dipped in 0.5% solution of *Pseudomonas fluorescens* for 30 minutes and are transplanted at a spacing of 90 × 60 × 75 cm in the paired row system in the main field.

3. Fertigation Techniques for Brinjal

For brinjal F_1 hybrid, a dose of 200:50:100 kg of NPK/ha is applied throughout the cropping period through split application. The split dose is applied once in every three days through fertigation.

Brinjal

Quantity of fertilizers Required per hectare : 200-250-100 kg of NPK/ha

Quantity of fertilizers through fertigation : 200-37.5-100 kg of NPK/ha

S.No.	Name of the Fertilizers	Quantity (kg/ hectare)
01.	Super phosphate	703
02.	19-19-19	79
03.	13-00-45	189
04.	12-61-00	37
05.	Urea	340

Application of Basal dose of 703 kgs of super phosphate during the last plough.

Fertigation Schedule for Brinjal

Stage of the crop	Ferilizer application interval	Fertilizer to be applied	No. of times to be applied	Quantity (kg/time)
First stage	From the day of planting to 10th day after planting (3,6 and 9th day)	19-19-19	3 times	31.17
		13-00-45	3 times	1.87
		Urea	3 times	8.53
Second stage	11th day to 40th day	12-61-00	10 times	1.85
		13-00-45	10 times	6.67
		Urea	10 times	10.68
Third stage	41st day to 70th day	19-19-19	10 times	3.95
		13-00-45	10 times	5.00
		Urea	10 times	10.00
Fourth stage	71st day to 150th day	12-61-00	27 times	0.69
		13-00-45	27 times	2.47
		Urea	27 times	3.96

1. Intercultivation

Micronutrient mixtures, Growth Regulators and bio fertilizers are applied at regular intervals.

2. Harvesting and Post Harvest Management

Harvesting of brinjal fruits is carried out at fully matured stage. Grading is done based on the size. The damages and pest affected fruits are sorted out and then packed in the crates for market.

3.1.3. Precision Farming Techniques For Banana

Banana grows well in well drained loamy soils and the best season differs with varieties and in different lands.

Wet Lands:

February to April: Poovan, Rasthali, Monthan, Karpooravalli and Neypoovan

April to May: Nendran

Selection of Planting Material

- The sword suckers of 1.5 to 2 Kg weight are selected for planting
- The suckers are selected only from the mother plots free from diseases, nematodes and other pests
- The tissue culture plants of 45 days old, well hardened and virus indexed plants with 5 to 6 leaves are selected.

Sucker Treatment

- The pseudo stem is cut leaving 20 cm from the corm.
- The roots and superficial damaged portion of the corm is trimmed
- The corms are dipped for 5 minutes in carbendazim at the rate of 0.1 per cent (One gram per litre of water) to avoid any diseases
- Pralinage is done with 40 gram of Carbofuran, 3 Gram granules per sucker (Dip the Corm in Slurry solution of 4 parts clay plus 5 parts of water and sprinkle Carbofuran to control the Nematodes)
- Alternatively, the Corms are dipped with 0.75 per cent Monocrotophos and shade dried for at least 24 hours and then planting is done.

Drip Irrigation System

- Installation of drip system is done with main, sub main pipes and the inline lateral tubes are placed at an interval of 1.50 meter (5 feet)
- The drippers in lateral tubes are placed at an interval of 60 cm and 50 cm spacing with 4 Liters per Hour (LPH) and 3.5 LPH capacities respectively

Main Field Preparation and Planting

- Ploughing is done at four times using chisel plough (One Time), Disc Plough (one Time) and Cultivator (Three Times)
- The raised beds are formed at 60 cm width along the lateral lines
- The pits of 45 cm^3size is dig out along the raised bed at 1.8 m intervals.
- The pits are filled with topsoil mixed with 10 kg Farm Yard Manure, 250 gram of Neem Cake, 20 gram of Furadon and 220 gram of Super phosphate

- Azospirillum and Phosphobacteria each at 20 gram per pit is applied during planting and at fifth month after planting
- At the time of planting 25 gram of Pseudomonas fluorescence is applied in each pit
- Pre emergence of weedicide of Fluchloralin at the rate of 2 litre per ha is sprayed through high volume sprayer
- Spacing practiced for Cavendish, G 9, Robusta are 5 feet x 7 feet. For other varieties, the spacing may be 5 feet x 8 feet. A total of nearly 3000 plants are accommodated in one ha area.
- Planting is done at the centre of the pit and cover the pit with top soil and pressed gently
- Sowing of Sunhemp on 45th day after planting and incorporated it after a month. This operation reduces nematode development.

Fertigation

- For Banana, a dose of 200:300 g of NK per plant is applied throughout the cropping period through split application and phosphorous alone is applied as basal dose at the time of planting. The split doses of N and K are applied once in every three days through fertigation.
- Water requirement is 25 litres per plant at normal density. Two suckers per pit require water upto 40 litres and Three suckers per pit requires 50 litres of water.

Fertigation Schedule

- The fertigation schedule prescribed under precision farming technique is furnished in the following Table 3.1.

Table 3.1 Fertigation Schedule for Banana under Precision Farming

Stage of the Crop	Duration (DAP)	Fertilizer Grade	Number of Splits	Quantity (Kg per Split / ha	Quantity (g/Split/ Plant)
I	01-70	13-00-45	14	29	9.7
		Urea	14	20	6.7
II	71 – 156	13-00-45	17	35	11.8
		00-00-50	17	11	3.5
		Urea	17	28	9.3
III	157 – 242	13-00-45	17	27	9.0
		00-00-50	17	7.3	2.4
		Urea	17	7.7	2.6
IV	243 – 264	00-00-50	05	36	12.0

Intercultivation

- Digging and Earthing up the plants are done after applying 20 grams of Furadon plus 0.5 kg Pungam oil cake at 3 Months after planting
- Digging followed by earthing up is done at fourth, sixth and eighth months after planting of banana suckers
- Desuckering and removal of diseased and dried leaves shuld be done for six times based on the intensity of sucker growth.
- Apply 20 gram of Furadon at fourth month after planting of banana
- Spray Zinc Sulphate (0.5 Per Cent), Ferrous Sulphate (0.2 Per Cent), Copper Sulphate (0.2 Per Cent) and Borax (0.1 Per Cent) at 3, 5 and 7th Month after Planting
- Remove male bud one week after opening the last hand
- Propping is done one month after shoot emergence
- Intercrops like Onion, leafy coriander, beetroot, Tomato can be grown. Inter cropping should be completed within 90 days of planting of Banana. Nematode repelling species like Sun Hemp or Marigold can be grown and incorporated with soil at maximum vegetative stage.

- One per cent of Urea and Potassium Sulphaate 1.50 per cent is sprayed after last hand opening and again sprayed at one after a month
- The fruit bunches are covered with ventilated (4 per cent) polythene sleeves (200x150 cm) with in 15 days of last hand opening

Harvesting

- The fruit bunches are harvested when the angularity of fruit disappears
- The bunch weight in Cavendish is 25 to 30 Kgs
- The bunch weight in Ney Poovan is 12 Kgs
- The bunch weight in Rasthali is 12 Kgs
- For uniform ripening, the fruit bunches are sprayed with Ethrel 5000 PPM (5 Ml per litre of water) with Sodium Hydroxide pellets

Plant Protection Measures

a. Pest Management

1. Rhizome Weevil (Cosmopolites sordidus)

- The adults are about 1.3 m long, shiny black with elongated and slightly curved snout and longitudinally striated elytra
- The female beetle bites a small hole in the rhizome, either on its exposed collar parts or underground parts to lay its eggs
- The larvae on hatching bore into the rhizome and feed by making tunnels in which the pupation occurs on maturity of the former.
- The adults can live upto 2 years and without food for 6 months
- The damage is done by both the adults and larvae. The adults feed during night on the psuedostems and the larvae on the rhizome whose growing parts may get destroyed resulting on the failure of the appearance of new shoots (suckers)

Management of Rhizome Weevil

- Use of uninfested suckers for the plantation
- Covering of the unexposed portion of the rhizome or cut portion of the pseudostem by a layer of earth to prevent entry of the weevils
- Treating the pits in which suckers are to be planted with Carbofuran 3G, 10 gram or Phorate 10 G5 gram per plant or Lindane 1.3 D 20 gram per plant

- Spraying around the collar with 0.05 per cent endosulfan or 0.04 per cent Monocrotophos has been found to prevent the attack

2. Pseudostem Borer (Odoiporus longicollis)

- Grubs or larvae are apodous, creamy white with dark brown head
- Adults are robust, reddish brown and black weevil
- Wilting of the plant, Bore holes and tunnels in the pseudostem are the main damages

Management of Pseudostem Borer

- Removal and destruction of the affected plants
- Spray Methyl Parathion 50 EC 2 ml per litre of water at the base of the plants
- Pseudostem Injection: Monocrotophos 36 WSC one ml, after diluting one ml in 4 ml of water at 45 days interval

3. Banana Aphid (Pentalonia nigronervosa)

- Dark nymphs and adults
- Winged adults are having black veined wings
- It is vector for bunchy top disease
- It seen in colonies on leaf axils and pseudostem

Management of Banana Aphid

- Spray methyl demeton 25 EC one ml per litre of water, dimethoate 30 EC one ml per litre of water, Monocroptophos 36 WSC one ml per litre @ 21 days interval
- Spray may be directed towards crown and pseudostem. The entire foliage need not be covered

4. Thrips

- Leaf Thrips – Helionothrips kadaliphilus; Panchaetothrips indicus
- Fruit rust Thrips – Chaetanaphothrips signipennis
- Flower Thrips – Thrips florum

Yellowing of Leaves, Rusty growth over fruits, Corky scab on fruits and flowers are the main problems created by Thrips

Management of Thrips

- Methyl Dimethon 25 EC – 2 Ml per litre of water
- Monocrotophos 36 WSC one ml per litre of water
- Disease Management in Banana

Panama / Fusarium Wilt (Fusarium oxysporum)

- Yellowing of leaf blades as a band along the margin and spread towards midrib
- Leaf starts wilting and the petiole gets buckled
- Leaf hangs between the pseudostem while the middle of the lamina is still green
- Longitudinal splitting of pseudostem above the soil level
- Red or Brown streaks appear in vascular tissue at the peripheral region
- Cut stem smells like rotten fish

Management of Fusarium Wilt

- Cultivation of resistant cultivars like Dwarf Cavendish, Poovan, Nendran, Moongil, Peyladen, Vamanakel, Raza vazhai
- Chipping of roots and dippling of rhizomes in 0.1 per cent Carbendazim for 5 minutes, dippling in clay solution and treating with 40 gram of Carbofuran
- Capsules filled with Carbendazim 60 mg or Pseudomonas fluorescens 60mg is applied at 2, 4 and 6 months after planting
- Removal of infected plants and application of lime 2 kg per pit

Bacterial Wilt / Moko (Ralstonia solanacearum)

- Yellow discoloration of inner leaf lamina close to the petiole, followed by wilting of inner leaves
- Pale yellow to dark discoloration of vascular bundles at the centre

Management of Bacterial Wilt

- Soil application of Pseudomonas fluorescens @ 2.5 kg per ha
- Soil drenching with Copper oxychloride 0.25 per cent plus Streptomycin sulphate @ 0.05 per cent

Sigatoka Leaf Spot (Mycosphaerella musicola)

- Light yellow or brownish green narrow streaks develop in the leaves
- Streaks enlarge in size developing into linear, oblong muddy brown to black spots
- Central portion becomes necrotic and spots become encircled by a yellow halo
- Rapid drying and defoliation of leaves occurs at larger stages

Management of Sigatoka Leaf Spot

- Spraying with Carbendazim at 0.1 per cent or Copper oxychloride 0.25 per cent or Chlorothalonil at 0.2 per cent

Anthracnose / Fruit rot (Gloeospsorium musarum)

- Fruit skin turns black, shrivels and becomes covered with characteristic pink acervuli
- Premature ripening and shrivelling of fruits

Management of Fruit Rot

- Spraying with Copper oxychloride 0.25 per cent or Carbendazim 0.1 per cent or Bordeaux mixture 1 per cent or Chlorothalonil at 0.2 per cent
- Post harvest dipping of fruits in Carbendazim 400 ppm or Benomyl 1000 ppm

Heart Rot / Head Rot (Erwinia carotovora)

- Middle tender leaf show rotting
- Pseudostem easily comes out from the corm portion
- Bacterial oozing from the cut edges of corm and pseudostem

Management of Head Rot

- Planting of resistant cultivar like poovan
- Spraying with Streptomycin 0.05 per cent Plus Copper oxychloride 0.25 per cent

Bunchy Top Virus (Vector: Pentalonia nigronervosa)

- Green streaks appear on secondary veins of leaf lamina, midrib and petiole
- Marginal chlorosis and curling of leaves
- Leaves turn brittle and petioles become shorter
- Several leaves unroll at a time

- Leaves become smaller, premature, crowns consist of stunted leaves showing a rosette or bunchy top appearance
- Opening bunches are constricted or choked by the Pseudostem and may split before normal emergence

Management of Bunchy Top Virus

- Removal of infected trees and suckers. Killing of tree is done by injecting 4 ml of 24 D diluted with water at 1:8 ratio
- Vector Management

3.1. 4 Precision Farming Techniques for Bitter Gourd

Bitter Gourd is a climbing vine grows well in well drained loamy soil. It prefers a pH of 6.5 to 7.0 and the temperature ranges from 25 to 30^0 C. If the temperature goes below 18^0 C, the vine growth slows down while above 36^0 C affects the female flower production and hence, the best season of sowing is June and July months.

1. Quality Seedling Production

Bitter gourd is a direct sown vegetable but polythene bag nursery is more advantageous to get early marketing and to avoid more gap filling.

A 200 gauge poly bags of 10 cm diameter and 10 cm height is used for sowing the seeds. The seeds of 1.8 kg are required to cover one hectare and the seeds are mixed with 500 g of *Azospirillum* and are shade dried for half an hour.

A potting mixture contains 1:1:1 ratio of topsoil, sand and well-decomposed farm yard manure is used as media and sowing is done by only one seed per hole. About 15 days old seedlings are transplanted to the main field.

2. Field Preparation and Drip Irrigation

The field selected for planting should be ploughed thoroughly by using Chisel Plough (once), disc plough (Once) and cultivator four times. Farm Yard Manure has to be applied at the rate of 25 tonnes per ha and 75 per cent of total recommended dose of 470 kg per ha. Single super phosphate is broadcasted before last ploughing. Azospyrillum and Phosphobacteria 2 kg per ha and Pseudomonas 2.50 kg per ha along with FYM 50 kg and Neem cake at the rate of 100 kg is applied before last ploughing.

Raised beds are formed in the well prepared field with the dimensions of 4 ft width (120 cm) at the rate of one metre height is formed along the laterals. Spread the lateral tubes in the centre of each bed. Spray pre emergent weedicide of Pendimethalin at the rate of one Kg a.i. per ha just before the planting of seeds. Planting or sowing is done at the holes made at 2 m distance.

Installation of drip system is done with main, sub main pipes and the inline lateral tubes are placed at an interval of 1.50 meter (5 Feet). The drippers in lateral tubes are placed at an interval of 60 cm and 50 cm spacing with 4 litres per hour and 3.5 litres per hour capacities respectively.

3. Fertigation

For bitter gourd a dose of 200:100:100 kg NPK/ha is applied throughout the cropping period through split application (Table 3.2). Fertigation is done once in every third day from date of sowing or transplanting.

Table 3.2 Fertigation Schedule for Bitter Gourd and Bottle Gourd under Precision Farming

Stage of the Crop	Duration	Fertilizer Grade	No. of Splits	Quantity (Kg/ Split)
		19-19-19	03	08.80
I Stage	01-10 (3rd, 6th, 9th DAS)	13-00-45	03	03.70
		Urea	03	09.80
		12-61-00	10	01.20
II Stage	11th to 40th DAS	13-00-45	10	06.70
		Urea	10	10.80
		19-19-19	10	04.00
III Stage	41st to 70th DAS	13-00-45	10	02.80
		Urea	10	10.60
		12-61-00	07	01.20
IV Stage	71st to 90 DAS	13-00-45	07	12.70
		Urea	07	14.70

1. Inter Cultivation

Staking is necessary to layout a horizontal spreading surface by providing eucalyptus poles, casuarina poles or bamboo stakes along with wire ropes so that the vine can be spread evenly.

A pandal or single line staking system is normally followed. The seedlings when starts producing tendrils should be staked with poles using banana fibre or threads so as to enable them to reach pandal.

Weeding is carried out at 2-3 times depending upon the weed growth. The earthing up is done to form the beds by 2 foot width and 1 foot height so as to facilitate easy drainage and anchorage to the growing vine. Ethrel spray of 250 ppm is done four times from 15th day after sowing at weekly intervals.

Micronutrients spray is done @ 1 gram per litre of water from 30th, 45th and 60th days after sowing.

Harvest

Harvesting is done from 45-50th day after transplanting or 60-65 days after sowing and harvest is done before the seeds are matured. Harvesting is continued once in 5-7 days and a total number of 8-10 harvests can be made. Average yield of 10-15 tonnes is recorded depending upon the varieties or hybrids.

Plant Protection

a. Pest Management in Bitter Gourd

1. Serpentine Leaf Miner (Liriomyza trifolii)

- Adult females of serpentine leaf miner thrust eggs into the epidermal layer of the leaves
- Minute orange yellow apodous maggots mine epidermal layers of the leaves and cause serpentine mines; drying and drooping of leaves
- The first larval and second stage of the leaf miner burrows into the mesophyll tissue
- The third stage larva concentrates its feeding towards the upper leaf surface
- When it is mature, it cuts a longitudinal slit in the leaf and leaves to pypate on the leaf surface or on the ground

- Damage caused by a single larva is minimal, however when large populations are present they are capable of destroying leaves and affecting the growth of the plants

Management of Serpentine Leaf Miner in Bitter Gourd

- Spray dimethoate 1.50 ml per litre or monocrotophos 1.50 ml per litre or malathion 1.50 ml per litre
- Spray NSKE 5 per cent

1. Aphids (Aphis gossypii)

- Colonies are found on tender shoots
- Suck sap from tissues; Plants remain stunted in growth
- In case of severe infestation plants may completely dry up and die away
- They also produce honey dew which favour the development of sooty mould

Management of Aphids in Bitter Gourd

- Spray dimethoate 1.50 ml per litre or Monocrotophos 1.50 ml per litre
- Release of first instar larvae of green lace wing Chrysoperla carnea @ 1000 per ha

2. White Fly (Bemesia tabaci)

- Eggs are pear shaped light yellowish in color
- On hatching the nymphs, which are oval, scale like and greenish white in color settle down on a succulent spot on leaves
- Adults are minute insects covered completely with white waxy bloom
- Reproduction is mostly oviparous, though parthenogenis also occur
- Nymphs and adults such the sap from the tender leaves
- The affected leaves become yellowish, wrinkle curl downward and are ultimately shed
- The honeydew exuded by these insects favour the development of sooty mould

Management of White Fly in Bitter Gourd

- Install yellow sticky traps smeared with sticky substances to attract and kill insects
- Spray Dimethoate 1.50 ml per litre or Monocrotophos 1.50 ml per litre or Malathion 1.50 ml per litre

3. Pumpkin Beetles (Rhaphidopalpa sp)

- The damaged roots and infested underground portion of stems start rotting due to secondary infestation by saprophytic fungi and the unripe fruits of such wines dry up
- Adult beetles feed voraciously on leaf lamina making irregular holes
- They prefer young seedlings and tender leaves and damage at early stage may even kill the seedlings

Management of Pumpkin Beetles in Bitter Gourd

- Spray Malathion 50 EC 2 ml per litre or Dimethoate 30 EC 2 ml per litre or Methyl Demeton 25 EC 2 ml per litre or Fenthion 100 EC 1 ml per litre

4. Spotted Leaf Beetle or Hadda Beetle (Epilachna Sp)

- Both adults and grubs of spotted leaf beetle cause considerable damage to leaves by scrapping away chlorophyll from epidermal layers of leaves which get skeletonised and gradually dry away
- Adults are active flyers and move about from plant to plant
- Grubs are stationary on the leaves and occur in large numbers
- Adults are hemispherical, pale brown and mottled with black spots
- Females lay elongate spindle shaped yellowish eggs in clusters of 40-50 on the lower surface of leaves
- Grubs are yellowish in color, stout with spines all over the body

Management of Spotted Leaf Beetle in Bitter Gourd

- Collect and destroy damaged leaves and egg masses on leaves
- Shake the plants to dislodge grubs, pupae and adults in a pail of kerosenated water early in the morning
- Spray Malathion 50 EC 2 ml per litre or Dimethoate 30 EC 2 ml per litre or Methyl Demeton 25 EC 2 ml per litre or Fenthion 100 EC 1 ml per litre

5. Blister Beetle (Mylabris pustulata)

- The adult beetles feed on pollen, petals of flowers and flower buds as a result fruit setting is adversely affected

Management of Blister Beetle in Bitter Gourd

- Hand collection and destruction of beetles during early morning hours when the beetles are comparatively less active
- Sparay Malathion 50 EC 2 ml per litre or Dimethoate 30 EC 2 ml per litre or Methyl demeton 25 EC 2 ml per litre or Fenthion 100 EC 1 ml per litre of water

6. Pumpkin Caterpillar (Diaphania indica)

- The eggs are laid on leaves
- On hatching the young caterpillars lacerate and feed on chlorophyll of foliage
- Later they fold and web together leaves and feed within the caterpillars also feed on flowers and bore into young developing fruits
- The caterpillars are elongated in shape and bright green in color, having two narrow longitudinal stripes dorsally
- Adults are medium sized moths and females have tuft of orange colored hair at their anal end

Management of Pumpkin Caterpillar in Bitter Gourd

- In early stages of infestation handpick and mechanically destroy the caterpillars
- Spray Malathion 50 EC 2 ml per litre or Dimethoate 30 EC 2 ml per litre or Methyl demeton 25 EC 2 ml per litre or Fenthion 100 EC 1 ml per litre of water

7. Thrips (Thrips sp)

- Both nymphs and adults feed gregariously on leaves, stems, flowers and fruits
- Feeding results in silvery scars on leaf surfaces especially alongside the midrib and veins
- Heavy infested plants are characterized by silvered or bronzed appearance of leaves and terminals and scarred and deformed fruits

Management of Thrips in Bitter Gourd

- Spray Dimethoate 30 EC 1.50 ml per litre or Monocrotophos 1.50 ml per litre or Malathion 1.50 ml per litre of water

8. Fruit Flies (Bactrocera Sp)

- The adult females punctures fruits by its ovipositor to lay eggs singly or in clusters of 5-12 inside the pulp and when full grown come out of the fruit, loop and unloop their elongated body to take a jump off the fruit, drop on the soil to burrow and pupate 2-8 cm deep
- Damage caused by the ovipositor heals up. No external symptom is left to identify attack of this pest. But as the maggots feed on the pulp the fruits rot and fall down
- Attack by this pest can be identified by oozing of brown resinous fluid from fruits, distorted and malformed fruits and premature dropping of fruits

Management of Fruit Flies in Bitter Gourd

- Collect and destroy the infested fruits and dried leaves
- Expose the pupae by ploughing and turning over soil after harvest
- In epidemic areas change the sowing dates as the fly population is low in hot dry conditions and its peak during rainy season
- Use attractants like citronella oil, eucalyputs oil, vinegar (acetic acid) dextrose and lactic acid to trap flies
- Use poison baiting in severe infestation; saturated sugar solution 5 ml plus Malathion 50 EC 5 ml plus 100 ml of fermented palm juice. Keep the bait in earthern lids placed at various corners of the field
- Use fly trap @ 50 nos per ha; keep 5 gram of wet fish meal in polythene bags with six holes (3mm diameter) two centimetre from the bottom of the bag. Add 0.1 ml of dichlorvos in cotton plug and kept it inside the bag. Dichlorvos should be added every week and fishmeal renewed once in 20 days.
- Spray Malathion 50 EC 1 ml per litre of water

9. Snake Gourd Semilooper (Plusia peponis)

- It is a specific pest on snake gourd
- The caterpillar cuts the edges of leaf lamina, folds it over and feeds from within the leaf roll
- Larva whitish greem, body has black warts of white longitudinal stripes and hump on its anal segment
- Adults are stout, dark brown with shiny brown fore wings

Management of Snake Gourd Semilooper

- Spray Malathion 50 EC 1 ml per litre of water

a. Disease Management in Bitter Gourd

1. Fusarium Wilt (Fusarium oxysporium f.sp.nireum, f.sp.melonis)

- Affected seedlings killed before their emergence
- Root rot noticed on older seedlings
- Plants show complete wilting
- Brown discolorations were seen inside the affected stem
- Whitish or pinkish fungus growth develops on the stem surface

Management of Fusarium Wilt

- Seed treatment with Trichoderma viride 4 gram per kg or Pseudomonas fluorescens 10 gram per kg
- Soil application of Pseudomonas fluorescens 2.5 kg mixed with 50 kg of FYM
- Soil drenching with Carbendazim 0.1 per cent

10. Root Rot (Pythium aphanidermatum, P.ultimum, P.irregulare)

- Dwarfing, yellowing and wilting of plants
- Roots of infected plants appear water soaked and flaccid
- Sunken, dark lesions are seen on fleshy roots
- Crown of the plant become girdled and the top portion withers easily
- Fruits on the ground become rotten

Management of Root Rot in Bitter Gourd

- Seed treatment with Trichoderma viride 4 gram per kg or Pseudomonas fluorescens 10 gram per kg
- Soil application of Pseudomonas fluorescens 2.5 kg mixed with 50 kg of FYM
- Soil drenching with Bordeaux mixture 1 per cent or Copper Oxychloride 0.25 per cent or Metalaxyl plus Mancozeb 0.2 per cent

11. Anthracnose (Collectotrichum lagenarium)

- Spots on the foliage begin as small yellowish or water soaked lesions and later turn to brown color

- Leaves break at the infected region
- Affected fruits become dark, shrivel and die
- Circular black and sunken cankers appear on fruit surface

Management of Anthracnose in Bitter Gourd

- Seed treatment with Thiram 2 gram per kg or Carbendazim 2 g per kg or Mancozeb 2g per kg
- Spraying with Carbendazim 0.1 per cent or Mancozeb 0.2 per cent or Copper oxychloride 0.25 per cent

11. Cercospora Leaf Spot (Cercospora citrulina, C.melonis, C.lagenariae)

- Small circular black spots with grey centre develop on foliage
- Fruits are reduced in size

Management of Cercospora Leaf Spot

- Spraying with Zineb 0.2 per cent or Difenconazole 0.05 per cent

12. Alternaria LeafSpot (Alternaria cucumerina, A.nigrescens, Macrosporium cucumerinum)

- Small circular to irregular spots with concentric rings appear on the leaves
- Leaves with numerous spots curl and shrivel
- An affected fruit become rotten, turn brown and gets mummified

Management of Cercospora Leaf Spot

- Spraying with Mancozeb 0.2 per cent

13. Downy Mildew (Pseudoperonospora cubensis)

- Yellow patches separated by dark green areas occur initially
- Lower surface of the infected area gets covered with faint purplish fungus layer
- Fruits are reduced in Size

Management of Downy Mildew

- Spraying with Metalaxyl plus Mancozeb 0.1 per cent or Chlorothalonil 0.2 per cent or Mancozeb 0.2 per cent or Fosetyl A1 2.20 kg per ha

14. Powdery Mildew (Erysiphe cichoracearum, Sphaerotheca fuliginea)

- White or greyish spots appear on upper surface
- Defoliation of leaves at later stage
- Infected plants produce small fruits

Management of Powdery Mildew

- Spraying with Dinocap 0.2 per cent or Carbendazim 0.1 per cent or Thiophanate Methyl 0.1 per cent or Tridemorph 0.1 per cent

15. Scab (Cladosporium cucumerinum)

- Light water soaked or pale green spots appear on leaves, stem and petiole
- Gradually the spot turns grey to white and become angular
- Veinlets in the affected region turn brown
- Leaves break at the affected region
- Slightly sunken, greyish spots occur on fruits
- Spots become darker and the infected tissue sinks to form a cavity

Management of Scab

- Spraying with Mancozeb 0.2 per cent or Chlorothalonil 0.2 per cent

16. Gummy Stem Blight and Black Spot (Didymella bryoniae)

- Seeds and seedlings rot at the early stage
- Greyish green circular spots appear on leaves and they later change to dark brown
- Canker occurs on stem, petiole and fruit stalk
- Affected stem decays and the plant start wilting
- Small circular dark green water soacked areas appear on fruits
- Lesions become dark as fungus fruiting body matures
- White fluffy mycelium develops on the fruit surface
- Numerous cracks originate from the centre
- Gummy exudates noticed on infected stem

Management of Gummy Stem Blight and Black Spot

- Spraying with Chlorothalonil 0.2 per cent or Mancozeb 0.2 per cent or Maneb 0.2 per cent or Benomyl 0.1 per cent

17. Choanephora Wet Rot (Choanephora cucurbitarum)

- Numerous small "Pins stuck in a pin cushion" like fungus growth develops on the fruit surface
- Initially they are white to brown in color later changes to purplish black
- Affected flowers, pedicels and immature fruits become translucent and rot

Management of Choanephora Wet Rot

- Fruit injuries should be reduced during the harvest

18. Fruit Rot (Pythium aphanidermaturm, Fusarium sp., Phytophthora sp., Rhizoctonia sp)

- Luxuriant woolly mycelia growth occurs on affected fruits
- Dark green water soaked lesions appear on fruits and they later develop into watery soft rots
- Tissues internally become watery and decaying matter emits bad odour

Management of Fruit Rot

- Soil drenching with copper oxychloride 0.25 per cent

19. Diplodia Fruit Rot (Diplodia natalensis)

- Wilting and blighting of stems and leaves
- Water soaked dark green lesions develop on stems and leaves
- Fruit decay appears around the stem
- The rind becomes slightly darken, water soaked and later turns brown
- Centre of the spots turn black, cracks and wrinkles

Management of Diplodia Fruit Rot

- Fruit must be harvested with long stem to prevent infection

20. Angular Leaf Spot (Pseudomonas syringae p. Lachrymans)

- Water soaked lesions develop on leaves

- They turn grey and form an exudates on the lower surface
- Angular holes appear on the spot
- Brown color rot appear on fruits

Management of Angular Leaf Spot

- Spraying with Streptomycin sulphate 0.05 per cent plus Copper oxychloride 0.25 per cent

21. Mosaic (Vec. Aphis craccivora, Myzus persicae)

- Yellow or light green mottling and marginal chlorosis of leaves
- Infected plants stunted in appearance
- Infected fruits mishappened, dwarfed, mottled or spotted

Management of Mosaic

- Removal of Infected plants
- Vector Management

22. Ring Spot (Vec. Xiphinema americanum)

- Yellowish green mottling and malformation of leaves
- Spots develop as haloes having definite pin point centre with water soaked appearance
- Infected plants stunted in appearance
- Fruit set and size are reduced

Management of Ring Spot

- Vector Management

3.1.5 Precision Farming Techniques for Bottle Gourd

Bottle Gourd is an annual vine trailing on the ground. This crop is a very hardy crop grows well in drought conditions. It is grown in wide range of soil with an optimum pH of 6-7.

1. Drip Irrigation System

- Installation of drip system is done with main, sub main pipes and the inline lateral tubes are placed at an interval of 1.50 meter (5 Feet)

- The drippers in lateral tubes are placed at an interval of 60 cm and 50 cm spacing with 4 LPH and 3.5 LPH capacities respectively

2. Field Preparation

- The field selected for planting is free from clots and brought into fine tilth using chisel plough (once), disc plough (once) and cultivator (four times)
- Farm Yard Manure at the rate of 25 tonnes per ha, 75 per cent of total recommended dose of single super phosphate (470 Kg / Ha), Azospirillum and Phosphobacteria each at the rate of 2 kg per ha, Pseudomonas 2.50 kg per ha along with FYM 50 kg and neem cake at the rate of 100 kg is applied before last ploughing
- Four foot width broad beds at one foot height are formed along the laterals using broad bed former

3. Seed Treatment

- A quantity of 3 kg seeds is required to sow one ha area
- The seeds are mixed with 500 gram of Azospirillum and the seeds are shade dried for half an hour
- The seeds are treated with Thiram 2g per kg of seeds or Carbendazim 2 gram per kg of seeds

4. Nursery Raising

- In Hi-tech horticulture the bottle gourd is planted by its 15 days old healthy seedlings raised in shade net houses. The seedlings are raised in protrays having 98 holes or in polythene bags
- Well decomposed cocopeat is used as media. Dibble the seed in each hole at the rate one per hole and keep the trays under shadenet house for about 15 days. In polythene bags pot mixture is used as a growing medium. Regular watering should be done twice in a day. About 15 days old seedlings are transplanted to the main field

5. Planting or Direct Sowing

- The lateral tubes are spread in the centre of each bed and irrigate the beds by operating the drip system continuouisly for 8-12 hours
- Spray pre emergence weedicide (Pendimethalin @ 1 kg a.i./ha just before planting)
- Planting the seedlings or sowing is done in the holes made at 60 cm distance

6. Fertigation

- For bottle gourd, a dose of 200:100:100 kg NPK per ha is applied throughout the cropping period through split application
- The fertigation schedule is followed as given in Bitter gourd is used every third day interval, after transplanting or direct sowing

7. Intercultivation

- Weeding and hoeing is done 2-3 times once in 15 days
- Staking is done during rainy season to avoid rotting of fruits and is used to avoid direct contact of fruits with soil

8. Harvest

- Harvesting is done when the fruits attain marketable size
- It starts from 50th day after sowing to 120 days depending upon the varieties or hybrids
- The average yield is 25 to 30 tonnes per ha

9. Plant Protection

- The plant protection measures given in respect of Bitter Gourd may be practiced for Bottle Gourd also.

3.1. 6 Precision Farming Techniques for Cabbage

1. Quality Seedling Production

Protected Nursery

- The cabbage seedlings are raised in shade net house. A nursery area of 5 cents with slanting slope of 2 per cent is required for the production of seedlings for one ha
- The nursery area is covered with 50 per cent shade net and the sides are covered with 40/50 mesh insect proof nylon net
- The raised beds of one metre width and convenient length are formed inside the nursery and above the beds the HDPV pipes are placed in 2 metre interval

Protrays

- The Protrays of 98 cells (54x27 cm) of 3.50 cm diameter and 0.8 mm thickness is normally used and is ideal for cabbage seedling production

- Around 600 Protrays are required for the production of 58333 seedlings which is required for one hectare at a spacing of 60 x 45 x 45 cm in three row planting

Growing Medium

- The Cocopeat is artificial soil less media and is used for the production of healthy and vigorous vegetable seedlings production
- The sterilized Cocopeat at the rate of 720 kg per ha is mixed with 10 kg of neem cake and Azospirillum and Phosphobacteria each at the rate of one kg. About 1.25 kg of the cocopeat medium is filled for each tray

Seed Treatment

- About 250 gram of hybrid cabbage seed is required for the production of seedlings for one ha. The seeds are treated in hot water at 50 degree Celsius for 30 minutes
- The seeds are mixed with Azospirillum at the rate of 100g per kg and are shade dried for half an hour. About 25 gram of Azospirillum is required for the seed treatment of 150 gram of Cabbage seeds

Sowing of Cabbage Seeds

- Sowing of Cabbage seeds in Protrays is done at the rate of one seed per cell. The seeds are covered with Cocopeat and is kept one over the other (8-10 Nos) and covered with polythene sheet for 5 days or till germination starts
- After 5 days when the seeds are germinated the Protrays are arranged on the raised beds inside the shade net nursery
- Watering is done by rose can every day (twice in a day) upto seed germination
- Panchakavya spray at 3 per cent (30 ml per litre) is done at 15 days after sowing
- 19:19:19 + MN at the rate of 0.5 per cent (5 gram per litre) solution are drenched using rose can or micro nutrient spray of 0.5 per cent is done at 18 days after sowing

The cabbage seedlings are ready for transplanting in 25 days and are transported to the main field through trucks without damage

Drip Irrigation System

- Installation of drip system is done with main, sub main and the inline laterals are placed at the interval of 1.5 meter

- The drippers at the interval of 60 cm and 4 Litres per hour or 50 cm dripper interval and 3.5 litres per hour are placed in the lateral system

Planting System and Population Maintenance in Cabbage

- The main field is ploughed four times using Chisel and Disc each at one time and cultivator two times
- Farm Yard Manure at the rate of 25 tonnes per ha is applied before the last ploughing
- Around 75 per cent of the total recommended dose of Super Phosphate of 588.93 Kg is applied as a basal dose
- Azospirillum and Phosphobacteria each at 2 kg per ha along with FYM 50 kg and Neem Cake of 100 kg is applied before last ploughing
- Raised beds of 120 cm width are formed at an interval of 30 cm and the laterals are placed at the centre of the bed

Planting Techniques

- Drip irrigation at field capacity is done for 8 to 12 hours depending upon the soil moisture level and the soil type
- Pre emergence weedicide spray of Fluchloralin at the rate of one litre a.i. per ha or Trifluralin 0.5 litre per ha through high volume sprayer is done just before transplanting. The soil should be moist while spraying
- Ropes are marked at the interval of 45 cm and planting is done at the spacing of 60x45x45 in the paired row system
- Gap filling is done at 7th day after transplanting

Fertigation Techniques

- Fertigation requirement for cabbage F1 Hybrid is 200:125:150 kg of NPK per ha is applied once in every three days throught the cropping period
- The fertigation schedule for cabbage is given in the Table 3.3

Table 3.3 Fertilizer Required for Cabbage under Precision Farming

Sl.No	Particulars of Fertilizers	Recommended Dose
01	Total Recommended Dose	200:125:150 Kg of NPK/ha
02	Water Soluble Fertilizers	200- 31.25 – 150 kg of NPK/ha
03	Basal Dose 75 per cent of 'P'	93.75 kg of P per ha
Sl.No	**Particulars of Fertilizers**	**Quantity (Kg/Ha)**
01	Super phosphate (75% of P as Basal Dose)	586
02	19:19:19 +MN	066
03	13-00-45	306
04	12-61-00	031
05	Urea	313

Table 3.4 Fertigation Schedule for Cabbage under Precision Farming

Stage of the Crop	Duration	Fertilizer Grade	Number of Splits	Quantity (Kg/ Split)
01	01-10 DAP (3rd, 6th and 9th DAP)	19:19:19	03	11.00
		13-00-45	03	06.50
		Urea	03	08.10
02	11th – 40th DAP	12-61-00	10	01.50
		13-00-45	10	10.00
		Urea	10	09.80
03	41st – 70th DAP	19:19:19	10	03.30
		13-00-45	10	08.60
		Urea	10	09.30
04	71st – 90 DAP	12-61-00	07	02.20
		13-00-45	07	14.30
		Urea	07	14.00

Intercultivation in Cabbage

- Carbofuran @ 33 kg per ha is applied at one week after transplanting to control nematodes
- Foliar spraying with Pseudomonas fluorescens 0.5 per cent (5 gram per litre) at 15 days interval is done for 6 times
- Weeding is done on 30 DAP depending upon the weed growth
- Micro Nutrient Mixture at the rate of 21 gram per litre (0.1%) is sprayed at 40th, 60th and 80th DAP

Harvesting and Post Harvest Management

- Harvesting is done when the heads attain the full size and are firm
- The harvested heads are packed in the collapsible crates and are transported to market without the damage

3.1. 7 Precision Farming Techniques for Cauliflower Quality Seedling Production

Protected Nursery

- The Cauliflower seedlings are raised in shade net house. A nursery area of 5 cents with slanting slope of 2 per cent is required for the production of seedlings for one ha
- The nursery area is covered with 50 per cent shade net and the sides are covered with 40/50 mesh insect proof nylon net
- The raised beds of one metre width and convenient length are formed inside the nursery and above the beds the HDPV pipes are placed in 2 metre interval

Protrays

- The Protrays of 98 cells (54x27 cm) of 3.50 cm diameter and 0.8 mm thickness is normally used and is ideal for Cauliflower seedling production
- Around 600 Protrays are required for the production of 58333 seedlings which is required for one hectare at a spacing of 60 x 45 x 45 cm in three row planting.

Growing Medium

- The Cocopeat is artificial soil less media and is used for the production of healthy and vigorous vegetable seedlings production
- The sterilized Cocopeat at the rate of 720 kg per ha is mixed with 10 kg of neem cake and Azospirillum and Phosphobacteria each at the rate of one kg. About 1.25 kg of the cocopeat medium is filled for each tray

Seed Treatment

- About 250 gram of hybrid Cauliflower seed is required for the production of seedlings for one ha. The seeds are treated in hot water at 50 degree Celsius for 30 minutes
- The seeds are mixed with Azospirillum at the rate of 100g per kg and are shade dried for half an hour. About 25 gram of Azospirillum is required for the seed treatment of 150 gram of Cabbage seeds

Sowing of Cauliflower Seeds

- Sowing of Cauliflower seeds in Protrays is done at the rate of one seed per cell. The seeds are covered with Cocopeat and is kept one over the other (8-10 Nos) and covered with polythene sheet for 5 days or till germination starts
- After 5 days when the seeds are germinated the Protrays are arranged on the raised beds inside the shade net nursery
- Watering is done by rose can every day (twice in a day) upto seed germination
- Panchakavya spray at 3 per cent (30 ml per litre) is done at 15 days after sowing
- 19:19:19 + MN at the rate of 0.5 per cent (5 gram per litre) solution are drenched using rose can or micro nutrient spray of 0.5 per cent is done at 18 days after sowing

The Cauliflower seedlings are ready for transplanting in 25 days and are transported to the main field through trucks without damage

Drip Irrigation System

- Installation of drip system is done with main, sub main and the inline laterals are placed at the interval of 1.5 meter
- The drippers at the interval of 60 cm and 4 Litres per hour or 50 cm dripper interval and 3.5 litres per hour are placed in the lateral system

Planting System and Population Maintenance in Cauliflower

- The main field is ploughed four times using Chisel and Disc each at one time and cultivator two times
- Farm Yard Manure at the rate of 25 tonnes per ha is applied before the last ploughing
- Around 75 per cent of the total recommended dose of Super Phosphate of 588.93 Kg is applied as a basal dose
- Azospirillum and Phosphobacteria each at 2 kg per ha along with FYM 50 kg and Neem Cake of 100 kg is applied before last ploughing
- Raised beds of 120 cm width are formed at an interval of 30 cm and the laterals are placed at the centre of the bed

Planting Techniques

- Drip irrigation at field capacity is done for 8 to 12 hours depending upon the soil moisture level and the soil type
- Pre emergence weedicide spray of Fluchloralin at the rate of one litre a.i. per ha or Trifluralin 0.5 litre per ha through high volume sprayer is done just before transplanting. The soil should be moist while spraying
- Ropes are marked at the interval of 45 cm and planting is done at the spacing of 60x45x45 in the paired row system
- 40 days old seedlings are dipped in Pseudomonas fluorescens 0.5 Per Cent for 30 minutes and are transplanted in the main field
- Gap filling is done at 7th day after transplanting

Fertigation Techniques

- Fertigation requirement for Cauliflower F1 Hybrid is 200:125:125 kg of NPK per ha is applied once in every three days throught the cropping period
- The fertigation schedule for cauliflower is given below (Table 3.5 & 3.6)

Table 3.5 Fertilizer required for Cauliflower under Precision Farming

Sl.No	Particulars of Fertilizers	Recommended Dose
01	Total Recommended Dose	200:125:125 Kg of NPK/ha
02	Water Soluble Fertilizers	200- 31.25 – 125 kg of NPK/ ha
03	Basal Dose 75 per cent of 'P'	93.75 kg of P per ha
Sl.No	**Particulars of Fertilizers**	**Quantity (Kg/Ha)**
01	Super phosphate (75% of P as Basal Dose)	586
02	19:19:19 +MN	082
03	13-00-45	243
04	12-61-00	026
05	Urea	326

Super Phosphate at the rate of 586 kg per ha is applied during last ploughing as a basal dose

Table 3.6 Fertigation Schedule for Cauliflower under Precision Farming

Stage of the Crop	Duration	Fertilizer Grade	Number of Splits	Quantity (Kg/ Split)
01	01-10 DAP (3rd, 6th and 9th DAP)	19:19:19	03	16.50
		13-00-45	03	02.30
		Urea	03	07.00
02	11th – 35th DAP	12-61-00	10	03.20
		13-00-45	10	14.00
		Urea	10	22.40
03	36th – 70th DAP	19:19:19	10	02.70
		13-00-45	10	10.40
		Urea	10	10.40

Intercultivation in Cauliflower

- Foliar spraying with Pseudomonas fluorescens 0.5 per cent (5 gram per litre) at 15 days interval is done for 6 times
- Weeding is done on 30 DAP depending upon the weed growth
- Micro Nutrient Mixture at the rate of 250 grams per ha (0.1%) is sprayed at 40th, 60th and 80th DAP

Harvesting and Post Harvest Management

- Harvesting is done when the heads curds attain the full size and are firm
- The harvested curds are packed in the collapsible crates and are transported to market without the damage

Pest Management in Cabbage and Cauliflower

1. Diamond Back Moth (Plutella xylostella)

- Diamond back moth is a major pest of all cruciferous vegetable crops and is of cosmopolitan in distribution. Adult is a small greyish moth having pale whitish narrow wings with inner margins yellow and three pale whitish triangular markings on hind margins on each forewing. At rest dorsal median patch of three diamonds shaped yellowish white spots clearly visible by joining both forewings. Hind wings have a fringe of long fine hairs. The eggs are oval and cylindrical, light lemon yellow and have a crinkly, sculptured surface. They are laid singly, usually on the under surface of the leaf. A single female may lay about 40-60 eggs. Larvae are 8-12 mm long pale, yellowish green in color, pointed at both ends with fine erect black hairs scattered over the body.
- The first instar larvae mine the epidermal surface of leaves producing typical white patches. Larvae, second instar onwards feed externally making holes on the leaves and soil them with excreta. Heavy infestations leave little more than the leaf veins. Pupation is inside the silken cocoon on the plant itself. It is active in cold season and completes 8-12 broods in a year.

Management of Diamond Back Moth

- Grow mustard as intercrop at 20:1 ratio to attract diamond back moths for oviposition. Periodically spray the mustard crop with insecticide to avoid the dispersal of the larvae

- Install pheromone traps at 12 per ha
- Spray Cartap hydrochloride at the rate of one gram per litre of water or Bacillus thuringiensis 2 gram per litre at primoridial stage (ETL 2 larvae / plant)
- Spray NSKE 5 per cent after primordial stage
- Release parasite Diadegma semiclausum at 50000 per ha, 60 days after planting
- Remove alternative hosts and crop residues to disrupt the development of diamond back moth
- Regulate irrigation with sprinkler to disrupt oviposition and dislodge the first instar larvae

2. Leaf Webber (Crocidolomia binotalis)

- It is a very serious pest of Cabbage and other Crucifers and commonly called as Cabbage cluster caterpillars. Eggs are laid in clusters of 40 – 100 on leaf surface. Young larvae feed gregariously and later web the leaves together and feed within the web. Due to gregarious feeding rotting of cabbage heads and cauliflower curds are common. Larvae are 18-20 mm in length and are pale violet brown with dorsal stripe. It pupates in a cocoon within the webbed up leaves and flowers or just below ground surface.

Management of Leaf Webber

- Remove webbed leaves and kill the larvae within
- Use light trap to attract and kill adults
- Spray phosalone 35 EC 2 ml per litre or Endosulfan 2 ml per litre or Cartap 50 WP 1 gram per litre of water
- Encourage the activitiy of Apanteles crocidolomiae

3. Cabbage Stem Borer (Hellula undalis)

- This is sporadic but occasionally serious and aborts head formation. Eggs are laid singly on the ventral side of leaves. Caterpillars first mine the leaves later feed on the leaves, shoots sheltered within silken passage and finally bore into the stems. They prevent head initiation causing multiple shoots or heads. Eggs are oval in shape and pinkish in color. Larvae are pale whitish brown in color with 4-5 purplish brown longitudinal stripes

Management of Stem Borer

- Collect and destroy larvae mechanically in the early stages of attack
- Spray phossalone 35 EC 2 ml per litre of water or endosulfan 2 ml per litre or Cartap 50 WP one gram per litre of water

4. Cabbage Butterfly (Pieris brassicae, Pieris rapae)

- This is a oligophagous pest having a limited host range of cruciferous crops. Eggs are flask shaped, yellowish in colour and laid in clusters on ventral side of leaves. Larvae are velvety bluish green in color with black spots and yellow dorsal and lateral stripes covered with white filaments. Larvae are gregarious in early stages and disperse as they grow. They skeletonize leaves and sometimes bore into heads of Cabbaage and Cauliflower. Adult is snow white butterfly having black markings on wings

Management of Cabbage Butterfly

- Collect and destroy the larvae mechanically in the early stages of attack
- Spray dichlorvos 2 ml per litre or endosulfan 2 ml per litre of water

5. Aphids (Brevicoryne brassicae, Lipaphis erysimi, Myzus persicae)

- Aphids are polyphagous soft bodied pear shaped insects feed on Cabbage and Cauliflower. The Cabbage aphid, Brevicoryne brassicae green peach aphid, Myzus persicae and the Mustard aphid Lipaphis erysimi are found on cole crops worldwide colonies of aphid are found on tender shoots and suck sap from plant tissues. In case of severe infestation plants may completely dry up and die away. Feeding damage from large numbers of aphids can kill seedlings and young transplants. On larger plants, feeding damage results in curling and yellowing leaves, stunted plant growth, and deformed heads. Contamination by dead aphids in the head or wrapper leaves can also be a problem. Dead aphids do not wash off easily and will cause a head to be unsuitable for fresh market sales. They also produce honew dew which favor the development of sooty mould. Cloudy and moist weather favor rapid multiplication

Management of Aphids in Cabbage and Cauliflower

- Remove and destroy damaged plant parts along with nymphs and adults
- Use yellow sticky traps 12 per ha to attract winged aphids

- Encourage the activity of Coccinellids, Chrysopids and Syrphids
- Spray Dimethoate 2 ml per litre or Methyl Dematon 2 ml per litre of water. Add wetting agent in spray fluid

6. Cut Worm (Agrotis spp)

- Cut worms are minor pests and commonly found in nurseries or at the time of transplanting. Eggs are laid in soil. Larvae are nocturnal in habit, cut seedlings at ground level and feed on tender parts, and destroy far more than they actually feed. Larvae are black colored with brown head and triangular spots at the spiracular area

Management of Cut Worm in Cabbage and Cauliflower

- Irrigate the field copiously and regulary
- Set up light trap in summer months
- Apply Chlorpyriphos 2 ml per litre in the Collar region during evening hours

7. Mustard Sawfly (Athalia lugens proxima)

- Adults of saw fly lay eggs singly along leaf margins by slitting the margin with the help of its sharp saw like ovipositor. The grub is green in color cylindrical and resembles a green caterpillar. The grubs feed on the leaves cutting small holes into lamina and ultimately leaves become skeletonized

Management of Mustard Sawfly

- Hand pick and destroy the grubs
- Spray dichlorvos 2 ml per litre or Quinalphos 2 ml per litre or Endosulfan 2 ml per litre of water

8. Painted Bug (Bagrada cruciferarum)

- It is a medium sized penatatomid bug with a black orange color markings with a large scutellum at least half as long as abdomen. Both nymphs and adults suck the sap from the leaves / shoots. The infested plants wilt and die. Nymphs almost resemble the adults but for the size and wings. Barrel shaped yellow colored eggs are laid on leaves stems and inflorescence. Freshly emerged nymphs are bright orange in color with bright red eyes

Management of Painted Bug

- Spray Malathion 50 EC 2 ml per litre of water

9. Flea Beetles (Phyllotreta downsei, Chaetocnema basalis)

- Adult flea beetles bite holes on foliage. Phyllotreta downsei is a small bluish black beetle with enlarged hind femur. Chaetocnema basalis is a tiny beetle 1.5-2.0 mm long ovate shiny black in color

Management of Flea Beetles

- Spray Malathion 50 EC 2 ml per litre of water

Disease Management in Cabbage

- The seeds are treated with bio control agent Pseudomonas fluorescens at the rate of 10 gram per kg of seed before sowing
- Pseudomonas fluorescens at the rate of 2.5 kg per ha along with 50 kg of Farm Yard Manure and applied before last ploughing
- Around 25 days old seedlings are dipped in Pseudomonas fluorescens 0.5 per cent for 30 minutes and are transplanted in the main field
- If alternaris leaf spot is noticed, foliar spray with Copper oxychloride 0.5 per cent or Captab 0.2 per cent or Mancozeb 0.2 per cent is done
- Metalaxyl + Mancozeb 0.1 per cent is applied as foliar spray to control downy mildew incidence in Cabbage
- If Black rot incidence occurs, Agrimycin 100 (0.1%) + Copper oxychloride 1.25 kg per ha is applied as foliar spray

Disease Management in Cauliflower

- The seeds are treated with bio control agent Pseudomonas fluorescens at the rate of 10 gram per kg of seed before sowing
- Pseudomonas fluorescens at the rate of 2.5 kg per ha along with 50 kg of Farm Yard Manure aand applied before last ploughing
- Around 25 days old seedlings are dipped in Pseudomonas fluorescens 0.5 per cent for 30 minutes and are transplanted in the main field
- If Black rot incidence occurs, Streptomycin 100 (0.1%) + Copper oxychloride 1.25 kg per ha is applied as foliar spray
- If Cauliflower mosaic virus symptom is noticed vector control (Aphids) is done

3.1. 8 PRECISION FARMING TECHNIQUES FOR TURMERIC

Ruling Varieties in Turmeric

- CO – 1
- BSR – 1
- BSR – 2
- ROMA
- SWARNA
- SUDHARSANA

Soil and Climate

- Turmeric is a rhizomatous crop prefers a friable well drained red loamy soil. The best season for sowing of rhizomes is May – June
- The growth of turmeric is good when the soil pH is 6 – 7. In saline soil or acidic soil the major and micro nutrients availability is poor so that the plants cannot absorb nutrients. Hence the nutrients deficiency symptoms are acute.

Drip Irrigation Systems

- Installation of drip system is done with main, sub main pipes and the inline lateral tubes are placed at an interval of 1.5 meter (5 feet).
- The drippers in lateral tubes are placed at an interval of 60 cm and 50 cm spacing with 4 litres per hour and 3.5 litres per hour capacities respectively

Field Preparation

- The main field is ploughed four times with Chisel and disc plough each one time and cultivator twice
- Farm Yard Manure at the rate of 25 tonnes per ha is applied before the last ploughing
- Around 75 per cent of the total recommended dose of Super Phosphate of 281 Kg is applied as a basal dose
- Azospirillum and Phosphobacteria each at 2 kg per ha along with FYM 50 kg and Neem Cake of 100 kg is applied before last ploughing

- Raised beds of 120 cm width are formed at an interval of 30 cm and the laterals are placed at the centre of the bed
- The beds are wetted for 8 – 12 hours through drip irrigation depending upon the soil moisture level

Rhizome Treatment and Planting

- Mother rhizome of about 2000 kg of rhizomes is required per ha
- The rhizomes are treated with 10 kg each of Azospirillum and Phosphobacteria
- The rhizomes are treated with Carbendazim for 30 minutes before planting
- The treated rhizomes are planted in the raised beds in three rows spaced at 45 cm. The spacing between the plants within the row is 15 cm.

Fertigation in Turmeric

- Fertigation is done for the recommended dose of 150:60:108 kg of NPK per ha is applied once in every three days throught the cropping period
- Around 75 per cent of the recommended dose of the Phosphorous (281 kg/ka) is applied as basal dose
- The fertigation schedule for turmeric is given below Table 3.7 & Table 3.8.

Table 3.7 Fertilizer Required for Turmeric under Precision Farming

Sl.No	Particulars of Fertilizers	Recommended Dose (Quantity (Kg/Ha)
01	Total Recommended Dose	150:60:108 Kg of NPK/ha
02	Water Soluble Fertilizers	150 - 15 – 125 kg of NPK/ha
03	75 per cent of 'P' as a Basal Dose	45 kg of P per ha
S.No	**Particulars of Fertilizers**	**Quantity (Kg/Ha)**
01	Super phosphate (75% of P as Basal Dose)	281
02	19:19:19 +MN	032
03	13-00-45	065
04	12-61-00	015
05	00 – 00 – 50	145
06	Urea	213

Super Phosphate at the rate of 281 kg per ha is applied during last ploughing as a basal dose

Table 3.8 Fertigation Schedule for Turmeric under Precision Farming

Stage of the Crop	Duration	Fertilizer Grade	Number of Splits	Quantity (Kg/Split)
		19:19:19	03	5.30
01	01-15 DAP (5th, 10th and 15th DAP)	13-00-45	03	5.80
		Urea	03	12.50
		12-61-00	12	00.60
02	16th – 75th DAP	13-00-45	12	04.00
		Urea	12	01.70
		12-61-00	12	00.60
03	76th – 135th DAP	00-00-50	12	05.40
		Urea	12	08.00
		19:19:19	21	00.80
04	136 – 240 DAP	00-00-50	21	03.80
		Urea	21	02.80

Intercultivation in Turmeric

- Weeding is done on 30 DAP depending upon the weed growth
- The raised beds are earthed up twice on 60th and 120th DAP
- Micro Nutrient Mixture at the rate of 375 gram each FeSo4, ZnSo4, Borax and Urea dissolved in 250 litre of water. Spraying is done thrice at 30th, 60th and 90th DAP

Harvesting

- Lodging of plants, yellowing and drying of leaves are the signs of maturity. It will start from nine months onwards
- The rhizomes are pulling out without any damage and stored the seed rhizome in sand media under partial shade
- For one ha, 7-9 tonnes of processed turmeric are obtained as yield

Plant Protection Measures for Turmeric

Pest Management in Turmeric

1. Rhizome Scale (Aspidiotus hartii)

- Minute, circular, light brownish to grey with a thin, pale membrane, yellowish to deep yellowish round insect inside the scale
- Plants become weak, pale and withered in the field, white coloured scales seen scattered on rhizomes and growing buds, shrivelling of rhizomes and buds.

Management of Rhizome Scale

- Soil drenching dimethoate 30 EC 2 ml per litre of water
- Soak the seed rhizomes in insecticide solution of either dimethoate 30 EC 1.75 ml per litre of water. Phosalone 35 EC 1.5 ml per litre of water, monocrotophos 36 WSC 1.5 ml per litre or dichlorvos 76 WSC 0.5 ml per litre for 15 minutes before storing or sowing

2. Thrips (Panchaetothrips indicus)

- Thrips asre small, slender and fast moving insects with fringed wings and the nymphs red yellowish in color
- They use their rasping and sucking mouthparts to scrape the leaf surface and suck up the exuding plant juices
- Tender foliage becomes spotted, pale and silvery streaks appear on the affected plants
- Females have a saw like structure that helps to make an incision in plant tissue for egg laying
- The nymphs grow in size and pass through four distinct stages, the last two of which are spent in the soil
- The females are able to reproduce without mating and therefore are not dependent on the presence of males of the species
- Consequently increases in the thrips population can occur very rapidly especially during the periods of hot dry weather

Management of Thrips

- Install yellow sticky traps smeared with sticky substances to attract and kill the insects
- Release first instar larvae of green lace wing Chrysosperla carnea at the rate of 10000 per ha and encourage other coccinelid predators
- Spray methyl demeton 25 EC 2 ml per litre or Dimethoate 30 EC 2 ml per litre or Endosulfan 2 ml per litre to control vector thrips
- Leaf Roller (Skipper) (Udaspes folus)
- Larva smooth, green with black head constricted neck
- Adult brownish black butterfly with eight white spots of different sizes on each forewing and a large white patch on each hind wing
- Leaves become folded or rolled longitudinally and defoliated; presence of greenish larva inside

Management of Leaf Roller

- Dimethoate 30 EC 2 ml per litre of water, Endosuslfan 35 EC 2 ml per litre, Quinalphos 25 EC 2 ml per litre of water

3. Rhizome Maggot (Formosina flavipes)

- Rhizomes and root tunnelled extensively rotting of rhizomes
- Methyl parathion – 50 EC one ml per litre

4. Shoot Borer (Dichocrocis punctiferalis)

- Larva pale greenish with pinkish tinge and fine hairs with dark head and prothoracic shield
- Adult medium sized moth, pale yellowish with small, black spots on the wings
- Pseudostem with holes plugged with excreta, dead heart panicles and spikes dry up above the point of infestation.

Management of Shoot Borer

- Dimethoate 30 EC one ml per litre or Malathion 50 EC one ml per litre of water

Disease Management in Turmeric

Rhizome Rot (Pythium aphanidermatum; P.graninicolum)

- Gradual drying of leaves along the margin
- Basal portion of shoot appeaers watery and soft
- Rhizome decomposes and turn into a decaying mass of tissue
- Rhizome become soft and turn brown

Management of Rhizome Rot

- Seed rhizome dipped in Zineb 0.3 per cent or Copper oxychloride 0.3 per cent or Metalaxyl 0.25 per cent for 30 minutes
- Soil drenching with Mancozeb 0.3 per cent or Captan 0.2 per cent or Copper Oxychloride 0.3 per cent or Bordeaux mixture one per cent or Metalxyl 0.25 per cent

Leaf Spot in Turmeric (Collectotrichum capscici)

- Elliptic or oblong spots develop on both the surface of the leaves
- Spots become greyish with numerous black acervuli arranged in concentric rings
- Yellow halo develops around the spots

Management of Leaf Spot

- Foliaar spraying with Zineb 0.25 per cent, Carbendazim 0.1 per cent or Copper chloride 0.25 per cent

Leaf Blotch (Taphrina maculans)

- Reddish brown spots appear on both the leaf surface. They develop into irregular big patches

Management of Leaf Blotch

- Spraying witht Bordeaux mixture one per cent or Copper oxychloride 0.25 per cent or Zineb 0.2 per cent or Mancozeb 0.2 per cent.

TNAU Experience

TNAU took an initiative to implement the turnkey project on precision farming during 2004-2007 in the farmers' field in Dharmapuri and Krishnagiri districts of Tamil Nadu in an area of 400 ha. Fertigation unit and water soluable fertilizers were distributed to farmers at free of cost.

Table 3.9 Yield of Major Vegetables under Normal Cultivation and Precision Farming

		Yield (Tonnes per ha)		
Sl.No.	**Crop**	**National Average**	**Precesion Farming**	**Percentage Increase in Precision Farming**
1.	Tomato	17.35	150	764.55
2.	Chilli	12.02	35	197.18
3.	Brinjal	10.46	200	1812.00
4.	Bhendi	6.28	16	154.78
5.	Tapioca	25.52	52	103.76
6.	Water Melon	12.71	60	372.07
7.	Musk melon	21.95	40	82.23
8.	Onion	11.32	21	85.51
9.	Cabbage	14.38	120	734.49
10.	Cauliflower	14.22	33	132.07
11.	Pumpkin	11.91	50	319.82
12.	Bitter Gourd	6.23	15	140.77
13.	Ridge Gourd	15.85	34	114.51
14.	Bottle Gourd	12.21	66	440.54
15.	Cucumber	6.48	20	208.64
16.	French beans	5.8	12	106.90
17.	Beet root	16.75	35	108.95
Average		13.03	56.41	332.94

Yield Improvement

As water and fertilizers were applied evenly to all the crops through fertigation, 25 to 50 per cent higher yield was registered. However, Table 3.9 revealed that the precision practices on an average could generate 333 present higher yield than the National average yield. Some of the examples are presented elsewhere.

Quality Improvement

By precision farming, 90 per cent of the harvested produces were of first grade. The produces were of good quality, size, weight, colour, attractiveness and taste and hence it fetched higher price in the market.

Water and Nutrients Saving

Nutrients and water were applied near the active root zone through fertigation which resulted in greater absorption by the crops. Fertilizer use efficiency through fertigation ranged between 80-90 per cent, which helped to save a minimum of 25 per cent of nutrients. Water was saved upto 40 per cent.

Extended Duration

The duration of the crop was extended by 2-3 months by ths precision farming system. Due to this reason, price fluctuation in the market was managed effectively.

SUMMARY

This chapter has given a highlight on Precision Farming Technology for different Principal vegetable crops. The technology has incorporated the needs of production which comprises preaparation of land, seeds, nursery establishment, planting, manuring, irrigation methodologies, plant protection and crop management techniques in detail for the benefit of farmers and the large scale producers of vegetables. Besides, the yield was also prescribed but it is not compared with the traditional way of producing the vegetables. However, the same has been presented in the ensuing chapter in which the non adopters of precision farming technology versus the adopters of precision technology was discussed and compared and highlighted the economic incentives ahead in respect of precision farming technology.

References

1. Mandal Debashis and Ghosh K.S., 2000, Precision Farming – The Emerging Concept of Agriculture for Today and Tomorrow, Current Science, 79 (12): Pp. 12 – 25.
2. Srinivasan, 2001, Precision farming in Asia: Progress and prospects, Geospatial Analysis Center, Regional Science Institute, Hokkaido, Japan. September.
3. http://www.aces.edu/anr/precisionag/GPS.php
4. http://www.smart-fertilizer.com/articles/drip-irrigation

CHAPTER IV

Methodological Approaches to Economics of Precision Farming

Dr. C. Sekhar and Dr. A. Vidhyavathi

Objectives

- To introduce the students on the applicability aspects of precision farming technologies to different sectors of agriculture
- To explain the Methodologies prevalent for appraisal of the Precision Technologies in India

Organization of the Chapter

- Introduction
- Application of Precision Farming in Different Sectors
- Economic Analysis of Precision Farming
- Methods and Tools of Economic Analysis
- Summary

4.0 INTRODUCTION

Precision farming or precision agriculture is one of the term currently used by the scientists of both biological and social disciplines. It is one of the word mostly used in agriculture and allied activities or allied disciplines to reap highest productivity with less input application after carefully following the prescriptions specific to technology which are recommended by the scientific forum for adoption in the field. It varies from sector to sector in agriculture. The allied sectors are Agriculture, Animal Husbandry, Dairying, Forestry, Fisheries and Horticulture. Let us see the specific technologies which are applicable to the concept of "Precision" in agriculture and allied sectors so that the students can have better understanding and differential concepts applied to different sectors and its outturn.

4.1 APPLICATION OF PRECISION FARMING IN DIFFERENT SECTORS

4.1.1 Agriculture

In Agriculture, the concept "Precision Agriculture" is mostly applicable to System of Rice Intensification, wherein, set of recommended technologies or practices to be carried out by the rice growers periodically without permitting elapses in time and quantum of input application and care. On doing so, the rice yield or productivity of rice is expected to be higher than the traditional practice of rice cultivation which could generate around 30 per cent better and higher yield over the conventional rice cultivation methods.

4.1.2 Animal Husbandry and Dairying

In Animal Husbandry sector, the concept of precision is applicable to Dairying, wherein, the prescribed quantum of input is fed to the animal for securing indented quantity of Milk in a day. To gain the desired meat quality, the culling animals are fed with the recommended food materials periodically and the quality of meat will be tested in the laboratories for gaining higher return per animal. Similarly, the principles are applied in poultry through contract farming approach. There are many success stories in contract farming with special reference to poultry. The case of Suguna Poultry Farms private limited, Coimbatore.

4.1.3 Forestry

The principle of "precision forestry" is that it uses modern tools and technology to get as much real information as it is possible to improve decision making process and to ensure current goals of forest management to regulate planting and felling.The best known and most frequently used tools of modern technology are remote sensing, navigation systems and geographical information systems. New trends are decision support systems and tools for tree identification and tools for wood material testing and measurement.

4.1.4 Fisheries

In fisheries, the practices of "Precision Aqua Farming" is practiced by regulating the resource management systems both in Inland and brackish water environment besides regulating or feeding regularly with the prescribed or recommended feed rate for the species chosen for aquaculture practices. Commonly this precision aqua farming is practiced for shrimp production. The resource management systems are well defined and monitored using the equipment like "***Xpercount***" which depicts the growth and number of aquatic population regularly. The concept of Precision Aqua Farming is being practiced following the concept of "Low Investment Higher Income".

4.1.5 Horticulture

The Concept of Precision Farming was widely practiced in respect of horticulture crops like vegetables. In Tamil Nadu, this was implemented as a 'turnkey project' in the District of Dharmapuri, one of the backward district. By determining the input per crop and regulation of resource on time, the technology could generate highest productivity in the dry tracts of Dharmapuri and becomes popular and then this technology could spread to other districts for adoption. Currently, the technology is widely in practice among the farm households with the technology advocacy from the technocrats of Tamil Nadu Agricultural University and the Department of Horticulture and Plantation Crops. The case studies in respect of precision technologies were appraised for its cost effectiveness with the traditional practices and are being documented in the forthcoming chapter on "Economics of Precision Farming".

Such an important technology and its effectiveness are to be assessed or appraised using the tools of economics to study or highlight the economic incentives associated with it. For that what type of tools or methods have to be employed is though known to many economists, the author feels that furnishing such methodologies as a separate chapter will be a guiding tool to the readers and students whom are pursuing higher education in agriculture and allied sectors. This book will serve as a hand book for the personnel engaged in transfer of agricultural technologies.

4.2 ECONOMIC ANALYSIS OF PRECISION FARMING

4.2.1 What is Economics?

Economics is the study of how people make decisions. A common economic decision that farmers make is whether doing something will return more than it costs. This decision is made in a business situation, where returns are measured in money terms or in any other currency term.

Economics is easy to explain in a business setting because one can use money as measures. But people make economic decisions happen every day, whether they are in business or not. For example, deciding how much nitrogen to put on your brinjal or maize this year, whether to buy a cosmetic or go out for a dinner, or whether to attend a farmer's meeting or spend time in dairying are all economic decisions. The concepts may be harder to grasp when costs and returns are not measured in money terms, but the concepts are still the same.

Economic decisions involve allocating resources to reach a goal. The goal may be to make as much money as you can, or to achieve highest productivity to secure award winner in the county or in the State during the current year. Whatever be the goal, the basic economic decision framework is the same:

- What are the costs?
- What are the appropriate resource mix advisable to the farming?
- During what time period one has to wait to generate return?
- What type of Effort that one should make in the production process?
- At what frequency, the input application and care has to be made?
- What quantum of benefits the producer is going to generate?
- Which marketing effort will fetch highest return per unit and
- What will be the price expected per unit of the output in the niche market?

These questions have to be answered. This section explores a few basic concepts from economics and shows how these concepts are used in decision making. Situations from agriculture are used to illustratethe applications of these concepts.

Economics is a social science that studies how people allocate their limited resources. If resources are not limited, then one need not have to make economic decision on how to allocate them. For example, air is not limited, so it is not necessary to decide how many breaths to take in the morning and many to save for your afternoon outdoor activities. Time, on the other hand, is very limited. The more time you spend working, the less time is available for outdoor non-agricultural activities. Money is also limited. The more you spend on food, the less is available for other goods and services. Hence the concept of allocation becomes important.

4.3 METHODS OF ECONOMIC ANALYSIS IN PRECISION FARMING

4.3.1 Ex-Ante and Ex-Post Analysis

Any technological evaluation for its efficacy, one should know and adapt the Ex-ante and Ex-post evaluation so that the technology is a viable one or not, one can assess. The main recommendations for the improvement of the methodology for ex-ante and ex-post evaluation of agricultural or irrigation projects which are operated in the Government of Tamil Nadu with the support of World Bank are summarised in the following 10 points. This approach is also can best be used in evaluating the technology like precision farming and its economic contribution to the farmer and the society as a whole.

1. Consider the Project Cycle

The "project cycle" has different stages like (problem identification, project preparation or formulation incorporating technical; institutional or managerial; social; commercial; financial and environmental aspects projects Appraisal, project implementation and project evaluation) from preliminary appraisal to ex-post evaluation. In the appraisal stage, one has to follow the comprehensive approach of evaluating the technology or the project which is in consideration. In this comprehensive approach, the ex-post evaluation should enable the ex-ante evaluation procedure to be fine-tuned through an ongoing feed-back process between the operating results of existing infrastructures and the assumptions used to evaluate new capital expenditure decisions.

2. Dynamic Approach to Ex-Ante Appraisal

The time dimension of the decision process must be integrated, managed and controlled. Consequently, appraisal cannot be made once and for all, but must adapt to such different stages, with more detailed analysis when the project or the technology is well defined. This called for the establishment of a clear design of the decision-making process for large irrigation projects like Tamil Nadu Irrigated Agricuture Modernization and Water Bodies Restoration and Management (TN IAM WARM) Project of Tamil Nadu being operated in several river sub basins with the aim of introducing SRI (System of Rice Intensification) technology for reaping higher productivity of rice in particular. Similarly other components have also been designed and refined in the steering committees.

3. Environmental and Social Impact Analysis.

The capability of Environmental and Social Impact Assessment (ESIA), when carried out at the very beginning of the project or technology appraisal, to influence the project technical solutions adopted in order to minimise the environmental and social costs to the society, as well as to organise the public debate on the technological impact and contribute to reach the consensus that should be extended to the whole assessment procedure of the technology in question or in usage recommendation.

4. Quantitative Risk Analysis.

A quantitative risk modelling a subsequent plan for risk management and the risk elimination are of paramount importance for improving the performance of Precision Technology projects approved by the Government of Tamil Nadu. Since the definition of probability distribution is the cornerstone of a meaningful risk analysis and, at the same time, is often the most challenging task, it is worth exploring possibilities to address this issue. Reference forecasting is suggested as an important tool for generating sound probability distributions. Here one should know what is Reference Forecasting? Which is presented below for better understanding.

4.1 Reference class Forecasting or Comparison class Forecasting is a method of predicting the future by looking at similar past situations and their outcomes. Reference class forecasting predicts the outcome of a planned action based on the actual outcomes in a reference class of similar actions to that being forecast. In the case of irrigation projects, or the technology projects, the experiences of precision farming implemented in other states or other developing countries can be taken as examples and the principal variables may be taken as a base for future evaluation in the context of Tamil Nadu

5. Monitoring the Project or Technology Development

In the progressive approach recommended along the whole project cycle, the continuous project or the technology implementation and its monitoring represents the starting point for a successful project cycle approach. The monitoring process should look at all factors that may potentially affect the project feasibility. This means that the checklist of the "monitoring" system that must be put in place should include not only investment costs, but also the socio-economic aspects of implementation, the extent of utility, the expected impact with respect to the demand for and supply of water and other nutrients while implementing the project and the environmental impact of the project.

6. Risk Management and Risk Elimination Strategies.

Scope of the "Risk management and Mitigation" is to identify possible strategic answers to the more critical items identified through the risk analysis. One should forecast the possible risks while implementation of the project or recommend risk planning to identify how the various risks can be managed and mitigated and one should identify the suitable person for managing such risks in the production environment or implementation scenario.

7. Systematically Perform Ex-Post Evaluation.

The Project Manager should recommend a systematic use of Ex-post evaluation on the lines of Ex-ante evaluation based on the base line information considered. The ex-post evaluation increases transparency by giving evidence to the effectiveness of the investments in relation to the reached ***Technical, Institutional, Social, Commercial, Financial, Economical and Environmental (TISCOFEE)*** objectives; provides elements to improve the ex-ante assessments of future interventions; collects relevant information about past projects to be used as reference class forecasting; and finally, by giving publicity to the real achievements of the projects, provides incentives for better and more accurate ex-ante analysis. Concerning ex-post evaluation one should recommend the following:

- Start the planning of the evaluation process at the stage of the project design together with the definition of the analytical framework;
- Maintain all the project documentation available;
- Identify the "after-without" the scheme to be compared with the "after" opening;
- Identify and quantify the discrepancies between the ex-ante appraisal and the ex-post results and assess as far as possible what caused the discrepancies, distinguishing between endogenous or exogenous factors.

- Concerning costs evaluation and undue optimism, a specific requirement of ex-post evaluations should be to investigate the methods used to obtain costs and the reasons behind the divergences between expected and actual costs.

8. Use Harmonised Models and Data.

Different infrastructure projects like irrigation, it should be made comparable on the basis of the assumptions set forth at least from the point of view of international projects or the technology. Given the relevance of demand forecast, recommend to define a common platform for the modelling tools supporting projects demand projections. A second area of harmonization is the availability of databases. Data plays a key role for projects appraisal as well for models. With a harmonised database, the evaluation framework at regional / national level could be considerably reinforced and become a common reference framework for developing partner countries.

9. Make Maximum use of Evaluation Results.

First of all, project manager recommends that presentation and dissemination of the outcome of the evaluation are as wide as possible, for two main reasons:

i. Democracy, and ii. By giving publicity to the real achievements of the projects, incentives for better and more accurate ex-ante analysis are provided.

iii. Secondly, recommend to use lessons that appear to follow from comparisons of outturn of the projects with forecasts to really enable the ex-ante evaluation procedure to be fine-tuned through an ongoing feedback process;

iv. To build up a database on costs and demand deviations to indicate how they varies in systematic ways across different types of schemes;

v. To use data from past to inform future estimates (e.g. reference forecasting) and

vi. To use past experience to derive the statistical distribution of parameter values as an important input to the risk analysis.

10. Establish a Dedicated Team.

We recommend the identification of a dedicated team in charge of collecting and capitalise ex-ante projects documentation, provide harmonised inputs for projects appraisal, collect information on projects monitoring and ex-post evaluation, establish minimum requirements for projects assessment and evaluation, and disseminate results. The dedicated team should have a progressive approach, starting from providing data already available, collecting information on past experience and promote project implementation monitoring.

While doing such ex-ante and ex-post analysis to assess the impact of the project or the technology, the following aspects should also be taken into account. They are

- The Concept of Time Value of Money
- Cost of Production
- Risk and Uncertainty
- Value of information
- Yield Gap Analysis
- Partial Budgeting Technique

4.3.2 The Concept of Time Value of Money

Many inputs, such as fertilizer and chemicals, are used as they are purchased, so evaluating their costs and returns is straightforward. However, evaluating returns from equipment that can last for several years is complicated because the costs and returns don't occur in the same year. The problem is in comparing the value of a money today with a value of money in the future. This involves the time value of money.

The central point to the time value of money is that a Rupee today is worth more than a Rupee tomorrow. The reason for this difference in value is that people normally would rather consume a good today than wait to consume it in the future, everything else being equal. Because of this preference to consume good now, people have to be paid to defer consumption to the future. If you put a Rupee in savings account for a year, the bank will return your Rupee of a dollar plus interest at the end of that year. Interest earned on savings, in effect, is payment for deferred consumption. Interest gives two aspects to the time value of money that is compounding and discounting.

Compounding is nothing but finding the future value of present income expressed in rupees or dollars. Discounting is nothing but finding the present value of future income generated from the project of long gestation period especially perennial crops like horticulture and forest trees.

4.3.3 Cost of Production

Under the head of cost of production, one should examine the opportunity costs, operational costs of the project or the technology, fixed costs involved because of the implementation of the project etc have to be individually analysed.

Not all resources that a farmer uses are purchased inputs, with costs determined in the market place. Some inputs, such as own land, fully paid for equipment and the farmer's own labor and management expertise, do not have a cash cost. Nevertheless, these inputs do have a cost associated with their use and they should not be treated as free resources. In each case, these resources have value in their current use as well as in alternative uses. The highest value in alternative uses is called the opportunity cost of the resource. For example, the opportunity cost of the farmer's time is what the farmer could receive in wages or salary by working for someone else. The opportunity cost of land is the income that could have been received if the land had been put to a different use.

Profits are maximized when the returns to the resource exceeds its opportunity costs; inother words, when the resource is earning more in its current use than in any other opportunity use. Every time a farmer drives a tractor across a field, two types of costs are incurred. Understanding the difference between these costs will help you to evaluate the benefits and costs of precision agriculture for your operation.

4.3.3.1 Operating Costs

Operating costs are the cash costs or out-of-pocket expenses of running the tractor. These costs are primarily fuel and lubrication, repairs and maintenance and labor costs (wages and benefits for the time operating, setting-up and taking-down).

4.3.3.2 Fixed Costs

Ownership costs are incurred whether the tractor is running or sitting in a shed. There are five of these costs:

- Depreciation
- Interest
- Taxes, Cess etc,
- Insurance
- Housing.

Depreciation is that portion of the tractor you use up in one year. Suppose you buy a tractor for Rs 10 lakhs that has an expected effective life of 10 years. If you use the tractor for 10 years, the cost of the tractor is spread out over those 10 years. In effect, you use 1/10th of the tractor each year, or the tractor can be considered to depreciate by *one lakh rupees /year.* (For tax reasons, you usually want to depreciate equipment and machinery as quickly as possible.)

Interest is the cost to borrow the money to buy the tractor, or the opportunity cost of your money if you use your own savings. Taxes, insurance and housing are usually a small percentage (2-5%) of the purchase price, but they do add to the cost of owning the tractor. Insurance and housing combined usually amount to no more than about 2-3% of the value of equipment. However, they should be added as a real cost when one considers a new piece of equipment or machinery. Precision agriculture systems often involve new and specialized equipments.

4.3.4 Risk and Uncertainty

Farming is a risky business. There are risks that the crop will become diseased, prices will fall, and foreign competition will push you out of business. An additional risk is that the new technology you just purchased is installed will not perform up to expectations, or that the company you bought it from will go bankrupt, leaving you without technical support. The following discussion of risk and uncertainty is presented to provide a framework for evaluating the risks associated with precision agriculture.

Every time you roll an unweighted, six-sided die, you face a 1/6 probability of rolling a1, or a2, or a...6. Suppose you receive a payoff, say Rupee of $ 1, if you roll a 6. Mathematically, the expected value of the game is:

$$\text{Rs 1 Prob.}\left[1/6\right] = \text{Re } 0.167.$$

In words: if you roll the die a lot of times, you would expect to receive the Rs 1 payoff once out every six rolls. On an average, for each roll you would receive Re. 0.167. Or, your expected payoff is Re 0.167.Now suppose it cost you Re 0.25 to play the described game. You can use economics to frame your decision; i.e., is the return greater than the cost?

$$\text{Rs 1Prob}\left[1/6j = \text{Re } 0.167 < \text{Re } 0.25\right.$$

In this case, the cost (Re 0.25) is greater than expected return (Re 0.167). If your goal is to maximize your return, you would not play the game. What if the game cost only Re 0.10? In that case, the return would be greater than the cost, and you would likely play the game.

4.3.5 The Value of Information

Precision agriculture has been defined as information technology applied to agriculture. Information has been an input in production for years. For example, farmers would know which fields were most productive, what would grow best in which field. Dairy farmers knew which cow would produce more milk if fed more, and which cow would just put on more weight. This intimate knowledge of the production process works well when fields are relatively small and a commercial herd has a score of cows. However, when production decision are made for fields that encompass a section and herds have thousands of cows, intimate knowledge has to be replaced by systematic information in order to have effective decisions made. Information as an input has unique characteristics.

First of all, data is not information. Data has to be screen and edited into a form which provides information. For example, 22 is data but provides no information. Knowing the temperature in Coimbatore, Tamil Nadu, India, is 22*C provides a context for the data, but still provides little information unless you know that 22*C is about 72*F.

4.3.6 The Yield Gap Analysis

All crops have a maximum potential yield, based on the genetic potential of the plant. Livestock have a similar potential genetic yield; e.g., the maximum yield of milk, eggs or meat from a given animal. This maximum is rarely achieved. The reason why can be as obvious as a disease or an insect infestation, or as hard to determine as the lack of water or nutrients at a critical stage of development. The difference between the maximum possible yield and the actual yield can be called the "yield gap". **Example:** The maximum possible Yield of Rice hybrid, CoRH2 is 4.50 tonnes; The Actual Yield obtained by a Farmer from CoRH2 Rice is 4.20 tonnes. The Yield gap is arrived at 0.30 tonnes.

4.3.7 Partial Budgeting Technique

The simple economics of a new technology is that if the returns from using the new technology are greater than the costs of the new technology, then use it. Otherwise, don't. The difficult part is that new technology is often very complex to implement, and determining costs and returns for a new system is rarely simple.

Partial budgeting is a method for comparing the costs and returns from a proposed change in a farm business. It is especially useful for evaluating a specific, limited change with what is currently being done. For example, a partial budget would be a good way to evaluate the costs and returns of a new combine. Suppose the new combine costs less to operate than the current combine, and is expected to harvest three more bushels of corn/acre due to less loss. The partial budget provides a consistent framework for comparing the lower operating costs and increased revenues from the new combine to the cost of buying the combine.

Both a strength and a weakness of partial budgeting is that it is limited to two alternatives. This means that partial budgeting is not a good way to determine which of three crops is the best one to grow, but a partial budget is an excellent way to evaluate whether a new technique or piece of equipment will benefit your operation. To use a partial budget to evaluate a proposed change requires that you are able toanswer four questions about that change (Table 4.1):

- What are the new or additional costs will be incurred?
- What are the current costs that will be reduced or eliminated?
- What are the new or additional revenue will be received?
- What are the current revenue will be reduced or lost?

Table 4.1 Partial Budgeting Format

A	B
Reduced Revenues	Additional Revenues
Additional Costs	Reduced Costs
Total Costs (Additional Costs + Reduced Revenues)	Total Benefits (Additional Revenues + Reduced Costs)
Net Change in Profits (B – A) = Rs. --------------------------	
Benefit Cost Ratio = ----------------------	

The Problem refers to the decision being evaluated, such as buy a new combine or keep the current one. Additional Costs are those costs that will be incurred with the new technique, method or enterprise. Recall that new equipment usually has two types of costs: operating costs and ownership costs. Both are important factors in accurately determining the profitability of an alternative. Reduced Revenues are current revenues that will be lost or reduced should the new alternative be adopted. Not all alternatives will have reduced revenues. Additional Revenues are those that will be received only if the new alternative is adopted. As with Reduced Revenues, not all alternatives will have Additional Revenues. Reduced Costs are those now being incurred that would longer be incurred if the new alternative is adopted. As with Additional Costs, both operating and ownership costs need to be considered. For example, replacing an old combine with a new combine means that both the operating and ownership costs incurred from the old combine will be eliminated.

Costs: The Additional Costs and Reduced Revenues are the costs of the new alternative. These can be considered the detriments of the new alternative.

Benefits: The Additional Revenue and the Reduced Costs are the benefits of the new alternative.

Net Benefits: If the Benefits are greater than the Costs, the new alternative has positive net benefits. Any alternative with negative Net Benefits should not be considered, as it will cost more than it will return.

Benefit/Cost Ratio: Looks at the relative values of the benefits and costs. An example can best explain why this ratio is important. Suppose you are evaluating two alternatives. Alternative 'A' has Rs 100,000 of Total Benefits and Rs 99,000 of Total Costs.

Alternative B has Rs 10,000 of Total Benefits and Rs 9,000 of Total Costs. Both alternatives have Net Benefits of Rs 1000, which at first glance may look profitable. However, the Benefit-Cost Ratio of Alternative A is 1.01, which means that the alternative is expected to return Re 0.01 for every Rs 1.00 spent on the alternative, or a 1% return on expenditures, while Alternative B has a Benefit-Cost Ratio of 1.10. Even though the Net Benefits are the same, the *B/C* Ratio showed that the return from Alternative B is much better. Both Net Benefits and the Benefit-Cost Ratio should be used to evaluate the results from a partial budget.

SUMMARY

The Chapater on Methodologies of assessing the Economics of Precision Farming has highlighted the essential tools and techniques which are used to appraise the Technique regarding its economic feasibility. On the lines, the important one is Partial Budgeting Technique which is always prescribed and practiced in the places of using new technology and its impact over the conventional technology which is capable of giving the additional return generation from that technology and hence the partial budgeting technique is the most ideal one in this case and another one is pre and post evaluation of adopting the technology by using the concept of Time Value of Money, one can also appraise for impact assessment.

References

1. http://www.precisionag.org/html/ch14.html
2. http://www.transport-research.info/sites/default/files/project documents/20121215_150809_37259_Deliberable_5.pdf

CHAPTER V

Economics of Precision Farming

Dr. C. Sekhar and Dr. A. Vidhyavathi

Objectives

- To infuse the ideas of Precision Farming Techniques in a Practical approach to the students
- The art of manning the technology by the farming community in their locale and
- To prescribe the constraints faced by them in adoptin of the Technologies like Precision Farming

Organization of the Chapter

- Introduction
- The Experiences of Precision Farming in Europe
- The Experiences of Precision Farming in India
- A Prelude to the Study Environment
- Methodology of Selection of Sample Farms
- Production Potentials of the Vegetables under Precision Farming
- Constraints Faced by the Adopters of Precision Farming Technique
- Summary

5.0 INTRODUCTION

Globally the area under agriculture and allied sector is shrinking very fast because of industrialization, urbanization due to burgeoning population and other developmental reasons. At the same time, Every Government is in a position to provide quality food, shelter and other related safety and security issues to the increasing population which called for enhancing the productivity of crops and livestock. For that improved technology or the Precision technology and its adoption to enhance the productivity many fold becomes paramount importance. In this context, every nation has motivated to adopt the Precision Agricultural Practices in certain crops and animals on test basis and the technology becomes contributive and hence the different nations has developed their own location specific, crop specific Precision Technologies and applied to different crops and animals for enhancing the productivity. Let us discuss the experiences one by one.

5.1 THE EXPERIENCES OF PRECISION AGRICULTURE IN EUROPE

A wide range of enabling technologies for PA are available. These technologies are used for object identification, geo-referencing, measurement of specific parameters, Global Navigation Satellite Systems (GNSS), connectivity, data storage and analysis, advisory systems, robotics and autonomous navigation. First implementation of PA practices already exist in arable, vegetable and dairy farming, but PA technologies can also be applied to other sectors. At the moment, a lot of progress has been made in PA development, and the PA market is fully embraced by the sector and investors, but the full potential of PA has not yet been harnessed.

5.1.1 Food Safety and Food Security

There are certain scenarios envisaged by the European Union. The Scenarios were whether optimistic (global sustained economic growth), pessimistic (recession, depression, end of globalisation) or disruptive (break-up of the European Union), food security and food safety were central. This is of course linked to the very essence of agriculture, which is to feed the entire humanity.

5.1.1.1 *Increasing Global Population and Low EU Agricultural Productivity Gains*

The most accepted scenario was based on the UN forecast of a world population reaching 9 billion people by 2050. The main question related to this scenario was *how the EU could contribute to feeding this growing population with low yield gains and declining agricultural land area*?

To achieve global food and nutrition security by 2050, agricultural global total factor productivity (TFP) – comparing the total outputs to the total inputs used for production of the outputs – will have to grow by an average rate of at least 1.8 per cent per year. According to the European Commission's Agriculture (DG AGRI) – based upon Eurostat data – TFP growth in EU agriculture has constantly remained below the percentage needed by the EU to contribute in a meaningful way to global food security. From 1995 to 2002, TFP grew by 1.6 per cent per annum in the EU-15. Thereafter, EU-15 TFP growth in agriculture dropped to just 0.3 per cent per annum (2002-2011).

To these low yield gains, we should add that, in the EU (also according to DG AGRI) there is a long-term decline in the number of holdings. Between 2005 and 2013, the average rate of decline was 3.7 % per year, resulting in the number of holdings being reduced by 1.2 million. The area of agricultural land also fell by 0.7 % over the same period due to increased forestry and urbanisation. Regardless of world demographics and global demand for agricultural commodities and food, it is obvious – if these trends persist – that EU agricultural productivity has to increase in order to maintain the same output.

5.1.1.2 *PA offers Technology Solutions for Producing More with Less Resources*

Beyond the sustainability issue, PA already offers technologies for producing more agricultural output with less input. For instance, sensor-based monitoring systems provide farmers with better information and early warnings on the status of crops, and improved yield forecasts. PA also plays a major role in animal husbandry.

A very good example is given by precision milking and feeding robots. The Netherlands, Germany and France are currently leading the shift towards automatic milking. Some 90 per cent of new equipment installations in Sweden and Finland, and 50 per cent in Germany include robotic milking.

Half of the dairy herds in north-western Europe will be milked by robots in 2025. Robotic milking generates about 120 data variables per cow per day such as: Animal Movements, Feed being distributed, Milk being produced, Quality of milk, Temperature, Coughs and other Cattle diseases. All these technologies noticeably improve the well-being of cows and lower their stress levels.

Dairy farms fully equipped with precision milking enjoy a substantial increase in yields. While the EU average annual milk production per cow is 6 915 kg, some precision milking demo-farms produce almost double that at 12 000 kg milk per year with the same agricultural input as traditional dairy farms. This is a clear example of what PA could deliver in terms of better yields with the same level of agricultural input.

5.1.1.3 *PA will Enhance Food Safety and Plant Health*

PA will contribute more and more to food safety. PA makes farming more transparent by improving tracking, tracing and documenting. Crop and livestock monitoring will give better predictions on the quality of agricultural products. The food chain will be easier to monitor for producers, retailers and customers.

It will also play a significant role in terms of plant health. Current technologies allow to monitor to different levels of resolution in precision farming. Grid level ranges from field monitoring (ca. 30 x 30 m) to plant level monitoring (ca. 30 x 30 cm). Forthcoming technologies will make leaf level (ca. 3 x 3 cm) and spots on leaves (ca. 0.5 x 0.5 cm) accessible to optical automated diagnostics. Diseases undetectable by traditional means will be prevented by automated optical sensing and intelligent planning options.

5.1.1.4 *Policy Options*

Irrespective of what the economic context might be in the next decades, PA will be needed by EU farmers to improve their yields on less available arable land. The strategic question here is:

Will the EU be one of the major global players for PA technologies?

Yet the EU has already taken some vigorous steps in addressing this challenge. The EU doubled its efforts with an unprecedented budget of nearly €4 billion, allocated to Horizon 2020 and the specific theme 'Societal Challenge 2', which partially relates to PA. Parallel to this, the EU has set

Fostering knowledge transfer and innovation in agriculture, forestry and rural areas as the first priority for rural development policy in 2014-2020. Rural development programmes will finance agricultural and forestry innovation through several measures which can support creation of operational groups, innovation services, investments or other approaches. In those two EU R&D funding tools, nine programmes include PA practices as an eligible priority. All stakeholders agreed that investments in research and development will be the key driving force for bringing about the agricultural jobs of tomorrow.

Accordingly, a substantial shift from the CAP (2021-2027) to enhance R&D in agriculture could be envisaged, especially in a period of persistent budgetary constraints during which other policy priorities are likely to supersede CAP priorities. More money could for instance be invested in cutting-edge technologies like biosensors, robotics, and spectrographic, imagery.

5.1.2 Environmental Sustainability of Farming in European Union

The potential of PA for cost saving can be illustrated by two examples discussed during the STOA (Science, Technology, Options Assessment) project workshop

By 2050, the global population will be in excess of 9.5 billion and we will require 70 to 100 per cent more agricultural output to meet the global demand for food.

Producing higher output by utilizing the lesser amount of resources through Precision Agriculture will be the driving force for sustainability meeting the needs of the European Union's environmental policies.

5.1.3 Key PA Technologies already in use with Positive Impacts on the Environment

PA uses not only satellite navigation and positioning systems but also a wide range of other technologies. These cover:

- Automated steering systems, which can take over specific driving tasks such as auto-steering, overhead turning, following field edges and overlapping of rows. Automatic steering systems reduce human errors. In addition, they contribute to effective soil and site management. Automated headland turns could, for instance, already save from 2 per cent up to 10 per cent fuel consumption.
- Geo-mapping, which is used to produce maps identifying, for instance, types of soils and levels of nutrients for particular fields.
- Sensors and remote sensing, with which data can be collected from a distance to evaluate soil and crop health, measuring parameters such as moisture, nutrients, compaction, and crop diseases. These sensors can be installed on mobile machines. EU farmers already make use of a wide range of sensors for capturing variations in properties of soils and crops, weather conditions and animal behaviour. Thermal, optical, mechanical and chemical measurements by sensors are applied to quantify crop biomass, plant stress, pests and diseases, soil properties, climatic conditions and animal behaviour.

- Agricultural robots of the future will be autonomous and able to reconfigure their own architecture to perform various tasks. They will offer an enormous potential for sustainability:
- They will ease the energy transition. Robots will be powered by electricity. The required electricity could be produced at the farm site.
- They can minimise soil compaction due to heavy machinery. Swarm robots will be lighter and able to intervene only where they are needed, staying permanently on the fields. *(note: Swarm robots are a group of simple robots, which can be coordinated in a distributed and decentralised way, in order to jointly execute more complex tasks)*
- Less work effort and resources input will be required, and robots will most likely provide greater output, as they already do in the dairy industry.
- Robots will optimise inputs used by farmers (fertilisers, pesticides, insecticides) and reduce the impact on soils and water tables.

5.2 A PRELUDE TO THE STUDY ENVIRONMENT

The main objective of this chapter is to document the experiences of the practitioners of Precision Farming Technologies besides documenting the area, production, productivity and export of select vegetables in India and to document the supporting infrastructure available in the State of Tamil Nadu. To document the same, the data were collected from the secondary sources of both Central and State Government reports. The reports taken into account for collecting the details on area, production, productivity and export of vegetables are furnished as follows.

- Agricultural Statistics at a Glance 2016 of Directorate of Economics and Statistics, Ministry of Agriculture, Government of India, New Delhi
- Food and Agriculture Organization Web Site
- Indian Horticulture Data Base 2016 of National Horticulture Board, Government of India, New Delhi
- Hort-Stat – 2016 of Government of Tamil Nadu, Chennai

The data gathered for different periods with respect to the area production and productivity of Vegetables, Export of Vegetables, assembling and trading centers for Vegetables and the importing institutions and states involved in vegetable export were gathered from the above reports and the data were analyzed using Descriptive Statistics like Mean and the current change in the data in respect of area, production and productivity of Vegetables over the base period was also documented to assess the technological and extension impact due to the Government Intervention through its planned programs in horticulture. The results of this analysis will be better utilized by the policy makers and the planners in horticulture and agriculture for further development of these sectors.

The productivity of Vegetables in the State was found to be marginally low when compared to the National Average Productivity and hence a comparison made with the intensely managed vegetable Gardens of Theni district under precision technologies was put into practice from the year 2009 and their experiences were also incorporated in this study as a case study approach as an illustration for easy implementation by other vegetable growers in Tamil Nadu.

Production of any agricultural commodity is mainly depended on geographical nature of a particular location. In this chapter, detailed information on location demographic details, physiology, soil type, rainfall, irrigation details, land utilization pattern, cropping pattern, crop rotation, land holding pattern, infrastructural facilities, financial and marketing facilities available in Theni district are presented. This information helps to understand the present study in proper perspective and for effective interpretation of results.

5.2.1 Location

Theni district is in the Southern part of Tamil Nadu. This district is surrounded by the Western Ghats, with it green stretches of cultivated lands and tea gardens. Silk cotton, soft towels, coffee seeds, cardamom, Garlic, are the main produce produced from the district. Theni district is the main route for the tourist bound from Madurai to Kochi via Bodinaickanur and Munnar and Madurai to Thekkadi wildlife sanctuary.

Periyakulam is a small town most picturesquely situated on the palm fringed banks of the Varaganathi, a perennial river with the great wall of the Kodaikanal hills. The Town consists of revenue villages namely, Thenkarai, Vadakarai and Keela Vadakarai. There are roads running westwards and northwards to the hills. The Town has been developing gradually but the extent has not increased on account of the fact that Periyakulam is purely agricultural town and is located far away from the railway communication. This is a century old Municipality located at the foot of the Kodaikanal hills.

Area of the Municipal limit is 2.11 sq km. This town is having 30 wards at present as per G.O. M.S. No. 189 MAWS dated 20.8.1996. Wards no.28 and 29 are reserved for SC/ST and wards No 9,10,11,20,22,23,27 and 30 are reserved for Women. Out of the wards allotted to Women, ward nos.8 and 25 are reserved for SC/ST Women.

The river Varaganathi runs in east-west direction and divides the town into equal half namely Vadakarai and Thenkarai. This river starts from the foot of the Kodaikanal hills and joins with Vaigai River.

5.2.2 Demographic Features

The demographic features of the district are furnished in Table 5.1. The Theni district had a population of 124, 5899 persons. The density of population was 434 inhabitants per sq.km (1120/sq mi). The district has a sex ratio of 991 females for every 1000 males. The literacy rate of Theni district during 2011 census was 77.26, higher than national literacy rate per cent. At gender wise, male and female literacy were arrived as 85.03 and 69.46 per cent respectively. Theni – Allinagaram and Bodinayakanur are the largest towns in terms of population, in this district.

Table 5.1 Demographic Features of Theni District (as per 2011 Census)

Sl. NO	Particulars	Theni District	Percentage
1.0	Total Population	1245899	100.00
1.1	Male Population	625683	50.22
1.2	Female Population	620216	49.78
2.0	Population Density per Sq.Km.	434	--
3.0	Sex Ratio (Number of Females to 1000 Male)	991	--
4.0	Literacy Rate (Percentage)	77.26	77.26
4.1	Male Literacy Rate	85.03	85.03
4.2	Female Literacy Rate	69.46	69.46

Source: http://www.census2011.co.in/census/district/46-theni.html

5.2.3 Physiographic Features

The 2,889 km^2 (1,115 sq mi) district lies at the foot of the Western Ghats between 9' 39' and 10' 30' North latitude and between 77' 00' and 78' 30' of East Longitude. Central location: 10°04′ N 77°45′ E 10.067°N 77.750°E. The district is bounded by Dindigul District to the north, Madurai District to the east, Virudhunagar District to the southwest, and Idukki district of the Kerala State to the west. The district is home to Theni, Periyakulam, Bodinayakanur, Cumbum, Uthamapalayam, Kombai, Gudalur, Chinnamanur, Andipatti, Thevaram, Pottipuram, Ramakrishnapuram, Lakshmipuram and several other small villages like Pathirakalipuram, SukkangalPatti, Uppukottai, Meenatchipuram, Melasindalaichery, Kottur or Pallavarayan Patti, a popular village in Tamil Nadu Jallikattu. It is diversified by several ranges and hills. A range of hills which runs parallel to Western Ghats from north to south separate it from the neighbouring state of Kerala.

5.2.3.1 *Soil Type*

Sandy loam, clayey and alluvial is the major soil type found in both the blocks of Theni District.

Table 5.2 Soil Type Prevalent in Theni District

Sl. No.	Soil Type	Location in which Soil Type is Available
01	Red Loamy Soil	Aundipatti, Theni, Periyakulam, Bodinayakanur and Uthamapalayam
02	Lateritic Soil	Aundipatti, Theni, Periyakulam, Bodinayakanur and Uthamapalayam
03	Black Soil	Aundipatti, Theni, Periyakulam, Bodinayakanur and Uthamapalayam
04	Red Sandy Soil	Aundipatti, Periyakulam and Uthamapalayam
05	Sandy Coastal Alluvial	---Nil---
06	Calcareous Soil	Aundipatti, Theni, Periyakulam, Bodinayakanur and Uthamapalayam
07	Clay Soil	Aundipatti, Theni, Periyakulam, Bodinayakanur and Uthamapalayam
08	Alluvial Soil	Cumbum Valley

(Source: Joint Director, Soil Survey and Land Use)

In respect of soil type, Cumbum valley is blessed with alluvial soil type. Red Loamy soil, Lateritic soil, Black soil, Red Sandy soil, Calcareous soil, Clay soil are the other soil types available in majority areas of Theni district revealed that the district is blessed with almost all soil types capable of accommodating any crops in its fold.

5.2.3.2 *Sources of Irrigation*

Sources of irrigation are important at district level to assess the nature of cropping pattern to be adopted by the farmers. In this regard, an effort has been taken to assess the sources of irrigation available in Theni district and the details are analyzed and presented in Table 5.3.

Table 5.3 Different Sources of Irrigation in Theni District (2015-2016)

Sl. No	**Name of the Block**	**Canals (nos.)**	**Length (km)**	**Wells (Nos) for Irrigation Alone**	**Tube Wells (Nos)**	**Wells used for Domestic Purpose**	**Major Small Minor**	**System Tanks <40**	**Tanks (nos.) >40 ha**	**Tanks (nos.)**
1.	Theni	17	42	3679	159	441	12	03	14	17
2.	Aundi patti	1	5	4015	117	616	00	04	17	21
3.	Myla dum parai	1	6	2088	79	700	00	02	10	12
4.	Periya kulam	31	51	5554	990	584	00	09	42	51
5.	Bodinaya kanur	34	43	2405	470	2325	00	28	00	00
6.	Chinna manur	5	18	2839	781	225	07	01	18	19
7.	Uthama palayam	3	11	3720	301	380	04	01	29	30
8.	Cumbum	15	50	3285	1120	480	06	00	00	00
9.	Total	107	226	27585	4017	5751	29	20	130	150

It could be observed from Table 5.3 that well irrigation is the major source for Theni District. Around 86 per cent of net irrigated area was under well irrigation followed by canals and tank irrigation accounted for 7.55 per cent and 3.44 per cent of net irrigated area. The irrigation intensity is arrived at 107. The Table 5.3 also revealed that the dependency on rain fed agriculture was very much higher in this district which may be witnessed from the fact that, only 48.28 per cent of gross area sown in Theni district was under irrigated agriculture.

Table 5.4 Blockwise Cropping and Irrigation Intensity in Theni District (2015-2016) Area in ha

Sl. No	Name of the Block	Gross area sown	Net area sown	Cropping intensity	Gross area irrigated	Net area irrigated	Irri gation inten sity
1.	Theni	11082	9831	113	8394	7143	118
2.	Aundipatti	17643	16632	106	9935	8924	111
3.	Myladum parai	16403	15151	108	6599	5346	123
4.	Periyakulam	19785	18574	107	12256	11046	111
5.	Bodinaya kanur	22421	19179	117	12507	9265	135
6.	Chinna manur	14011	11908	118	9355	7428	126
7.	Uthama palayam	12472	10208	122	8526	6412	133
8.	Cumbum	16535	14595	113	10731	8792	122
9.	**Total**	**130352**	**116078**	**112**	**78303**	**64356**	**122**

As shown in the Table 5.4 the cropping intensity and irrigation intensity is more than 100 per cent in the blocks of Theni district indicating that farmers are able to cultivate and irrigate area more than once in a cropping year.

5.2.3.3 Climate

Theni district falls under third agro- climatic zone which is designated as Western Zone of Tamil Nadu. Normally, semi arid tropical monsoon type of climate is prevailing in the plains. However hilly areas are recorded low temperature and fairy heavy rainfall.

The mean maximum and minimum atmospheric temperatures are 38 degree Celsius and 24 degree Celsius respectively. Heat becomes intense in the months of April and May and shoots up unless interfered by rains.

5.2.3.4 Rainfall Pattern

The season wise rainfall is depicted in Table 5.5. From this, it could be observed that the normal annual rainfall received was 829.80 mm of which the contribution by North-East Monsoon was much higher (46 per cent) followed by South West Monsoon (21 per cent) followed by summer rain (27 per cent).

Table 5.5 Season wise Distribution of Rainfall (mm) in Theni District (2015 & 2016)

Sl. NO	Season	Normal Rainfall	Actual Rainfall		Percentage to Total
			2015	2016	
1	South West Monsoon	170.80	306.34	153.95	20.58
2	North East Monsoon	382.40	435.07	131.47	46.08
3	Winter	53.90	15.47	07.08	06.50
4	Summer	222.70	358.95	136.17	26.84
	Total	829.80	1115.83	428.67	100.00

(Source: Office of the Assistant Director of Statistics, Theni)

With this rainfall, the people of Theni District had the habit of cultivating almost all the crops both under rainfed and irrigated. Theni District is famous for fruits and vegetables and spices production and distribution.

5.2.4 Land Use Pattern

Analyzing the existing land use pattern prevalent in Theni District will give an overall understanding on the sources of supply of raw materials to the markets. Cumbum, Uthamapalayam and Chinnamanur are the nodal centre for sources of supply of vegetables to Theni District. In this respect the land use pattern was analyzed and the results are presented in Table 5.6.

Table 5.6 Land use Pattern in Theni District (2015-2016)

Sl. NO	Particulars	Area (ha)	Percentage to Total
01	Forests	103717	31.99
02	Barren and Uncultivable Land	43320	13.36
03	Land Put to Non-Agricultural Uses	24613	7.59
04	Cultivable Waste Land	2865	0.88
05	Permanent Pastures and other Grazing Lands	315	0.10
06	Land Under Miscellaneous Tree Crops and Groves	1225	0.38
07	Current Fallow Land	6716	2.07
08	Other Fallow Land	25385	7.83
09	Net Area Sown	116078	35.80
10	Area Sown More Than Once	14273	4.40
11	Geographical Area According to Village Papers	324231	100.00
12	Total Cropped Area	130351	40.20

(Source: Office of the Assistant Director of Statistics, Theni)

The land utilization pattern of Theni district is given in Table 5.5. The total geographical area of the district is 324231 hectares. The net area sown is arrived at 116078 ha (35.80 per cent) with area sown more than once was 3.21 per cent. Forest is accounted for 103717 ha (31.95 per cent) of total geographical area of the district. The forest area of Theni district is almost equal the norms of National Forest Policy of 1988 which prescribed one third of the total geographical area.

5.2.5 Crop Rotations

The major crop rotations normally prevalent in Theni district under different land use situations are presented below:

Periyar Project Areas

1. Paddy – Paddy – Pulses
2. Paddy – Sugarcane (3 years rotation)

Manjalaar & Sothuparai Project Areas

1. Paddy – Cotton / Paddy
2. Paddy – Sugarcane (3 years rotation)

Tank fed Areas

1. Paddy – Cotton / Gingelly / Pulses

Well Irrigated Areas

1. Groundnut / Sunflower
2. Cotton – Groundnut
3. Paddy – Sugarcane (3 years rotation)
4. Cotton – Chillies
5. Cotton – Bajra / Cholam Paddy – Cotton
6. Paddy – Banana
7. Cholam / Bajra – Groundnut
8. Maize – Vegetables
9. Coconut

Rainfed Areas

- Cholam, Maize, Bajra, Ragi, Samai, Pulses, Cotton, Groundnut, Gingelly and Castor.
- I Crop: Paddy (Oct – Feb) II Crop: Pulses (March – May) Single Crop: Sugarcane / Banana (December – January) (June – July)

5.2.6 Distribution of Land Holdings

Land is an important asset to the farmers in raising crops based on the moisture regime and nature of the soil. The farmers of Tamil Nadu are classified as marginal, small, medium and large farm category based on their size of holding. It is important to know how many marginal and small farmers in the zone of study are available so as to draw certain policy recommendations. In this respect, the details on size of holding available in Theni district are analyzed and the results are presented in Table 5.7.

Table 5.7 Land Holding Pattern of Theni District (As per Agricultural Census 2010-2011)

Sl. NO	Category	Size of Holdings		Percentage to Total	
		Number	Area (ha)	Number	Area
1	Marginal	88919 (69.42)	39792.57 (30.21)	69.42	30.21
2	Small	26415 (20.62)	36033.79 (27.36)	20.62	27.36
3	Semi Medium	9298 (7.25)	24911.18 (18.91)	07.25	18.91
4	Medium	401.71 (0.31)	16510.31 (12.53)	00.31	12.53
5	Large	10.94	144465.17		
	Total	128081	131713.02		

(Source: Agricultural Census, ADS, Theni)

Land holding pattern of Theni district is presented in Table 5.7. It is found that about 128,081 farmers cultivated in an area of 1,31,713 hectares in the district. Mean size of holding with the farmers of Theni District is arrived at 1.03 ha. Marginal farmers formed major part of the cultivators (69.42 per cent) cultivating only 30.21 per cent of area. Small farmers constituted 20.62 per cent cultivating 27.36 per cent of area. Thus numerically marginal and small farmers together constituted 90.04 per cent cultivating only 57.57 per cent of total area of the district. The rest of the area (42.43 per cent) has been cultivated by 9.96 per cent of other farmer (large category) implying skewed distribution of land holdings. The overall average size of land holdings was 1.03 ha.

5.2.6.1 *Cropping Pattern*

The area under important crops cultivated in Theni District was shown in Table 5.8. From the Table 5.8, it could be seen that the major crops in the study area are cereals, pulses, fruits, coconut and groundnut. Among the cereals, Cholam, Cumbu and Paddy had the largest share. The maximum area under Cholam and Cumbu is mainly due to the nutritional requirements of the dwelling population due to average rainfall received in Theni District.

5.2.6.2 *Area under Important Crops in Theni District*

Table 5.8 Area Under Important Crops in Theni District (2015-2016)

Sl. No	Name of the Crops	Area (Ha)	Production (Tonnes)	Productivity (Kgs Per Ha)
01	Paddy	14400	61662	4282
02	Cholam	13005	28738	2210
03	Cumbu	2840	4859	1711
04	Ragi	34	73	2131
05	Redgram	2083	2452	1177
06	Greengram	253	74	291
07	Blackgram	330	86	260
08	Groundnut	1899	3743	2971
09	Gingelly	604	262	434
10	Sunflower	100	159	1589
11	Cotton (Bales of 170 kg of lint)	1364	3570	445
12	Sugarcane (in terms of Cane)	7658	769797	101000
13	Chillies	189	96	506
14	Onion	558	5376	9635
15	Banana	4577	194517	42499

(Source: Director of Economics and Statistics, Chennai)

The area under fruits and vegetables is arrived at 5326 ha which is one third area of paddy crop raised in Theni District. Currently, awareness is found to be increased to the farmers on raising cash crops like fruits and vegetables and spices in this belt.

5.2.7 STATUS OF LIVESTOCK POPULATION

Livestock is often called as Mobile Banks of the farmers. If a farmer owns a big herd of cattle, he will be called as Wealthy Farmer and respected in the village in a different style. It is also prevalent among the Thoda Tribes. A Tribe is respected in Nilgiris based on the number of Buffaloes he owns, maintains and take care of the animals. It acts as ready cash to the farmers on emergency situation. Even the animals can be called ATM (Any Time Money) provider. Hence, assessing the livestock population is an important one and act as a proxy for judging the wealth of the farmers. It is noticed that total livestock population in Theni District is 369,283. It could be noticed that the cattle had the largest share (37 per cent) in total livestock population.

Table 5.9 Details on Livestock Population in Theni District

Sl. NO	Name of the Animals	Number of Animals	Percentage to Total
		Animal Population	
01	Cattles	137665	37.28
02	Buffaloes	6397	01.73
03	Sheep	123256	33.38
04	Goats	101965	27.61
	Total Number of Animals	369283	100.00
		Poultry Population	
05	Poultry Birds	1259899	100.00

(Source: Office of the Assistant Director of Statistics, Theni)

Sheep and goats accounted for the next higher share in the district. It is observed that the poultry population in Theni District is found to be on the higher side which is arrived at 12.60 lakhs. From this it could be concluded that higher demand is prevailing for the poultry and hence number of poultry units found to be on the increasing side. In order to meet the requirements of poultry, maize, cholam cultivation becomes major crops in the district.

5.2.8 INFRASTRUCTURE FACILITIES

Both institutional and non-institutional agencies are playing a vital role in meeting the requirements of the farmers. The major institutional agencies in the district are Primary Agricultural Cooperative Banks, District Central Co-operative Banks, Scheduled Commercial Banks and their branches. The co-operative societies serve as a useful agency for lending the required credit as well as a source of supply of agricultural inputs like fertilizers and pesticide to the farmers. Besides these, scheduled commercial banks are also providing loans to the farming community.

Table 5.10 Details of Financial Institutions in Theni District

Sl. NO	Type of Institutions	Number of Institutions	Percentage to Total
1.	Primary Land Development Banks	04	3.45
2.	Co-operative Apex Banks	00	00.00
3.	District Central Co-operative Banks	00	00.00
4.	Primary Agricultural Credit Societies	81	69.83
5.	Employees Co-operative Societies	21	18.10
6.	Co-operative Marketing Societies	02	01.72
7.	Co-operative Stores	8	06.90
	Total	116	100.00

(Source: Office of the Assistant Director of Statistics, Theni)

The district is provided with a good net works of roads and railway lines connected to important places in the state. NH 45 is Connecting Theni to Chennai via Trichy in North. State High way is connecting Theni and Coimbatore in the Western Direction. The details of road and rail infrastructure are presented in Table 5.11.

Table 5.11 Details of Road and Rail Infrastructure in Theni District

Sl.No	Details of Infrastructure	Length in Kms	Percentage to Total
01	National Highway	162.70	06.83
02	State Highway	725.77	30.47
03	Corporation and Municipal Roads	301.85	12.67
04	Panchayat Union and Panchayat Roads	1132.93	47.57
05	Town Panchayat Roads	-	00.00
06	Forest Roads	58.30	02.45
	Total	2381.55	100.00
	Railways		
01	Meter Gauge Lines	41.09	100.00
02	Broad Gauge Lines	Nil	00.00
	Total	41.09	100.00

Among the road infrastructure and type of roads, the panchayat union and the village panchayat roads alone accounted for 47.57 per cent to the total road infrastructure followed by state highways accounted for 30.47 per cent to the total. Theni District had only 06.83 per cent of National Highway roads and the corporation road had shared only 12.67 per cent to the total road infrastructure.

In respect of Railway lines, A Meter gauge line is accounted for 100 per cent to the total railway line in Theni District (41.09 km).

5.2.9 MARKETING FACILITIES

Theni market committee is operating 21 regulated markets around Theni district for crops like Cotton, Chillies, Paddy, Pulses, Groundnut, Banana, Maize, Jaggery, Cane, Silk Cotton, Jowar, Tamarind, Coconut, Garlics etc. The markets are located in Theni, Thirumangalam, Melur, Cumbum, Chinnamanur, Usilampatti, Bodinaickanur, Madurai, Dindigul, Oddanchatram, Natham, Gopalpatti, Palani, Batlagundu, T. Kallupatti, Vadipatti, Vadamadurai, Aundipatti, Vedachandur, Uthamapalayam and Periyakulam.

5.2.10 STATE AND CENTRALLY OPERATED SCHEMES FOR HORTICULTURE IN THENI DISTRICT

There are various State and Central schemes being operated in Theni district to develop the horticulture in the district through the scheme inputs are supplied in subsidized rates. Table 5.12 gives the details about the schemes implemented in this district.

Table 5.12 Details of the Horticultural Schemes Implemented in Theni District

S. No	Name of the Scheme	Inputs Supplied	Subsidy (per cent)
A	State Schemes		
1.	Integrated Horticulture Development Scheme	Vegetable Seeds, Planting Materials	50.00
2.	National Agriculture Development Programme – Precision Farming	Seeds, Fertilisers	-

S. No	Name of the Scheme	Inputs Supplied	Subsidy (per cent)
3.	National Agriculture Development Programme – High tech productivity Enhancement Programme	Vegetable Seeds	50.00
4.	National Agricultural Insurance Scheme	-	-
B	State and Central Shared Schemes		
1.	National Horticulture Mission	Seeds, Planting Materials	50 -75
2.	National Mission on Micro Irrigation	Drip Irrigation Components	75 – 100
3.	Externally Aided Schemes – TN IAMWARM	Seeds, Planting Materials	50.00

There are 3 schemes which are funded by the state government and the other three schemes are shared by the State and Central Governments. Shared schemes are National Horticulture Mission for seeds and planting material has the subsidy of 50 to 75 per cent. In respect of Drip Irrigation components, the subsidy given is around 75 to hundred per cent. The World Bank Sponsored IAM WARM Project is providing only 50 per cent subsidy for the seeds and planting materials.

5.2.11 METHODOLOGICAL APPROACH

The farmers who are cultivating Vegetables in Theni District formed the base for the study. In the first stage, the Theni District was selected as it has higher number of farmerspracticing the Precision Farming Technologies in the Western part of Tamil Nadu. In the second stage the blocks in which the precision farming technologies was higher in number and had two to three years of experience were choosen. This condition was fulfilled by two blocks viz., Periyakulam and Theni Blocks and hence these two were selected. In the Third Stage, the villages in which the select vegetables which are common in day to day affairs in the villages like Brinjal, Tomato and Chilli cultivated farmers under the precision farming technologies were selected following a three stage random sampling technique.

5.2.11.1 Selection of Sample Farmers

The number of farmers growing vegetables under precision farming technologies and the list of villages and the farmers who are purposively selected are presented in Table 5.13.

Table 5.13 Details of Sample Farmers Selected for the Study

Sl. No	Name of the Villages	Number of Precision Farmers	Number of Farmers Selected	Percentage to Total
01	Lakshmipuram	30	00	00.00
02	Vadaveeranaickenpatti	80	00	00.00
03	Jallipatti	18	18	54.54
04	Bomminaickenpatti	22	06	18.18
05	Koduvilarpatti	20	08	24.24
06	Upparpatti	17	01	03.04
07	Kottur	20	00	00.00
08	Seelayampatti	20	00	00.00
	Total	227	33(14.54)	100.00

(Figures in Parentheses Indicate Percentage of Samples Selected to the Total)

Accordingly, a sample of 33 farmers hailing from four villages of Periyakulam and Theni Blocks were selected. The selected sample farmers from the total population for the study are accounted to be at 14.54 per cent.

From the sample farmers, the details of land use, crop management for its production and productivity, marketing practices adopted by the precision farmers, cost of production and the constraints associated with the vegetable cultivation are assessed using the structured and pretested questionnaire exclusively designed for the purpose and the same were analyzed and documented with respect to the objectives enshrined in the study by adopting a descriptive statistics like mean and percentage analysis.

5.2.11.2 Concepts and Issues Related to Cost of Production

Cost of Production is one of the important area which acts as a scale to which the vegetables should be priced above so as to get considerable profit. It involves both fixed costs and variable costs to be incurred by the producers of the farm vegetables. In this regard, availing a discussion on the concepts related to those costs and the cost of production will be a worthy one.

5.2.11.2.1 *Fixed Cost*

For the present study, fixed cost is conceptualized as the sum of costs of fixed inputs such as land, building, van and pump sets for a specific time period.

5.2.11.2.2 *Variable Costs*

In the present study variable cost such as cost of raw materials, labour charges, processing, drying and packing charges in addition to the interest in working capital, depreciation on buildings and implements all these costs involved in the production of Vegetables that could vary proportionately with the increase and decrease with the output of Vegetables.

5.2.11.2.3 *Cost of Production*

Shukla (1966) classified the costs into Cost A_1, Cost A_2, Cost B and Cost C.

Cost A_1	included costs and hired expenses actually incurred less rent
Cost A_2	covered Cost A_1 plus rent paid for leased in land
Cost B	included Cost A_2 plus rental value of owned land plus interest on fixed Capital minus land revenue on owned land
Cost C	covered cost B plus imputed value of family labour.

Kapoore (1968) defined the cost concepts as follows:

Cost A_1	Consisted of hired human labour, value of hired and owned bullock labour, value of hired and owned machine labour, value of owned and purchased seed, value of insecticides and pesticides, value of owned and purchased manures, value of fertilizers, depreciation, irrigation charges, land revenue, cess and other taxes, interest on working capital and miscellaneous expenses.

Cost A_2	Cost A_1 plus rent paid for leased in land
Cost B	Cost A_2 plus imputed renal value of owned land. (Less land revenue paid) plus imputed interest on owned fixed capital (excluding land)
Cost C	Cost B plus imputed value of family labour

According to Dewett (2009), Cost of Production includes the following elements

- Wages of Labor
- Interest on Capital
- Rent or Royalties paid to the owners of Land or other property used
- Cost of Raw Materials
- Replacement and Repairing Charges of Machinery items
- Depreciation of Capital Goods and
- Profits of the Manufacturer or the Producer of the Commodity sufficient to induce the Entrepreneur / Agripreneur to carry on the Production of the Commodity

In the present study, the concept used by Dewett (2009) is adopted for working out the cost of production of Vegetables like Brinjal.

5.2.11.2.4 ***Returns***

5.2.11.2.4.1 *Gross Returns*

In the present study gross income is defined as the total value of output produced per annum taking into account the market price as received by the farmers in the village for the Vegetables.

5.2.11.2.4.2 *Net Return*

In the present study, net return is defined as the difference between the gross income and the total cost incurred by the Producers / Traders of Vegetables.

5.3 EXPERIENCES OF PRECISION FARMING IN INDIA – THE CASE OF TAMIL NADU

In the preceeding section, detailed discussions were made on trends in area production and productivity of select vegetables and the state wise details on production and productivity which revealed that many of the states are not able to realize higher potentials of hybrid varieties. Instead, the states are realizing the low productivity barring few other states particularly West Bengal and Karnataka are ahead of achieving the productivity. To achieve the higher productivity of vegetables per ha, the experiences of farmers who are practicing the Precision Farming Techniques and its documentation is vital to highlight the merits of those technologies and hence a detailed discussion is made under the following sub headings

5.3.1. Area under Precision Horticulture in Theni District

5.3.2. Details of Crops Cultivated under Precision Farming in Theni District

5.3.3. Area under Different Crops during Pre and Post Practice of Precision Farming Techniques

5.3.4. Tomato Cultivation under Precision Farming Technology

5.3.5. Brinjal Cultivation under Precision Farming Technology

5.3.6. Cultivation of Chillies under Precision Farming Technology

5.3.7. Details of Price Realized by the Farmers for their Output Realized from Precision Farming

5.3.8. Cost of Production of Brinjal in Precision Farming Farms of Theni District

5.3.9. Economics of Guava Orchards under Precision Farming

5.3.10. Awareness on Drip Irrigation System among the Practitioners of Precision Farming

5.3.11. Details of Socio Economic Changes among the Adopters of Precision Farming

5.3.12. Perception of the Sample Farmers on Precision Farming Techniques

5.3.13. Perception of the Sample Farmers on their Exposure Visits

5.3.14. Details of Entrepreneurial Initiatives Made in the Sample Farms

5.3.15. Constraints Faced by the Practitioners of Precision Farming Techniques

The discussions on the enlisted headings will give a better understanding on the potentials associated with the precision farming techniques in the sample farms of Theni District.

5.3.1 Area under Precision Horticulture in Theni District

The success stories of Precision Farming Techniques which was launched first of its kind in Dharmapuri District on Turn Key Basis proved that careful and prescribed applications of technology at appropriate hours will ensure better results and provide better living conditions to the farming communities. After realizing the potentials of precision farming techniques, the Government of Tamil Nadu gave clear direction through the Directorate of Horticulture and Plantation Crops to spread the technology to almost all the districts. Accordingly, this technology is in practice in Theni District every year. Hence, the details of area under precision horticulture were analyzed and the results are presented in Table 5.14.

Table 5.14 Area under Precision Horticulture in Theni District

Sl.No	Name of the Villages	Name of the Block	Area in Ha	Percentage to Total
01	Lakshmipuram	Periyakulam	30.00	13.22
02	Vadaveeranaickenpatti	Periyakulam	80.00	35.24
03	Jallipatti	Periyakulam	18.00	07.93
04	Bomminaickenpatti	Periyakulam	22.00	09.69
05	Koduvilarpatti	Theni	20.00	08.81
06	Upparpatti	Theni	17.00	07.49
07	Kottur	Theni	20.00	08.81
08	Seelayampatti	Theni	20.00	08.81
	Total	02	227.00	100.00

Table 5.14 revealed that the Precision Horticulture was practiced in 8 villages of Theni District. Among the Eight villages, 4 belong to the Periyakulam Block and the other 4 belonged to the Theni Block of Theni District. Periyakulam Block had the precision horticulture in 150 ha. Theni Block had the precision horticulture in 77 ha. From the year of 2009, the precision farming techniques were practiced in the

Periyakulam Block of Theni District and hence higher area under this Block. Among the four villages in Periyakulam Block, Vadaveeranaickenpatti is the village had 80 ha under precision farming horticulture with the initiatives of the Horticultural Officers and the Assistant Agricultural Officers of Department of Horticulture and Plantation Crops, Vadugapatti. In this respect, analyzing the area under precision farming in the sample villages becomes important.

5.3.2 Details of Crops Cultivated under Precision Farming in Theni District

The details of crops cultivated under precision farming in Periyakulam and Theni Blocks of Theni District is analyzed and the details are presented in Table 5.15.

Table 5.15 Details of Crops Cultivated under Precision Farming in Theni District

Sl.No	Name of the Villages in which Vegetables Cultivated	Name of the Vegetable Crops Cultivated	Area in Ha	Percentage to Total
		Periyakulam Block		
1.1	Lakshmipuram	Banana	20.00	13.33
	Lakshmipuram	Turmeric	10.00	06.67
1.2	Vadaveeranaickenpatti	Banana	80.00	53.33
1.3	Jallipatti	Brinjal	09.20	06.13
	Jallipattti	Chillies	08.80	05.87
1.4	Bomminaickenpatti	Tomato	06.00	04.00
	Bomminaickenpatti	Bhendi	04.00	02.67
	Bomminaickenpatti	Banana	12.00	08.00
	Sub Total (1)		150.00	66.80
		Theni Block		
2.1	Koduvilarpatti	Tomato	07.50	09.74
	Koduvilarpatti	Bhendi	07.50	09.74
	Koduvilarpatti	Banana	05.00	06.49
2.2	Upparpatti	Banana	15.00	19.49
	Upparpatti	Tomato	01.00	01.30
	Upparpatti	Bhendi	01.00	01.30

Sl.No	Name of the Villages in which Vegetables Cultivated	Name of the Vegetable Crops Cultivated	Area in Ha	Percentage to Total
2.3	Kottur	Banana	20.00	25.97
2.4	Seelayampatti	Banana	20.00	25.97
	Sub Total (2)		77.00	33.20
	Total (1+2)		227.00	100.00

Table 5.15 revealed that the Periyakulam Block is cultivated with Banana, Bhendi, Tomato, Chillies, Brinjal and Turmeric in an area of 150 ha. In Theni Block of Theni District, Tomato, Bhendi and Banana are the crops cultivated under 77 ha which are respectively accounted for 67 per cent and 33 per cent to the total area under precision farming. Among these crops, Banana alone cultivated under 112 ha in Periyakulam Block and 60 ha in Theni Block. Put together, the area under Banana is arrived at 172 ha which is accounted for 75.77 per cent to the total area under precision farming in Theni District. However, only the crops like Tomato, Brinjal and Chilli alone were taken into discussion for the study as these vegetables are routine in the kitchen of all type of consumers. However, it is important to analyze the area under different crops during pre and post implementation of precision farming technologies in the sample farms of Theni District.

5.3.3 Area under Different Crops during Pre and Post Practice of Precision Farming Techniques

The area under different crops during pre and post implementation of precision farming technologies in the sample farms of Theni District is very important to highlight the use of hybrid varieties in the farm lands of precision technology adopters and to delineate the advantages of precision technologies. These details are analyzed and the results are presented in Table 5.16.

Table 5.16 Area under Different Crops during Pre and Post Practice of Precision Farming Techniques

Sl. No	Name of the Crops Cultivated	Before Precision Farming		After Precision Farming	
		Area Under Crops in Ha	Percentage to Total Size of Holding	Area Under Crops in Ha	Percentage Change
01	Chillies	00.30	15.38	00.40	33.33
02	Tomato	00.30	15.38	00.40	33.33
03	Brinjal	00.25	09.62	00.50	100.00
04	Bhendi	00.20	07.69	00.20	00.00
05	Cotton	00.80	30.77	00.80	00.00
06	Maize	01.00	30.77	01.00	00.00
07	Cumbu	00.25	09.62	00.25	00.00
08	Total Area under Crops	03.10	119.23	03.55	14.52
	Size of Holding	02.60	100.00	02.60	00.00
	Cropping Intensity		119.23		136.54 (17.31)

(Figures in Parentheses Indicate Percentage Change in the Cropping Intensity)

Table 5.16 revealed that the average size of holding available with the sample farms was arrived at 2.60 ha. It remains same during pre and post scene of adoption of Precision Farming Techniques in their farm lands. But the total area under crops is found to be increased from 3.10 ha to 3.55 ha leaving the cropping Intensity of 119 per cent during the pre implementation of precision farming and the same was rose to 136.54 per cent during the post practice of precision farming techniques. Since the Department of Horticulture and Plantation Crops are fixing the target of precision farming only one ha per farmer, the allocation of land by the farmers were resorted to the crops like Brinjal, Tomato, Chillies and Bhendi in an area of 0.50 ha, 0.40 ha, 0.40 ha and 0.20 ha respectively.

5.3.4 Tomato Cultivation under Precision Farming Technology

The Department of Horticulture and Plantation Crops of Government of Tamil Nadu is providing the inputs like seeds and the water soluble fertilizers and the technology support to the farming community and hence most of the farmers are registering willingly with the Department under National Agricultural Devleopment Project (NADP) and other schemes available with the Department of Horticulture and Plantation Crops and they are providing subsidy for one ha per farmer. The possible seeds supplied by the Department are Tomato, Brinjal, Bhendi and Chillies. In this respect, the details of Tomato cultivation under precision farming technology are assessed and the details are presented in Table 5.17.

Table 5.17 Technology Adoption by the Tomato Cultivating Farmers under Precision Farming Technique

Sl. No	Details of Technology	Before Implementation of Precision Farming	After the Implementation of the Precision Farming
01	Selection of the Variety / Hybrid	Variety	Hybrids
02	Season	No Much Importance for the Season	Timely Sowing of Tomato
03	Method of Sowing	Raised Bed Nursery	Pro Tray Nursery
04	Plant Population Maintained	Higher Population	Optimum Population
05	Spacing Practiced	150 X 45 CM	180 X 60 CM
06	Quantity of Seeds Used	100 Grams	75 Grams
07	Quantity of Fertilizers (WSF) used (Total NPK)	Rs 8000/-	Rs 15000/-
08	Weed Control Practices	Manual Weeding	Weed Management Practices Adopted
09	Yield Obtained (Tonnes / Ha)	24 Tonnes	56 Tonnes
Percentage Increase in the Yield			133.33

Table 5.17 revealed that the Tomato cultivation was analyzed for its status of pre and post implementation of precision farming technology with the farmers. The Parameters taken into account are selection of variety, season of sowing, method of sowing, plant population maintained, spacing adopted, quantity of seeds used, quantity of water soluble fertilizers used, weed control practices and the yield realized per ha. Under precision farming techniques, almost all the parameters are varying. Notably, inrespect of selection of variety, the hybrids are used. The quantity of seeds used per ha found to be reduced. Spacing practiced was enhanced. Application of water soluble fertilizers in different split doses was found to be increased. All these timely applications of inputs have resulted in 56 tonnes of Tomato per ha which is found to be increased to the tune of 133 per cent over the traditional method of cultivation practiced by the farmers in the sample villages and hence they feel that the Precision Technology is much needed one for the farmers and they believe that this technology is a boon to them.

5.3.5 Brinjal Cultivation under Precision Farming Technology

In the farm lands the Brinjal is cultivated under 0.50 ha under precision farming techniques. The details of achievements made in Brinjal cultivation are analyzed and the results are presented in Table 5.18.

Table 5.18 Technology Adoption by the Brinjal Cultivating Farmers under Precision Farming Technique

Sl. No	Details of Technology	Before Implementation of Precision Farming	After the Implementation of the Precision Farming
01	Selection of the Variety / Hybrid	Variety	Hybrids
02	Season	No Much Importance for the Season	Timely Sowing of Vegetables
03	Method of Sowing	Raised Bed Nursery	Pro Tray Nursery
04	Plant Population Maintained	Under Population	Optimum Population
05	Spacing Practiced	60 X 30 CM	90 X 60 CM
06	Quantity of Seeds Used	100 Grams	75 Grams

Sl. No	Details of Technology	Before Implementation of Precision Farming	After the Implementation of the Precision Farming
07	Quantity of Fertilizers (WSF) used (Total NPK)	Rs 7000	Rs 15000
08	Weed Control Practices	Manual Weeding	Weed Management Practices Adopted
09	Yield Obtained (Tonnes / Ha)	31.500 Tonnes	84.000 Tonnes
Percentage Increase in the Yield			166.67 Per Cent

Table 5.18 revealed that the parameters stipulated for precision cultivation is also practiced by the farmers in respect of Brinjal cultivation. Here, the optimum plant population is maintained per ha. Notably, the quantity of seeds used per ha is found to be less. Adoption of hybrid for cultivation is practiced. Spacing was enhanced. Application of Water Soluble Fertilizers was found to be increased which are fed to the plants in split doses as and when needed based on the prescriptions of the officials of the Department of Horticulture and Plantation Crops. All these jointly enhanced the productivity of Brinjal to the tune of 84 tonnes which is accounted to be 167 per cent over the conventional method of cultivation of Brinjal which could generate the productivity of only 31.50 tonnes. But in the data published by the Ministry of Agriculture, Government of India, Tamil Nadu statistics in respect of vegetable cultivation is included in other states. It might be due to poor area under vegetable cultivation in Tamil Nadu.

The Yield realized from the Brinjal farm was arrived at 84 tonnes per ha. If this output is sold at the rate of Rs 7000 per ton, gross revenue of Rs 5.88 lakhs was realized from per ha farm. From the above information, the cost of production of Brinjal is arrived at Rs 3.75 per kg. When we compare the selling price of Rs 7 per kg, the revenue generation is 87 per cent more than the cost of production of Brinjal per kg and hence one can easily enter into the process of implementing the precision farming technologies in their farm.

5.3.6 Cultivation of Chillies under Precision Farming Technology

Similar to Tomato and Brinjal cultivated under Precision Farming techniques by the sample farmers, how the chilli is being cultivated and the yield obtained are analyzed and the results are presented in Table 5.19.

Table 5.19 Technology Adoption by the Chilli Cultivating Farmers under Precision Farming Technique

Sl. No	Details of Technology	Before Implementation of Precision Farming	After the Implementation of the Precision Farming
01	Selection of the Variety /Hybrid	Variety	Hybrids
02	Season	No Much Importance for the Season	Timely Sowing of Vegetables (Dec – Jan)
03	Method of Sowing	Raised Bed Nursery	Pro Tray Nursery
04	Plant Population Maintained	More Population	Optimum Population
05	Spacing Practiced	60 X 30 CM	90 X 60 CM
06	Quantity of Seeds Used	100 Grams	75 Grams
07	Quantity of Fertilizers (WSF) used (Total NPK)	Rs 8000/-	Rs 15000/-
08	Weed Control Practices	Manual Weeding	Integrated Weed Management
09	Yield Obtained (Tonnes / Ha)	16.800 Tonnes	28.350 Tonnes
Percentage Increase in the Yield			68.75

Table 5.19 revealed that the parameters stipulated for precision cultivation is also practiced by the farmers in respect of Chilli cultivation. Here, the optimum plant population is maintained per ha. Notably, the quantity of seeds used per ha is found to be less. Adoption of hybrid for cultivation of Chilli is practiced. Spacing was enhanced. Application of Water Soluble Fertilizers was found to be increased which are fed to the plants in split doses as and when needed based on the prescriptions of the officials of the Department of Horticulture and Plantation Crops. All these jointly enhanced the productivity of Chilli to the tune of 28.350 tonnes which is accounted to be 69 per cent over the conventional method of cultivation of Chilli which could generate the productivity of only 16.80 tonnes.

5.3.7 Details of Price Realized by the Farmers for their Output under Precision Farming

Production of output of Tomato, Brinjal and Chilli are realized under precision farming technique was considerable and hence one should find the ways and means of distributing the same to the consumers at appropriate price per unit of the produce is very much important. The output realized from the precision farming technique was found to be quality produce and hence it should command higher price per unit of the produce. In this context, the details of price prevailed per unit of the produce was assessed and the same are presented in Table 5.20.

Table 5.20 Details of Price Realized by the Farmers for their Output under Precision Farming

Sl. No	Name of the Crop	Before Implementation of Precision Farming (Price Rs/Kg)	After Implementation of the Precision Farming (Price Rs/ Kg)	Percentage Change in the Price
01	Brinjal	05.00 - 06.00	07.00 – 08.00	33.33
02	Chillies	05.00 - 06.00	08.00 - 10.00	66.67
03	Onion	07.00 - 08.00	10.00 - 12.00	50.00
04	Tomato	03.00 – 04.00	07.00 – 08.00	100.00
05	Gourds	04.00 - 05.00	07.00 - 08.00	60.00
06	Sugarcane	06.00 – 08.00	12.00 – 14.00	75.00
07	Cotton	05.00 – 06.00	10.00 – 12.00	100.00
08	Maize	06.00 – 07.00	07.00 – 08.00	14.29

Table 5.20 revealed the details of percentage change in the price realized for the precision output in the sample farms. Brinjal found to generate 33.33 per cent increase in the price per kg when compared to the output realized from the traditional cultivation practices because of quality, color and free from infestation. Similarly, the Chillies found to generate 66.67 per cent increase in the price for the precision output when compared to the output realized from the traditional method of cultivation of chilli. Tomato found to generate cent per cent increase in the price because of its uniform color, appearance and free from infestation and has higher keeping quality. All these contributed to fetch higher price per unit of Tomato.

To sum up, the output realized from the precision farming technique found to attract higher price per kg when comparing the same with the output realized from the traditional method of cultivation. In this context, the cost of production of precision output is important to assess the level of comparative advantage in revenue generation from the precision output. For that, Brinjal alone was taken into account for arriving the cost of production as all other crops attract similar investment per ha.

5.3.8 Cost of Production of Brinjal in Precision Farming Farms of Theni District

The output drawn from precision farming is found to be qualitative one and hence a doubt arises among the traditional farmers and others that the investment for precision farming might consume higher than the traditional method of cultivation. Hence a detailed analysis on cost of production of Brinjal under Precision farming technique was analyzed and the results are presented in Table 5.21.

Table 5.21 Cost of Production of Brinjal in the Precision Farming Farms of Theni District

S. No	Details of Expenditure	Number of Units	Price Per Unit	Expenditure / Acre
	1. Annual Fixed Cost			
1.1	Annual Fixed Cost (Rental Value of Owned Land per Ha)	One Ha	108325.00	108325.00
1.2	Annual Fixed Cost per Crop	One Crop	54163.00	54163.00
1.3	Cost of Gunny Bags @ Rs 40 per Bag assuming the life of 5 years	50.00		0400.00
1.4	Cost of Plastic Trays for Establishing Nursery assuming the Life of 5 Years	50.00	100.00	1000.00
1.5	Cost of Drip and Its Accessories assuming the life of the asset is 10 years	One Unit	70000.00	2000.00
	Total Annual Fixed Cost			165888.00
	2. Production Cost			
2.1	Ploughing of Land @ Rs 750 per Plough	03.00	750.00	2250.00

S. No	Details of Expenditure	Number of Units	Price Per Unit	Expenditure / Acre
	1. Annual Fixed Cost			
2.2	Clearing the Weeds and Grasses using the women labor	05.00	200.00	1000.00
2.3	Land Preparation and Channel Formation @ Rs 400 per Man Day	05.00	400.00	2000.00
2.4	Cost of Brinjal Seed @ 75 Grams per Ha plus seed Treatment	75.00	66.67	5000.00
2.5	Planting of Seedlings @ 20 women days per Ha	20.00	200.00	4000.00
2.6	Water Soluble Fertilizers (NPK)	50.00	300.00	15000.00
2.7	Application of Farm Yard Manure @ Rs 4000 per Tractor Load	04.00	4000.00	16000.00
2.8	Weeding Two Times @ 20 Women Days per Weeding	20.00	200.00	4000.00
2.9	Application of Pesticides and Chemicals	024	380.00	9120.00
2.10	Harvesting Cost @ 2 Man Days and 5 Women Days per Picking @ 48 Pickings	336.00	257.00	86352.00
2.0	Production (Variable) Cost (II)			144722.00
	3. Post Harvest Expenses			
3.1	Sorting and Packing Charges in Gunny Bags @ 5 Man Days	05.00	400.00	2000.00
3.2	Transportation Charges to Kerala and Bangalore Wholesale Merchants @ Rs 100 per Bag	25.00	100.00	2500.00
3.3	Sub Total Cost (III)			4500.00
Total Cost				
4.1	Annual Fixed Cost (I)			165888.00
4.2	Variable Cost (II)			144722.00
4.3	Post Harvest Expenses			4500.00

S. No	Details of Expenditure	Number of Units	Price Per Unit	Expenditure / Acre
		1. Annual Fixed Cost		
4.0	Total Cost			315110.00
Cost of Production of Brinjal				
5.1	Output Realized per Ha in Tonnes	84.00	7000.00	588000.00
5.2	Cost of Production per Tonne of Brinjal			03751.309
5.3	Cost of Production per Kg of Brinjal			03.75

Table 5.21 revealed that the total cost incurred towards establishing one ha brinjal farm consumed 3.15 lakhs. Out of this, the annual fixed cost consumed is arrived at Rs 1.66 lakhs and the variable expenses in establishing Brinjal farm was arried at 1.45 lakhs and the post harvest expenses was arrived at Rs 4500/- The items under annual fixed cost included were rental value of owned land, plastic trays for raising nursery seedlings, Gunny bags and Drip Irrigation systems.

5.3.9 ECONOMICS OF GUAVA ORCHARDS UNDER PRECISION FARMING

5.3.9.1 Establishment Cost Involved in Guava Orchards

One of the delicious and mostly preferred fresh fruit is Guava by the consumers next to Banana. Analyzing the costs and returns generated from Guava Orchards will be a piece of information to the producers of Guava so as to decide their selling price even on the market glut situations and hence the details of cost involved in establishing the Guava Orchards are assessed and the details are presented in Table 5.22

Table 5.22 Cost of Establishment of Guava Orchards

Sl.No	Particulars of Establishment	Value of Input in Rs per Ha	Value of Labour Use per Ha	Total Cost per Ha
01	Field Preparation	7630	000	7630 (11.50)
02	Digging of Pits	000	4684	4684 (07.06)

Sl.No	Particulars of Establishment	Value of Input in Rs per Ha	Value of Labour Use per Ha	Total Cost per Ha
03	Cost of Planting Material	9390	1880	11270 (16.99)
04	Cost of Manures	3098	2717	5815 (08.77)
05	Cost of Fertilizers	5846	2875	8721 (13.15)
06	Cost of Irrigation	000	18000	18000 (27.14)
07	Cost of Weeding	000	3145	3145 (04.74)
08	Plant Protection Chemicals	1967	5090	7057 (10.64)
Total Establishment Cost		27931	38391	66322 (100.00)

Table 5.22 revealed that the total establishment cost involved in establishing the orchard is arrived at Rs 66322 per ha. Among this, the cost of irrigation alone consumed the cost to the turne of Rs 18000 which is accounted for 27 per cent to the total establishment cost followed by the planting materials consumed around 11270 per ha which is accounted for 17 per cent to the total establishment cost.

The cost of fertilizers and manures during the time of establishment alone consumed Rs 14536 which is accounted for around 26 per cent to the total establishment cost. The cost towards plant protection chemicals is accounted for 11 per cent to the total. From that one could infer that the planting materials and irrigation are the major cost under the head of establishment cost of Guava Orchard

5.3.9.2 Variable Cost involved in Establishing the Guava Orchard

The details of variable cost involved are nothing but the maintenance costs involved for the guava orchard. The average variable cost involved per annum is worked out during the bearing period is presented in the Table 5.23.

Table 5.23 Average Variable Cost Involved in Guava Orchard during the Bearing Period

Sl.No	Particulars of Cost	Value of Input in Rs per Ha	Value of Labour Use per Ha	Total Cost per Ha
01	Ploughing	00.00	1500	1500 (00.70)
02	Farm Yard Manure	2465	1623	4083 (01.83)
03	Fertilizer	9357	3156	12513 (05.62)
04	Weeding	1875	1565	5440 (02.43)
05	Pruning and Training	00.00	3173	3173 (01.42)
06	Plant Protection Chemicals	3126	2825	5951 (2.67)
07	Irrigation	00.00	10410	10410 (04.67)
08	Harvesting and Loading	00.00	159392	159392 (71.57)
09	Interest on Working Capital @ 10%			20246 (09.09)
Total Variable Cost per Year				222708 (100)

Table 5.23 revealed that the total variable cost involved in the maintenance of Guava orchard is arrived at Rs 2.23 lakhs per annum. Among the total variable cost, the cost of harvesting and loading alone accounted for around 72 per cent to the total variable cost. The next important cost under the head of maintenance is fertilizer and irrigation which are respectively accounted for 5.62 per cent and 4.67 per cent to the total variable cost. The reason behind the higher harvesting cost is mainly due to plucking the fruits from the tree, grading the fruits into different size classes and loading the same in transport vehicles for the consuming ends.

5.3.9.3 Costs and Returns from Guava Orchard

The details of costs and returns and the cost of production are analyzed and the resultls are presented in Table 5.24. In arriving the total cost, the fixed expenses involved per annum is arrived at Rs 3200 and the annualized cost on establishment is arrived at 4378/-

Table 5.24 Cost and Returns Realized from Guava Orchards per Annum

Sl. No	Particulars of Cost and Returns	Amount in Rupees per ha
01	Total Fixed Cost	3200.00
02	Annualized Cost of Establishment	4378.00
03	Total Variable Cost	222708.00
04	Total Cost of Cultivation	230286.00
05	Cost of Production per Kg	14.74
06	Gross Income	546875.00
07	Yield in Tonnes per Ha	15.625
08	Price Received by the Producer per Kg	35.00

The total cost of cultivation of Guava is arrived at Rs 2.30 lakhs per ha. The yield of Guava fruits per ha is arrived at 15625 kgs which is 50 per cent higher yield than the state average yield of Guava. The cost of production incurred by the farmer per kg of output of Guava is arrived at Rs.14.74. The selling price of the Guava per kg in the local market is arrived at Rs 35/ which is 137 per cent higher than the cost of production. From this data, one could understand that the cultivation of guava could yield higher returns with appropriate maintenance.

5.3.10 Awareness on Drip Irrigation System among the Sample Farms

The sample farmers have been asked to express their opinion on post adoption of precision farming techniques after the expiry of the schemes in vogue. For that a mixed opinion could be received from the respondents who are practicing precision farming techniques. The details are analyzed and the results are presented in Table 5.25.

Table 5.25 Awareness on Drip Irrigation System among the Sample Farms

Sl. No	Opinion of Farmers	Number of Farmers	Percentage to Total
01	Willing to Continue the Drip Irrigation in Future Without Even Subsidy	06	18.18
02	Willing to Continue the Drip Irrigation in Future if 75 Per cent subsidy is Given	14	42.42
03	Willing to Continue the Drip Irrigation in Future if atleast 50 Per Cent Subsidy is Given	09	27.27
04	Unwilling to Continue the Drip Irrigation if Subsidy is not given due to the Higher Cost of Investment	04	12.13
	Total	33	100.00

Table 5.25 revealed that only 18 per cent of the practitioners of precision farming expressed their opinion of willing to continue the Drip Irrigation in Future without even subsidy. Around 42.42 per cent of the respondents revealed that they are willing to continue the Drip Irrigation in future if 75 per cent subsidy is extended to them. Around 27.27 per cent of the sample farmers have expressed that they are willing to continue the Drip Irrigation in future if atleast 50 per cent of the subsidy is given. However, 12 per cent of the respondents categorically rejected the option of continuing but expressed that they are unwilling to continue the Drip Irrigation if subsidy is not given due to the Higher cost of Investment in Drip infrastructure. Besides it needs expertise also for close monitoring to enhance the longevity of the drip infrastructure. Hence the subsidy should be extended for precision farming.

5.6.11 Details of Socio Economic Changes among the Adopters of Precision Farming

Adoption of any technology creats some impacts interms of social and economical among the practitioners of particular technology and hence effort has been taken to assess the details of social and economic changes created due to the project among the sample farmers and the details are analyzed and the results are presented in Table 5.26.

Table 5.26 Details of Socio Economic Changes Made Due to the Implementation of Precision Farming Technique in the Sample Farms

Sl. No	Details of Changes Made	Number of Farmers Reported	Percentage to Total
01	Improvements Made in the Existing House	07	21.21
02	Purchase of New Consumer Durables – Two Wheelers	14	42.42
03	Increase in the Frequency of Intake of Non Vegetarian Items	33	100.00
04	Better Linkage Developed with the Department of Horticulture and Plantation Crops	29	87.88
05	Enhancement in the Awareness Related Participation with the Technical Departments and TNAU	32	96.97
06	Increased Investment in the Farm and Savings	30	90.91

Table 5.26 revealed that the details of socio economic changes occurred among the sample farmers are Improvements were made in the existing house, purchase of new consumer durables particularly the two wheelers, increase in the frequency of intake of non vegetarian dishes, better linkage developed with the Department of Horticulture and Plantation Crops and enhancement in the consultation, Enhancement in the awareness related to participation with the technical departments and TNAU and Increased the Investment in the Farm and in Savings.

Among these changes, increase in the frequency of taking non vegetarian dishes found to be reported by all the sample farmers followed by 97 per cent of the farmers reported that Enhancement in the awareness related participation with the Technical Departments and TNAU and Increased investment in the farm and in savings was reported by 91 per cent of the sample farmers. Better linkage developed with the

Department of Horticulture and Plantation Crops was reported by 88 per cent of the sample farmers. From this one could understand that the precision farming technique has imparted notable changes among the lives of farming community.

5.6.12 Perception of the Sample Farmers on Precision Farming Techniques

Amidst the developments imparted by the Precision Farming Techniques, one should be interested in listening to the perception of the users of the technology need to be analyzed and hence the details are analyzed and the details are presented in Table 5.27.

Table 5.27 Perception of the Practitioners of Precision Farming Techniques in Theni District

Sl. No	Particulars	Perception of the Practitioners of Precision Farming (Per Cent)			
		Excellent	Very Good	Good	Average
01	Adoption of Recommended Crops in the Farm Lands	23	06	04	00
02	Timely Supply of Inputs	24	05	04	00
03	Supply of Inputs are Quality	19	12	02	00
04	Technical Support from Extension Functionaries from Horticulture	21	08	02	02
05	Installation of Drip and Fertigation at Field	18	11	03	01
06	Advice on Pest and Disease Control	27	06	00	00
07	Exposure Visits Undergone	31	02	00	00
08	Yield of Crops	29	02	02	00
09	Quality of the Output Realized	25	08	00	00
10	Marketing Arrangements	00	05	18	10

Sl. No	Particulars	Perception of the Practitioners of Precision Farming (Per Cent)			
		Excellent	Very Good	Good	Average
11	Selling Price of the Produce of Precision Farming	00	11	09	13
12	Loan Facility Arranged with Commercial Banks / Primary Co-Operative Banks	00	04	10	19

Table 5.27 revealed that the perception of sample farmers on Precision Farming Technology and its usage level was ranked as Excellent followed by very few farmers alone have reported that the technology was contributing good to the farmers. If one could examine the individual parameters, adoption of recommended crop variety scored the excellent and very good scale and only very few reported that the recommended variety is only good. Next parameter is timely supply of inputs which is also ranked as excellent according to majority of the sample farmers. Because of adoption of particular variety, the farmers felt that the yield performance was found to be Excellent when compared to the traditional varieities in practice.

Farmers were dissatisfied with the marketing arrangements, selling price of the produce derived from the precision farming technique and the loan facility arranged with the commercial banks and primary cooperative banks found to gain the perception only to the level of 'Average' category. Hence, the Department of Horticulture and Plantation Crops should establish marketing arrangements through user / producer groups and the same should be made available at higher price per unit of the produce they produce. Loan facility is another problem. Poor farmers are unable to invest even small money to take up this technology and hence loan tie up with the commercial banks / primary agricultural cooperative banks need special attention to gain higher success from the technology.

5.6.13 Perception of the Sample Farmers on their Exposure Visits

The success of precision farming technology is not only on the prescription and adoption of technology but also based on the exposure visits and training attended by the farmers / practitioners of precision technology. Hence, it is important to assess their perception on the exposure visits and the training organized by the Department of Horticulture and Plantation Crops and the details are analyzed and the results are presented in Table 5.28.

Table 5.28 Perception of the Sample Farmers on their Exposure Visits to Different Places

Sl. No	Particulars of Places Visited	Perception of the Sample Farmers (Per Cent)			
		Excellent	Very Good	Good	Average
Training					
1.1	Training on Precision Farming at Jallipatti	20	09	04	00
Exposure Visits					
2.1	Coimbatore – TNAU for Precision Farming Technique	29	04	00	00
2.2	Dharmapuri for Learnings on the Practice of Precision Farming	30	03	00	00
2.3	Hosur – Roja Flower Factory	23	08	02	00
2.4	Bangalore – Safal Market	19	10	03	01
2.5	Bangalore – Gardens and Nurseries	33	00	00	00
2.6	Chennai – Koyambedu Wholesale Fruits and Vegetable Markets	25	06	02	00
2.7	Chennai – Koyambedu Wholesale Flower Market	18	12	01	02

The Horticultural Officer, Department of Horticulture and Plantation Crops, Vadugapatti has organized training and the exposure visits to the farmers at Jallipatti Village. Table 5.28 revealed that the training organized was rated as excellent by the 60.61 per cent of the respondents and 27 per cent of the respondents ranked the training as very good. In respect of the exposure visits organized by the Department, Visit to Tamil Nadu Agricultural University, Coimbatore is an excellent one according to

87.88 per cent of the sample farmers. The visit to the Precision Farming Practitioners field at Dharmapuri was rated as an Excellent one by 90 per cent of the respondents.

Visit to gardens and commercial nurseries were rated as excellent by almost all the farmers. But the visit to safal market was rated as excellent one according to only 57 per cent of the respondents due to the poor understanding on the usefulness of the safal market and the Visit to the Chennai – Koyambedu Fruits and Vegetable Market and its usefulness was rated as excellent according to 76 per cent of the sample farmers.

To sum up, the perception on the technologies, training and the exposure visits to the sample farmers were rated as an Excellent one which is followed by the ranking as Very Good which has been rated by only few respondents. Hence the effort taken up by the Department of Hortiuclture and Plantation Crops is found to be an excellent one according to the respondents and hence the support facility towards these activities should further be extended in a refined manner to the farming community.

5.6.14 Entrepreneurial Initiatives Made in the Sample Farms

The exposure visit to different places including commercial nurseries would have mooted some innovative ideas among the sample farmers and hence the details of entrepreneurial initiatives taken up by the farmers are analyzed and the details are presented in Table 5.29.

Table 5.29 Details of Entrepreneurial Initiatives Made in the Sample Farms

Sl.No	Details of Initiatives Made	Number of Farmers	Percentage to Total
01	Production and Distribution of Vegetable Nursery	14	42.42
02	Establishment of Commercial Horticulture Nursery	02	06.06
03	Distribution of Drip Accessories	01	03.03
04	Production and Distribution of Siddha Medicines	01	03.03
05	Distribution of Farm Machineries and Sprayers on Hire Basis	02	06.06
	Total Samples	33	60.60

Table 5.29 revealed that the sample farmers were taking up the entrepreneurial initiatives are Production and Distribution of Vegetable Nursery, Establishment of Commercial Horticulture Nursery, Distribution of Drip Accessories to the farmers in their villages, Production and Distribution of Siddha Medicines and Distribution of Farm Machineries and Sprayers on Hire Basis are the prominent initiatives. Among these, the production and distribution of vegetable nursery was established by 42 per cent of the respondents. Establishment of Commercial Horticulture Nursery was prevalent only among six per cent of the sample farmers. Similarly, the distribution of farm machineries and sprayers supply on hire basis was established by only six per cent of the sample farmers. The distribution of Drip accessories and production and distribution of Siddha Medicines were respectively prevalent among the respondents whom are accounted for three per cent to the total.

To sum up, the main entrepreneurial initiatives made in the sample farmers were production and distribution of Vegetable nurseries and establishment of horticulture nurseries among the sample farmers. If the farmers were trained in other non farm technologies and value addition issues, the entrepreneurial initiatives could further be enhanced among the farmers. Hence the Department of Horticulture and Plantation Crops and Tamil Nadu Agricultural University should have joint venture approach in supporting the farmers in terms of provision of value addition technologies and building the entrepreneurial initiatives.

5.6.15 Constraints Faced by the Practitioners of Precision Farming Technology

Amidst all these technological infusion and adoption efforts, the success comes out after crossing over certain constraints. Hence, the details of constraints crossed over to enhance the productivity using the technology need to be documented. In this context, the details of constraints faced by the farmers are analyzed and the results are presented in Table 5.30.

Table 5.30 Constraints Faced by the Practitioners of Precision Farming in the Sample Farms

Sl. No	Particulars of the Constraints	Number of Farmers Reporting	Percentage to Total
01	Cost of Precision Input is High	33	100.00
02	Quantum of Labor Use is Comparatively Higher than the Traditional Practice	25	75.76
03	Absence of Know How for Maintenance of Drip System	28	84.85
04	Turbulent Situation in Sharing of Water from the Common Property Resources in the Joint Family / Divided System	05	15.15
05	Need for an Intensive Care in Adoption of Precision System	14	42.42

Table 5.30 revealed that the constraints faced are Cost of Precision Input is comparatively higher according to cent per cent of the sample farmers. Followed by absence of Know how for maintenance of Drip Systems among the sample farmers were accounted for 85 per cent. The quantum of labor use is comparatively higher than the traditional practice was reported by 76 per cent of the sample farmers. Any new technology which is productivity friendly should ignore labor intensiveness. But the precision farming technology found to attract higher labor for different operations which inturn promote employment opportunities in the rural pockets. However, availability of labor is another constraint. The drip accessories need an intensive care system according to 42 per cent of the respondents. Hence, maintenance of drip accessories related training is an important one. In this respect, the Department of Horticulture and Plantation Crops can initiate Drip Maintenance related training as one of the component in the training program for its successful implementation.

SUMMARY

The success of precision farming technology is not only on the prescription and adoption of technology but also based on the exposure visits and training attended by the farmers / practitioners of precision technology. The perception of sample farmers on Precision Farming Technology and its usage level was ranked as Excellent followed by very few farmers alone have reported that the technology was contributing good to the farmers. Hence the Department of Horticulture and Plantation Crops and Tamil Nadu Agricultural University should have joint venture approach in supporting the farmers in terms of provision of value addition technologies and building the entrepreneurial initiatives.

References

1. Government of India (2016)."Agricultural Statistics at a Glance 2016", (Directorate of Economics and Statistics, Ministry of Agriculture, Government of India: New Delhi)
2. Food and Agriculture Organization Web Site
3. Government of India (2016). "Indian Horticulture Data Base 2016", (National Horticulture Board, Government of India: New Delhi)
4. Government of Tamil Nadu (2016). "Hort-Stat – 2016", (Department of Horticulture and Plantation Crops: Chennai)

CHAPTER VI

Economics of Precision Agriculture: The Case of System of Rice Intensification (SRI)

Dr.M.Anjugam and Dr. C. Sekhar

Objectives

- To expose the Stakeholders on the Status of Rice Production in India and Tamil Nadu
- To identify the Precision Technologies used in Agriculture for enhancing productivity of Rice
- To analyze the impact of Precision Technologies applied in Agriculture and to transform the same to others for adoption.

Organization of the Chapter

- Status of Global Rice Production
- Status of Rice Production in India
- Status of Rice Production in Tamil Nadu
- Agricultural Economy of Tamil Nadu
- System of Rice Intensification (SRI) as the Precision Technique
- Critical Steps to be followed for Adoption of SRI
- Crop Response
- Significant Impact Events on Rice in India
- Area Under SRI

- Knowledge and Experience in Adoption of SRI
- Application of SRI Principles to Other Crops
- Economics of SRI in Tamil Nadu
- Summary

6.1 STATUS OF GLOBAL RICE PRODUCTION

So far, we have discussed the precision technological advantage in respect of fruits and vegetables. Being rice is a staple food, how the precisin technology is being implemented in certain farms are important so as to take up the same in spirit.

Rice is an important crop and consumed widely across the globe as a staple food. During 2014-15 and 2015-16, the world rice production has marginally decreased by 1.67 per cent (from 480 million tonnes to 472 million tonnes), trade by 7.14 per cent (from 42 million tonnes to 39 million tonnes) and consumption by 0.23 per cent (from 476 million tonnes to 475 million tonnes). According to the March 2015-16 statistics, China ranks first globally producing 143 million tonnes followed by India producing 107 million tonnes against the world total of 472 million tonnes. India's rice production has witnessed an increased Compound Annual Growth Rate of 1.43 percent during 2005-06 and 2016-17; however, area under rice cultivation has not increased substantially during the same period.

India was the largest exporter of rice in 2015-16 due to opening up of non-basmati rice export followed by Thailand, Vietnam and Pakistan. India's rice export was 10.41 million tonnes during the year 2015-16. Globally, the major rice exporters are Thailand, Vietnam, India, United States, Burma and Pakistan. India is still the largest exporter of rice.

6.2 STATUS OF RICE PRODUCTION IN INDIA

In India, rice is also a major staple food after wheat and is one of the world's largest producers of white and brown rice, accounting for around 20 per cent of world rice production. India is the second largest producer and consumer of rice in the world. The country produces rice in a large quantity and rice production is estimated at a record of 108.86 million tonnes which is higher by 2.21 million tonnes than the previous record production of 106.65 million tonnes achieved during 2013-14. Rice is grown widely across the nation in more than 20 states and in an area of over 400 lakh hectares. Out of these states, the top 10 rice producing states account for more than 80 per cent of total rice production in India viz., West Bengal, Uttar Pradesh, Madhya Pradesh, Punjab, Tamil Nadu, Bihar, Chhattisgarh and Odisha. With the green revolution attained post-independence, India has been able to produce enough rice to meet the requirements of its growing population. The agricultural production methods in the country have been modernized in order to obtain better and higher yield.

6.3 STATUS OF RICE PRODUCTION IN TAMIL NADU

Tamil Nadu ranks fifth in total rice production in the country. Paddy productivity in Tamil Nadu has always been the second highest in the country, next only to Punjab. The production of rice in Tamil Nadu was 4.05 million tonnes for the year 2013-14. The Minimum Support Price for paddy for the year 2013-14 was Rs.1310 (Common) and Rs.1345 (Grade-A).Rice is the major constituent accounting for 85.20 per cent of the total food grains production in the state. Tamil Nadu is one of the leading rice growing states in India, has been cultivating rice from time immemorial as it is endowed with all favorable climatic conditions suitable for rice growing. In Tamil Nadu, 90 per cent of the farmers belong to small and marginal category and their operational holdings account for 56 per cent of the total areas. The Gross Cropped Area in Tamil Nadu is around 58.43 lakh hectares of which the Gross Irrigated Area is 33.09 lakh hectares which is 57 per cent and the balance 43 per cent of the area are under rain fed cultivation.

In spite of rice being an important ingredient of household food basket, the yield level has been low and uncertain in India. The operational holding size is shrinking and the land and water resources are also being degraded. Apart from these, the demand for rice is also in the increasing trend. With all such problems, the food security can be assured only by increasing the rice production. The conventional method of rice cultivation will not be able to meet out the increasing demand, the main reason being low productivity. These issues can be met out only using resources efficiently by adopting new precision technologies with more emphasis on increased productivity. One such technology is the System of Rice Intensification (SRI). The SRI technology was investigated thoroughly after laying several demonstrations at field and found effective and hence, the Tamil Nadu Agricultural University has started popularizing such technology as the one for precision agriculture. This technology alone can increase the rice productivity with a comprehensive package of practices involving less seed, water, chemical fertilizers and pesticides.

6.4 AGRICULTURAL ECONOMY OF TAMIL NADU

Tamil Nadu Economy is an agricultural economy which depends mostly on agriculture. Around 60 per cent of the agricultural households either in one way or other depend on agriculture and its contribution. Around 85- 90 per cent of the farmers are marginal and small farmers in terms of number and area. Around 13 per cent of the state income generation is obtained from agricultural sector. The net sown area is found to be 62.59 lakh ha which is accounted for 37.50 per cent to the total geographical area. The area under paddy crop alone is estimated at 20 lakh ha with a productivity of only 2777 kgs. per ha. The paddy production is accounted for 75 per cent of total food grain production.

6.5 SYSTEM OF RICE INTENSIFICATION (SRI) – A PRECISION TECHNOLOGY

The SRI methodology was developed by Fr. Henrie de Laulanie, a French Jesuit who assembled the set of simple, but synergistic practices that now constitute SRI. SRI changes the management of rice plants and of the soil, water and the nutrients that support them in simple but specific ways. SRI is not a set of technology but rather derives from certain insights about how rice plants can be induced to become more productive particularly by eliciting greater root growth (TNAU, 2009).

The system of rice intensification involves cultivating rice with as much organic manure as possible, with young seedlings planted singly at wider spacing in a square pattern, with intermittent irrigation that keeps the soil moist and frequent inter-cultivation with weeder that actively aerates the soil.

6.6 CRITICAL STEPS TO BE FOLLOWED IN ADOPTION OF SRI

The success of SRI entirely depends on the adoption of critical management steps. The steps involved are delineated below.

- Age of the Rice Seedling
- Number of Seedlings per hill
- Square Planting Technique
- Water Management Technique
- Usage of Mechanical Weeder (Cono-Weeder)

6.6.1 Age of the Seedling

Young seedlings if carefully planted into an aerobic soil environment, not submerged in hypoxic soil, have more vitality and resilience, and more growth potential as can be explained in terms of phyllochron analysis (Stoop et.al. 2002). So young seedling of two weeks (14 Days) duration is recommended.

6.6.2 Number of Seedlings per Hill

This intuitive notion is contradicted by the known fact that when any plant is crowded together, the root growth is inhibited, something that applies to rice as much as any other plant species. Hence, planting of single seedling should be adopted under this technology.

1. Recommended Seed Rate is at 7.50 kgs / ha
2. Seed treatment should be done with Pseudomonas 10 g per kg of paddy seed
3. Azophos Biofertilizer at the rate of 75 grams per kg of seed should be used
4. Soaking of seeds for 24 hours
5. Incubation for 24-36 hours
6. Preparation of raised seed beds @ 1 X 5 M and 20 beds are required per ha
7. Application of finely powdered organic manure before and after spreading of paddy seeds in nursery
8. Powdered DAP may be applied @ 95 grams per bed in total of 1.90 kg should be used.
9. Spread Polythene sheets over the beds evenly. Old poly-sacks can also be used
10. Fill the Soil evenly over the sheets up to 4 Cm.
11. Uniformly spread 375 grams of seeds in each 5-sq. meter nursery bed
12. Watering is to be given using the Rose can. It is an advisable one
13. Cover the seed bed using the locally available mulching materials like straw
14. Nursery area is reduced to 250 Square Meter
15. Thin spreading of seeds to avoid congestion is recommended
16. Deep ploughing is to be carried out

17. Perfect leveling of floor must be done to raise the crop
18. Formation of drainage channels for every 2 meters
19. Running of markers to draw lines between drainage channels
20. Transplanting 14 days old seedlings with 2-3 leaves
21. Pulling the seedlings along with seed and mud without disturbing the roots
22. Planting at 25cm x 25cm spacing
23. Planting only one seedling per hill
24. Planting in shallow depth
25. Maintaining thin film of water while transplanting
26. Maintaining the field in saturated condition always without water stagnation
27. Weeding with Cono-weeder at 10 days interval
28. Incorporation of weeds into soil enhances nutrient enrichment

6.6.3 Square Planting Technique

High density among plants inhibits their growth and performance, although spacing is a parameter to be optimized rather than maximized so as to get the greatest number of large and fertile panicles per unit area. Square planting with 25 cm X 25 cm ensures optimum space for efficient utilization of resources. Besides, it eases operation of weeder usages.

- SRI is entirely a new concept and certain critical elements are new to the farmers who are in conventional mode
- Square Planting is one such element to be practiced
- Rolling marker is a Unique tool in easing the square planting
- It reduces the drudgery in transplanting
- Eases the labor and the process gets simplified
- Squared impressions laid in the field and single seedlings are planted

6.6.4 Water Management Techniques

Hypoxic conditions cause rice plant roots to degenerate. Continuous flooding early in the growth cycle will diminish the size and health of plants roots systems, making them less able to tolerate water stress later in the cycle. Plants with truncated roots cannot access the residual soil moisture in lower horizons that is accessible to plants which have large and functioning roots systems to maintain their growth and productivity. Hence, alternate wetting and drying is advocated.

6.6.5 Recommended Water Management for SRI

- Agriculture is the single largest consumer of water in the state using 75 per cent of the state's water and rice is the principal food crop of the state and majority of area is under irrigation
- It has been found that in the most intensively cropped areas under rice, where ground water is often used for irrigation, water table has been falling at the alarming rate of one meter per year or more
- Water management is one of the critical steps in SRI and provision of aerobic environment in rice fields is the core point in SRI
- Irrigate to 2.5 cm depth (after hairline crack formation up to panicle initiation and after that after disappearance of ponded water
- Around 40 to 50 per cent of water saving is visible from planting to harvest of paddy

6.6.6 Mechanical Weeder Usage

Square planting eases the cono / rotary weeder operation in two directions. And thereby weed management could be effected efficiently. The greatest part of the space between plants is cleaned by the "teeth" on the small rotating wheels of such weeder and weeds germinating near the plants will be covered by the churned-up mud, which will give the rice enough time to grow in advance of the weeds. Hence, mechanical weeding is recommended four times from 10 days after planting

- Usage of weeder is one of the critical step in SRI. The Cono-Weeder should be used at 10 days interval from the transplanting day
- Three laborers are enough to weed one acre. Weeds are trampled and on decay the nutrients are ploughed back to the soil.

- Soil is frequently disturbed which has beneficial physic-chemical and biological results in soil.
- Root pruning triggers the tillering that results in bursting out of tillers
- Water level should be properly monitored for usage of weeders
- The left-out weeds should be hand weeded.

6.7 CROP RESPONSE

Figure 6.1 Wider Spacing for Weeder Operation

- Weeding permits higher root growth.
- Higher tillering of even shy tillering genotypes;
- Higher plant height;
- Increase in number of productive panicles per square meter;
- Higher number of grains per panicle
- Higher percentage of filled grains per panicle
- Resistance to Lodging
- Leaves remain green up to harvest
- Resistance to adverse effects of drought
- Higher nutrient use efficiency
- Higher biochemical activities
- Higher grain yield and straw yield

6.7.1 Profitability

- Higher Out turn of Milled Rice
- Higher net profit

6.7.2 Other Benefits

- All genotypes (high yielding varieties, hybrids, land races) respond well
- Reduced rodent damage in the field

6.7.3 Advantages of SRI

6.7.3.1 *Nursery*

- Reduction in nursery area from 800 square meters to 100 square meters (to plant one ha)
- Drastic reduction in seed rate from about 60 kg per ha to 7.50 kg per ha
- Lesser pest problems in the nursery
- Reduction in nursery duration from 21-30 days to 14 days

6.7.3.2 *Transplanting*

- Square planting to enable the crisscross use of weeder
- Reduction in planting density from 50 – 66 hills per square meter to 16 hills per square meter
- Reduction in number of seedlings per hill from 2- 3+ to One

6.7.3.3 *Weeding*

- No requirement of herbicide
- One laborer with a rotary weeder can cover an area of about 0.13 ha in a day
- Reduction in labor requirement for weeding from 30- 60 per ha to 25 per ha
- Multiple benefits of using weeder (weed control; less labor for weeding; incorporation of top dressed fertilizer; disturbance to the soil; pruning of older roots increased crop growth)
- Timely removal of weeds facilitating latter absorption of applied nutrients
- Small farmers can do the weeding by themselves.

6.8 SIGNIFICANT IMPACT EVENTS ON RICE IN INDIA

Rice is being a staple food to Indians, the rice related events are taking place in the country. The series of events taken place have been documented and presented below in Table 6.1 for the benefit of the readers to understand the significant developments.

Table 6.1 Significant Rice Events happened in India

Sl. No	Year	Name of Events
01	1906	Introduction of Single Seedling Planting in Madras Presidency
02	1911	Introduction of *Gaja* Planting in Madras Presidency
03	1912	Beginning of Systematic Study on Rice Production in Agriculture School established at Coimbatore
04	1946	Establishment of the Central Rice Research Institute at Cuttack
05	1951	Introduction of the Japanese method of rice cultivation
06	1965	Launching of All-India Coordinated Rice Improvement Project (ICAR)
07	1969	Introduction of IR8 variety, surge in expansion of rice area in Punjab
08	2000	First Evaluations of SRI, followed by Trials in other States of India
09	2006	First National SRI Symposium held in Hyderabad
10	2007	Annual Symposium on SRI held in Agartala
11	2007	SRI Technique has been incorporated as one of the Major activity in the TN IAM WARM Project operated in the State of Tamil Nadu with World Bank Funding support
12	2008	Symposium on SRI held at Tamil Nadu Agricultural University, Coimbatore

6.9 AREA UNDER SRI

SRI has greater potential for tillering and root growth; earlier arrival in the main field extends time for tillering. Seed requirements are reduced. Wider spacing in this technology facilitates no competition for nutrients, water and space. Inter-cultivation with Cono- weeder is also made possible. Water saving is achievable up to 40 per cent.

Table 6.2 SRI on Rice Area, Production and Productivity in Tamil Nadu

Year	Rice Area (Lakh ha)	Area under SRI (Lakh ha)	Productivity (kg/ha)	Total production (Lakh Tonnes)
Base year(2002-06)	17.54	-	2667	47.36
2007-08	17.89	4.20	2817	50.40
2008-09	19.32	5.38	2682	51.83
2009-10	18.46	6.49	3070	56.65
2010-11	19.06	8.50	3039	57.92
2011-12	19.05	10.01	3915	74.59
2012-13	14.93	-	2712	40.50

(Base year-5 year Average (normal); Season and Crop Report, Directorate of Economics and Statistics, Government of Tamil Nadu, Chennai).

6.10 KNOWLEDGE AND EXPERIENCE IN SRI

In 2004, Government of Tamil Nadu implemented a pilot project funded by World Bank in Adavinayinar river basin (Part of Tamiraparani basin) in which enhancing water-use efficiency was a key objective. Tamil Nadu Agricultural University (TNAU) was involved in this project as one of the constituent department in the Tamil Nadu Irrigated Agriculture Modernization and Water Bodies Restoration and Management (TN IAM WARM) Project sponsored by World Bank and SRI was one of the principal technology to be demonstrated to farmers both by TNAU and the Department of Agriculture. Positive results from this pilot project paved the way for the implementation of a five-year project, Irrigated Agriculture Modernization and Water bodies Restoration and Maintenance (IAM WARM) by World Bank in 63 river basins from 2007, with a budget allocation of Rs.27420 million. SRI was included in this project as a major thrust in the agricultural component.

6.10.1 SRI Initiatives in Tamil Nadu

- SRI experiments were conducted during 2003-2004 at AC&RI, Killkulam, TNAU. Results revealed that on an average, grain yield of 3,892 kg/ha which was 28 per cent higher than the conventional farm was realized and 53 per cent less irrigation water required for SRI farms over conventional farms.

- Evaluation of two on-farm trials in Tamiraparani basin in south Tamil Nadu revealed that mean grain yields under SRI was 1.57 tonnes per ha and around 31 farmers recorded grain yields of over 8 tonnes per ha under SRI (Thiyagarajan, 2006). This is again supported with a case study. The results of case study are presented elsewhere.
- TNAU has recommended SRI as a technology in Tamil Nadu to increase rice productivity and save irrigation water.
- The State Department of Agriculture held demonstration trials in all rice-growing areas of the state during 2004.
- It was planned to bring an additional area of 7.5 lakh ha under SRI cultivation in Tamil Nadu during 2008-09 (Policy Note, 2007-08).

SRI was introduced in Tamil Nadu in 2007. Since its introduction, the area under SRI is on the increasing trend with 10.01 lakh ha in 2011-12. The yield levels are also consistently increasing due to large scale adoption. The favorable monsoon environment and SRI acreage led and SRI acreage led to the increase in rice productivity.

The SRI efficiently uses the scarce resources and is accessible to the poor farmers. **SRI is initially labor-intensive**. It needs 50 per cent more man-days for transplanting and weeding. Once the right skills are learnt and implemented, the labor costs will be lesser.

On the other hand, the traditional method of paddy cultivation requires large quantities of inputs, the other hand, the traditional method of paddy cultivation requires large quantities of inputs, like water, fertilizers and pesticides, being used indiscriminately. However, in spite of many advantages, farmers have difficulties in adopting SRI as it requires high precision in handling of important resources.

Figure 6.2 More Number of Panicles per Hill in SRI

The average grain yields obtained by farmers in Conventional and SRI cultivation in various river basins assisted by the World Bank-IAMWARM Project in Tamil Nadu are presented in Table 6.3.

Table 6.3 Performance of Paddy under SRI Technique in River Sub Basin Commands

Sl. No	Name of the River Sub Basins	Grain Yield in Tonnes per Ha		Maximum Grain Yield in SRI (Kgs)	Percentage Increase in SRI over Conventional
		Conventional	SRI		
1	Aliyar	3484	4488	6475	28.82
2	Arjunanadhi	5488	6375	7880	16.16
3	Kottakkaraiyar	4044	5023	7500	24.21
4	Manaimuthar	4185	5596	7500	33.72
5	Pambar	4050	4924	6740	21.58
6	South Vellar	4100	6005	7500	46.46
7	Upper Vellar (Plate 1)	5250	7140	9987	36.00
	Average	4372	5650	7655	29.23

Average yield with conventional methods ranged from 3.5 tonnes per ha to 5.3 tonnes, while the SRI yields ranged from 4.5 tonnes per ha to 10.0 tonnes per ha . The increase in yield ranged from 16.16 to 46.5 per cent, with an average of 29.23 per cent. Field reviews have concluded that not all of the recommended SRI practices have been fully used by the farmers, or used as recommended, so these results are from what should be considered as 'Partial SRI.'

The Extension Functionaries of Department of Agriculture, Government of Tamil Nadu has practiced this "Partial SRI" because of several constraints. This problem of assessing what represents full or correct SRI is a widespread one for this innovation, since SRI use, with its set of half a dozen practices, is a matter of degree more than of kind. However these are discussed as a case study in the Tamirabarani Sub basin by conducting exclusive study to assess the efficiency of the SRI technology.

Figure 6.3 Vice Chancellor, TNAU is in Discussion with SRI Farmers

6.10.2 Other State Experiences in SRI

6.10.2.1 *Andhra Pradesh*

- In AP, SRI method assessment project was jointly undertaken by WWF-ANGRAU in rabi season
- 250 paddy farmers spread over 11 districts were targeted for implementation
- Study results revealed that 50 per cent of water and 95 per cent of seed is saved in SRI (WWF, 2005).
- Plants look healthier and yield was higher in SRI cultivation.
- Led to popularization of the program with the budget outlay of 4 crores to take up SRI demonstration plots in each village of the state to train the farmers, supply of implements like cono-weeders and markers at 50 per cent subsidy to the farmers (The Hindu, dated November 16, 2005, AP).

6.10.2.2 *West Bengal*

- West Bengal - PRADAN- an NGO studied the experiences of farmers adopting SRI technique during 2004 Kharif season.
- SRI plots produced an average of 32 per cent higher paddy yields even with the partial adoption of SRI practices.
- 59 plots in Balrampur, the average paddy output from SRI farms was 6,282.65 kg/ha (49.8 per cent higher) compared to 4,194.13 kg per ha in conventional fields.
- Average straw output was 5.15 tonnes per ha in SRI Fields compared to 3.46 tonnes per ha in conventional fields.

- In Jhalda block, the increase was only 11.9 per cent for a number of reasons including drought, only one weeding and transplantation of older seedlings.

6.10.2.3 *Gujarat*

- Gujarat - SRI method yielded 5.81 tonnes per ha with 46 per cent less water usage while the conventional practice yielded 5.84 tonnes per ha of grain yield.
- JDM Foundation in Punjab - SRI practice is the answer to Punjab's water problems as it can lead to water savings of up to 60-70 per cent during the paddy-growing season (Sambu Prasad, 2006).

6.10.3. SRI Initiatives in the Rain-fed Lands

Although SRI was developed for irrigated transplanted rice cultivation, the principles of SRI have been extended and adapted to rain-fed rice cultivation with some modifications. In Eastern India, SRI principles have been applied for both transplanted and direct-seeded rice under rain-fed conditions. For transplanted rice, the practices are similar to irrigated SRI except for adjustments in water management which depends upon rainfall. Keeping the paddy fields under standing water (when there is sufficient rainfall) is avoided with proper drainage. With direct seeding, lowering the seed rate, and wider spacing of rows (in the case of line sowing), thinning of the plant population, and inter-cultivation have been followed. The main objective is to reduce labor requirements. In Dharwad region of Karnataka, AME Foundation has been spearheading SRI in rain-fed areas. AMEF (Dharwad) has developed a 4-row seed drill to sow seeds at 10 inch intervals instead of the 6-row seed drill with 6-8 inch rows already being used by farmers.

6.11. APPLICATION OF SRI PRINCIPLES TO OTHER CROPS

The term 'SRI' is more appropriately used as an adjective rather than as a noun. We should avoid the reification of SRI, i.e., making it into a thing when it is better thought about in terms of characteristics and qualities, framing issues and choices as matters of degree rather than of kind. Water management must be done differently in upland compared to lowland areas, but SRI principles for managing water can be made relevant for upland farming. Spreading of SRI should be approached more as a problem solving exercise than as usual top-down 'extension' (Uphoff, 2008). These developments have taken place in most places due to the enthusiasm of farmers and NGOs assisting them, spurred by their observations on the effects of SRI, without the intervention of any established research systems. The System of Wheat Intensification

(SWI) is becoming popular in Himachal Pradesh, Uttarakhand, and Uttar Pradesh. SWI has been recognized as the 'Innovation of the Week' on the World watch Institute's 'Nourishing the Planet' blog in May 2011. In Africa, SWI is being successfully tried in Mali and Ethiopia also. SRI concepts and methods have been extended to other crops as well. The Sustainable Sugarcane Initiative (SSI) for sugarcane is being tried in five states and on a fairly large scale in Uttar Pradesh.

SRI principles of lower seed rate, limited water use, and inter-cultivation are being applied to crops like finger millet (Ragi), redgram, mustard, pigeon pea, and other crops in various states of India. Intuitive farmers and NGOs are terming the new method of cultivation as the 'System of Crop Intensification' (SCI), the 'System of Root Intensification', or the System of Sustainable Crop Intensification (SSCI). An SRI Secretariat promoted by SDTT and hosted by the "LIVOLINK FOUNDATION" at Bhubaneswar, Odisha, has taken the initiative to document farmers' practices of applying the SRI principles for growing crops like wheat, finger millet, rapeseed / mustard, brinjal, tomato and chilli.

6.12 ECONOMICS OF SRI IN TAMIL NADU

With this background, a study was taken up in Tamil Nadu with the following specific objectives to highlight the effectiveness of the technology in relation to Socio-Economic Benefits. They are;

1. To identify the reasons for adoption of SRI technique by the farmers,
2. To assess the economic feasibility of SRI technique over the conventional paddy cultivation, and
3. To identify the constraints in adoption of SRI technique and to suggest the suitable policy measures to enhance rice productivity in Tamil Nadu.

6.12.1 Sampling and Methodology

Tamil Nadu is one of the developed state in India in terms of adoption of agricultural technologies and is fulfilling the self sufficiency in food production through concerted efforts. That too precision agriculture is one of the technology in which Tamil Nadu has gained much expertise. To document the experiences of Tamil Nadu, a sample survey and its test results were made as a paradigm to the followers across the country. To have a detailed study, the most popular zone of Tamil Nadu in terms of Agricultural Diversification is Western Zone. Hence, the Western Zone of Tamil Nadu was selected for the study. One block per district was selected namely Pollachi and Sathyamangalam blocks in Coimbatore and Erode districts, respectively. Structured interview schedule was used to collect the information on cost, income, reasons for adoption and the constraints in adoption of SRI technology. The required information was collected from 30 farmers who adopt the SRI technique (adopters) and 30 farmers who are not adopting the technique (non-adopters). The data pertained to the crop year 2007-08. Average and percentage analysis was used wherever possible. Reasons for adoption of SRI technique and the constraints encountered by the farmers were identified by using Garrett Ranking Technique.

Partial budgeting

Partial Budgeting is normally used to estimate the changes in cost and returns due to new technology like SRI in the place of conventional method of planting of paddy. If the net gain is positive, it gives incentives to the farmer to decide on adopting the SRI technique.

6.12.2 Results and Discussion

Under this head, the extent of adoption of SRI technique by the farmers in the study environment, varietal adoption, comparision of SRI Technique with that of traditional method of cultivating the rice, factors influencing the adoption of SRI Technique and the costs and returns realized in the practices are highlighted for the benefit of the stakeholders.

6.12.2.1 *Adoption of SRI Practices*

Nearly 67 per cent of farmers adopted SRI technique from the training offered by the Dept. of Agriculture followed by Tamil Nadu Agricultural University under TN IAM WARM Project. In addition, 60 per cent of farmers adopted SRI only in the past 6 months (one season) and 20 per cent of the farmers adopted this technique for the past 2 years. Moreover, 60 per cent of farmers adopted SRI technique in the 2nd season (November to February) since no assured irrigation from the canal during the 1st season (Kuruvai).

6.12.2.2 *Varietal Adoption in the Study Area*

The major ruling paddy varieties adopted are ADT45 (30.46 per cent) and ASD16 (24.49 per cent) in case of non –adopters and ADT45 (26.09 per cent) and ADT43 (25 per cent) in case of adopters. CORH3 hybrid is also being cultivated under SRI farms as revealed in the following graph on varietal adoption in the Tamirabarani sub basin.

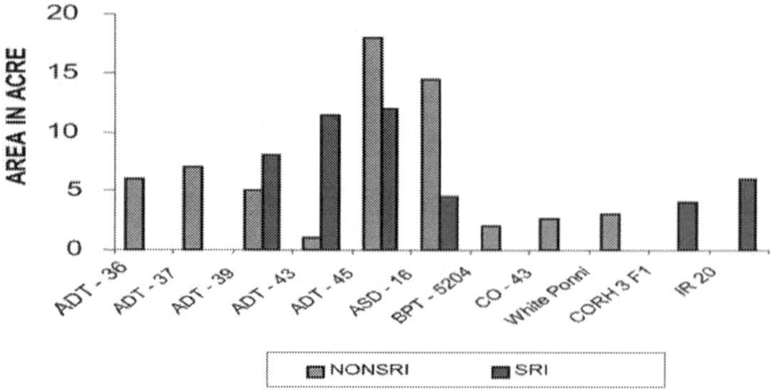

Figure 6.4 Varietal Adoption in the Study Area

6.12.2.3 *Comparison of SRI Farmer's Practices over Conventional Method of Paddy Cultivation*

The comparison of package of practices followed in paddy cultivation under conventional method, SRI practices recommended and the farmer's practices made in respect of SRI were compared and the details are presented in Table 6.4.

Table 6.4 Comparison of SRI Farmer's Practices over Conventional Paddy Cultivation

Input/ Activity	Conventional Paddy	SRI technique – Recommended	SRI – Farmer's Actual Practices
Seed	20-30 kg per acre	2 kg per acre	21 kg per acre
Spacing	Closer spacing 15X10, 20X10, 20X20	25 X 25 cm	Not all of them practiced
Transplanting	Seedlings about 30 days old	8-14 days old	20-25th day
No. of seedlings per hill	2-3 seedlings	Only one seedling	1-2 seedlings
Fertilizers	Application of chemical fertilizers, insecticides and herbicides	Preference given to organic fertilization, insecticides and herbicides	Mostly Chemical fertilizers are used

Input/ Activity	Conventional Paddy	SRI technique – Recommended	SRI – Farmer's Actual Practices
Plant protection	Application of insecticides and herbicides, 2-3 times spray	Insecticides, pesticides are not necessary	One time spray is followed when it is required
Weeding	Manual Weeding, Herbicide application	Non chemical means of weed control by Cono-weeder (2 times)	Cono-weeder (2 times) + Manual weeding (2 times)

Source: Primary Survey

Table 3 revealed that the farmers are actually not practicing the recommended technology in spirit. There is violation of the technology follow up. It might be mainly due to the reluctance in switching over to the newer technology. It might be due to risk averting nature that prevailed among the farmers. Recommended seed rate is only 2 kg per acre. But they use 21 kgs per acre. It is an increase of ten time's higher input application.

Square planting is recommended with the spacing of 25 cm. But the farmers are practicing line planting as advocated in the Department of Agriculture. It had visualized the technology gaps. In all the practices with respect to input application, the farmer's practices were varying with respect to the recommended technology and hence variation in yield could be visualized. The details we have discussed in **Table 2** itself.

6.12.2.4 *Factors Influencing Adoption of SRI Technique*

Factors influencing adoption of SRI technique was assessed using Garrett Ranking Technique (Table 6.5) and the results revealed that minimum seed requirement, low nursery duration and availing subsidy from the department were the major factors influencing the adoption of SRI by the famers.

Table 6.5 Factors influencing Adoption of SRI technique

S.No.	Reasons for adoption	Score	Rank
1	Minimum seed requirement	74	I
2	Low nursery duration	57	II
3	To avail subsidy	52	III
4	Low nursery cost	51	IV
5	Less cost of cultivation	43	V
6	Low labour requirement	37	VI
7	Motivation by Extension officials	35	VII

Source: Primary Survey

Among all the reasons, as revealed in **Table 4**, the Minimum Seed Requirement found to be the principal reason among the farmers of Tamirabarani river sub basin and hence the adoption is predominant. If the farmers are really want to minimize the cost of production, one should follow the recommended practices by the Tamil Nadu Agricultural University, Coimbatore.

6.12.2.5 *Operational Cost of Cultivation*

The operational cost of cultivation was worked out to Rs.33610 and Rs.36153 per ha for adopters and non –adopters, respectively. Though the operational cost of cultivation is non-significant, there are other inputs which is significantly differ with SRI adopters and non –adopters (Table 6.6). The cost incurred in nursery preparation, seed, irrigation, weeding and plant protection chemicals were found to be statistically significant.

Table 6.6 Operational Cost of Cultivation among SRIAdopters and Non Adopters

Operational cost	SRI Adopters		Non –adopters		'T' value
	Amount (Rs./ha)	% to total	Amount (Rs./ha)	% to total	
Nursery preparation	687.86	2.05	1083.39	3.00	1.51 ***
Main field preparation & Transplanting	9612.03	28.60	10521.08	29.10	0.98

Operational cost	SRI Adopters		Non -adopters		'T' value
	Amount (Rs./ha)	% to total	Amount (Rs./ha)	% to total	
Seed	300.85	0.90	1446.85	4.00	26.19*
Fertilizers	3642.51	10.84	4024.59	11.13	1.25
Irrigation	1599.97	4.76	1889.74	5.23	1.66**
Weeding	4769.27	14.19	2866.38	7.93	-3.30*
Plant protection chemicals	641.80	1.91	985.43	2.73	2.07**
Harvesting and threshing	11588.91	34.48	12511.26	34.61	0.64
Interest on working capital	766.34	2.28	824.34	2.28	
Total Operational Cost	33609.54	100	36153.06	100	1.20

Note: *Significant at 1 %, ** Significant at 5%; *** Significant at 10%

Source: Primary Survey

6.12.2.6 *Comparison of Yield and Return*

The average yield was worked out to 67.23 quintals and 54 quintals per ha for adopters and non-adopters respectively (Table 6.7). It showed significant difference between adopters and non adopters i.e. 13.23 quintals higher in SRI farms and is found to be statistically significant. Similarly, gross income and net income over operational cost was also found to be statistically significant at one per cent level.

Table 6.7 Yield and Income of SRI Technique over Conventional Paddy Cultivation (Rs/Ha)

Sl. No.	Particulars	SRI Adopters	Non-Adopters	'T' value
1.	Yield: Main product (Quintals)	67.23	54.00	4.42 *
2.	Price realized by the farmer (Rs./ Quintal)	842.67	811.33	
3.	Value of Main Product (Rs.)	57034.93	43643.48	
4.	By Product (Qtls)	50.83	48.95	

Sl. No.	Particulars	SRI Adopters	Non-Adopters	'T' value
5.	Price Realized by the Farmer (Rs.)	159.33	124.33	
6.	Value of Bi-Product	7871.07	5471.44	
7.	Gross Income (3+6)	64906.00	49114.92	4.22*
8.	Operational Cost	33609.54	36153.06	
9.	Net Income over Operational Cost (7-8)	31296.46	12961.86	4.05*
10.	Output – Input Ratio	01.93	01.36	

Source: Primary Survey

The Output – Input Ratio was worked out to be 1.93 in case of adopters whereas it was only 1.36 in case of non –adopters **(Table 6)**. It revealed that the SRI Technique is capable of fetching higher out turn per unit of investment.

6.12.2.7 *Success Stories of SRI Techniques across India*

A number of innovative farmers in our country has adopted SRI Method of cultivation of Rice and accomplished commendably high yields over 20 tonnes per ha against the national average of 2 tonnes per ha which is presented in Table 6.8.

Table 6.8 Innovative Farmers and their Achievement in Rice Yield per Ha

Sl. No	Name of the Farmer	Name of the Village	Name of the State	Yield of Rice in Tonnes per Ha
01	Mr. Sethumadhavan	Alanganallur, Madurai District	Tamil Nadu	24.00
02	Mr. Sumant Kumar	Naalanda	Bihar	22.40
03	Mr. Solaimalai	Andaman, Madurai District	Tamil Nadu	20.60
04	Mrs. T. Amalarani	Vasudevanallur, Thirunelveli District	Tamil Nadu	18.00
Average Yield among the Innovative Farms				21.25

(Source: Anon, 2014)

Table 6.8 revealed that the success was made because of the enthusiastic support received from the Scientists and the extension functionaries from the State Department of Agriculture. The Farmers are adviced to adopt such a promising precision technology to get higher yield and at the same time in conserving the natural resource like water (Anon, 2014)

6.12.2.8 *Economic Impact*

The partial budgeting analysis revealed that (Table 6.9) the net gain in adoption of SRI technique over conventional paddy cultivation was Rs.13725 per ha. The cost was reduced in all the activities particularly in seed and transplanting, the cost reduction was found to be more. On the other hand, yield was increased by 13 quintals per ha which was found to be more efficient. The results indicated that SRI technique is found to be economically beneficial for the farmer when compared to the conventional method of cultivating paddy in their holdings and hence the group leaders of the village can always have an eye on newer technologies and its release to infuse the technique to their fellow farmers to reap higher output per unit of land resources.

Table 6.9 Analysis of Economic Impact of SRI using Partial Budgeting Technique

Credit (Gain)	Amount (Rs.)	Debit (Loss)	Amount (Rs.)
A. Added Return		**C. Added Cost**	
Yield: Main product: 13 qtls per ha @ Rs.842 /Qtl	11140	Weeding	1903
By-product:2qtl/ha@ Rs.150/quintal	100		
Sub total	11240	Sub total	1903
B. Reduced Cost		**D. Reduced Return**	
Nursery preparation	396		
Main field preparation & Transplanting	909		
Seed	1146		
Fertilizers	382		
Irrigation	290		

Credit (Gain)	Amount (Rs.)	Debit (Loss)	Amount (Rs.)
Plant protection	344		
Harvesting	922		
Sub total	***4388***	***Sub total***	...
Total Gain (A+B)	15628	Total Loss (C+D)	1903
Net Gain	13725		

Source: Primary Survey

6.12.2.9 *Constraints in SRI Adoption*

The constraints in adoption of SRI technique was assessed using Garrett Ranking technique (Table 6.10) and the results revealed that difficulties are prevalent in use of Cono-weeder by the farmers followed by lack of support from extension officials and health problem due to operation of Cono-weeder were identified as the major constraints in adoption of SRI by the famers.

Table 6.10 Constraints in Adoption of SRI by Farmers

S. No.	Constraints	Scores	Rank
1.	Difficulty in use of Cono-weeder	58.06	I
2.	Lack of technical support from the extension officials	57.67	II
3.	Health problem due to operation of cono-weeder	47.06	III
4.	Lack of awareness among the labourers about SRI technique	37.22	IV

From the study, results concluded that power operated Cono-weeder usage by using either battery or diesel may be activated and supplied to the farmers for efficient operation of Cono-weeder so as to increase the efficiency of the laborers. Skill oriented training should be given to the agricultural laborers regarding SRI practices. This will definitely help the farmers to adopt this technique in a larger way to increase the productivity and net income at farm level and also this technique will be a better solution to meet the food demand. Overall, SRI is a precision technology in which the recommended practices were not implemented by the fellow farmers due to hesitation and reluctance which called for periodic training to them in adopting the technology in full spirit.

6.12.3. SUMMARY

The Sysmtem of Rice Intensification in Tamil Nadu and other places revealed that the rice productivity is increased to many fold when comparing the traditional way of cultivating rice in their fields. This was proved by some case studies conducted in different states of India and Tamil Nadu. For that the input requirement were also less. If the farmer is able to take up the practice of cultivating rice with SRI technology, an absolute increase of 30 per cent yield is visible and hence the technology need to be spread almost in all the places of possibilities.

References

1. Anon (2014), "Innovations in Cultivation of Rice", ***Kisan World,*** 41(8): P.12.
2. Barah, B.C. (2009) Economic and Ecological Benefits of System of Rice Intensification (SRI) in Tamil Nadu, ***Agricultural Economics Research Review***, **22**:209-214.
3. Basavaraja, H., S.B Mahajanashetti, P. Sivanagaraju (2008) Technological change in paddy production: a comparative analysis of traditional and SRI methods of cultivation, ***Indian Journal of Agricultural Economics***, **63**(4):630-640.
4. Government of Tamil Nadu (2008) "***Policy Note on Agriculture 2007-08***", (Ministry of Agriculture: Chennai)
5. Government of Tamil Nadu (2016) "In the Last Five Years, Production Surpassed 100 Lakh Tonnes Four Times" ***The Hindu (Chennai Edition),*** Date: 20.08.2016
6. The Hindu (2015) Precision Agriculture in AP, The Hindu Dated November 16, 2015
7. Shambu Prasad (2006) "***System of Rice Intensification in India - Innovation History and Institutional Changes***", WWF Project on Dialogue on Water, Food and Environment
8. Thiagarajan, T.M. (2006) "Adopting SRI in Tamil Nadu, India", ***LEISA Magazine***, 22(4): December 2006
9. Thiagarajan, T.M. and B. Gujja (2013) "***Transferring Rice Production with SRI – Knowledge and Practice***" (AGSRI: Hyderabad) pp. 01-205

Web References

- http://blogs.worldwatch.org/ nourishing the planet/tag/system of-wheat-intensification/

CHAPTER VII

Precision Aquacultural Practices in India

Dr. C. Sekhar and Dr. A. Pounraj

Objectives

- To introduce the users of the book to the Precision Aquacultural Practices to develop expertise
- To highlight the economic incentives associated with the Precision Aquacultural Practices which are practiced by the successful entrepreneurs

Organization of the Chapter

- Introduction
- Aquaculture Scenario in the World
- Aquaculture Scenario in India
- Evolution of Rice Fish Culture Systems in India
- Impact of Aquaculture Practices
- The Positive Aspects of Aquaculture
- The Possible Conflicts in Aquacultural Practices
- Precision Technologies for Inland Fish Culture
- Best Aquacultural Practices and its Certification
- Summary

7.1 INTRODUCTION

Aquaculture is defined as the rearing of fish in artificial or natural bodies of water by manipulation of the environment with the aim of increasing production beyond natural limit. Fish production currently generates less than 3 per cent in aquaculture where as with appropriate promotional strategies it could match capture fisheries output and cost effectiveness (Jamu and Ayinla, 2003) is also visible.

Aquaculture -- also known as fish or shellfish farming -- refers to the breeding, rearing, and harvesting of plants and animals in all types of water environments including ponds, rivers, lakes, and the ocean. Researchers and aquaculture producers are "farming" all kinds of freshwater and marine species of fish, shellfish, and plants. Aquaculture produces food fish, sport fish, bait fish, ornamental fish, crustaceans, mollusks, algae, sea vegetables, and fish eggs.

Aquaculture includes the production of seafood from hatchery fish and shellfish which are grown to market size in ponds, tanks, cages, or raceways. Stock restoration or "enhancement" is a form of aquaculture in which hatchery fish and shellfish are released into the wild to rebuild aquarium trade, and growing plant species used in a range of food, pharmaceutical, nutritional, and biotechnology products. Wild populations or coastal habitats such as oyster reefs. Aquaculture also includes the production of ornamental fish for the entertainment purpose.

"Precision Aqua Farming" is practiced by regulating the resource management systems both in Inland and brackish water environment besides regulating or feeding regularly with the prescribed or recommended feed rate for the species chosen for aquaculture practices. Commonly this precision aqua farming is practiced for shrimp production. The resource management systems are well defined and monitored using the equipment like "***Xpercount***" which depicts the growth and number of aquatic population regularly. The concept of Precision Aqua Farming is being practiced following the concept of "Low Investment Higher Income".

Marine Aquaculture refers to the culturing of species that live in the ocean. U.S. marine aquaculture primarily produces oysters, clams, mussels, shrimp, and salmon as well as lesser amounts of cod, yellowtail, barramundi, seabass, and seabream. Marine aquaculture can take place in the ocean (that is, in cages, on the seafloor, or suspended in the water column) or in on-land, manmade systems such as ponds or tanks. In India, cage culture is practiced for raising oysters, clams, shrimp, crabs etc.

Recirculation aquaculture systems that reduce, reuse, and recycle water and waste can support some marine species.

Freshwater or Inland Aquaculture produces species that are native to rivers, lakes, and streams. U.S. freshwater aquaculture is dominated by catfish but also produces trout, tilapia and bass. In Indian waters, mostly the Catla, Rohu, Mrigal, Common Carp, Grass Carp, are cultivated in the same water. This kind of practice is otherwise called as Composite Fish Culture. Whereas, Tilapia, Murrel and Catfish are cultivated individually as monoculture in the fresh water environment. Freshwater aquaculture takes place primarily in ponds and in on-land, manmade systems such as recirculation aquaculture systems.

Brackish Water Fish Farming is a system of aquaculture that focuses on the production of quality fin and shell fish that are found in the creeks, lagoons, and estuaries through rational rearing. It has a capacity of bridging through wide gap between fish demand and supply. Fish production from aquaculture is seen as the only means to bridge the widening gap between domestic fish supply from depleting return from capture fisheries and demand. To augment production from the aquatic resources, in the form of shrimps, crabs, bivalves and fishes from the low lying, barren, unproductive or marginally productive coastal saline lands, swamps and other brackish water bodies are to be brought under coastal aquaculture. Modern coastal aquaculture is an offshoot of the traditional aquaculture and it is largely confined to shrimp aquaculture (Harikumar and Rajendran, 2007).

Aquaculture is viewed as alternative option for increasing fish/shrimp/prawn production. The technological development in the inland fisheries paved the way for making use of inland water resources spread over the different districts of the State for aquaculture development. Aquaculture currently enjoys the distinction of being one of the fast growing food production sectors in the State. But following the wide spread out breaks of viral diseases, the rate of development has declined sharply. Serious concerns were voiced at this time about the future of this sector. The State government is now in the process to formulate a 'Master plan for aquaculture' (Harikumar and Rajendran, 2007).

While the shrimp-oriented aquaculture industry in India recorded exceptional growth for the last three decades in spite of its high exposure to risk and uncertainties, the farming /culture of various other species has not picked up to the expected level enabling the optimum use of potential areas suitable for aquaculture (Sathidas et.al., 2009).

7.2 AQUACULTURE SCENARIO IN THE WORLD

Aquaculture remains a growing, vibrant and important production sector for high protein ood. The reported global production of food fish from aquaculture, including, crustaceans, molluscs and other aquatic animals for human consumption, reached 52.5 million tonnes in 2008. The contribution of aquaculture to the total production of capture fisheries and aquaculture continued to grow, rising from 34.5 percent in 2006 to 36.9 percent in 2008. In the period 1970–2008, the production of food fish from aquaculture increased at an average annual rate of 8.3 percent, while the world population grew at an average of 1.6 percent per year. The combined result of development in aquaculture worldwide and the expansion in global population is that the average annual per capita supply of food fish from aquaculture for human consumption has increased by ten times, from 0.7 kg in 1970 to 7.8 kg in 2008, at an average rate of 6.6 percent per year.

Production from aquaculture is mostly destined for human consumption. Globally, aquaculture accounted for 45.7 percent of the world's fish food production for human consumption in 2008, up from 42.6 percent in 2006. In China, the world's largest aquaculture producer, 80.2 percent of fish food consumed in 2008 was derived from aquaculture, up from 23.6 percent in 1970. Aquaculture production supplied the rest of the world with 26.7 percent of its food fish, up from 4.8 percent in 1970 (SOFIA, 2010). Despite the long tradition of aquaculture practices in a few countries over many centuries, aquaculture in the global context is a young food production sector that has grown rapidly in the last 50 years or so. World aquaculture output has increased substantially, from less than 1 million tonnes of annual production in 1950 to the 52.5 million tonnes reported for 2008, increasing at three times the rate of world meat production (2.7 percent from poultry and livestock together) in the same period. In contrast to world capture fisheries production, which has almost stopped growing since the mid- 1980s, the aquaculture sector maintained an average annual growth rate of 8.3 percent worldwide (or 6.5 percent excluding China) between 1970 and 2008. The annual growth rate in world aquaculture production between 2006 and 2008 was 5.3 percent in volume terms.

The growth rate in the rest of the world (6.4 percent) from 2006 to 2008 was higher than that for China (4.7 percent). The value of the world aquaculture harvest, excluding aquatic plants, is estimated at US$98.4 billion in 2008. The actual total output value from the entire aquaculture sector should be significantly higher than this level, because the value of aquaculture hatchery and nursery production and that of the breeding of ornamental fishes are yet to be estimated and included. If aquatic plants are included, world aquaculture production in 2008 was 68.3 million tonnes, with an estimated value of US$106 billion.

Rice and Shrimp Farming in Vietnam

According to Sammut 2017, Improved Management Practices (IMP) are helping rice-shrimp farmers improve their yields despite high salinity. The experience sharing is furnished as follows.

The People have farmed rice in the Mekong Delta for centuries. More recently, the farmers near the coast have diversified their rice systems by also growing shrimp, either concurrently or in rotation with the rice. This makes the most of the available resources during the dry season, when salinity is too high to grow rice.

Rice is grown on the platform during the wet season when salinity is usally low. Shrimp are grown in the surrounding ditch and other water compartments on the farm in the dry season when salinity is high, but can be grown during the wet season if the shrimp are acclimatized. In recent years, even in the wet season salinity levels in soil and water have been too high for rice, causing crop losses or very low yields.

The project on Improving the Sustainability of Rice-Shrimp Farming Systems in the Mekong Delta, Vietnam has been helping the farmers find solutions. The project is exploring ways to improve yields of both rice and shrimp.

For Example: Rice yields improved with modified rice platform techniques, changing the timing of planting and using salt-tolerant rice varieties. Fertilizer costs have also been reduced by using waste from the shrimp farming.

At Tan Bang Commune in Camau Province, where some of these practices have been tested, trial farms are getting rice yields upto five times higher than nearby farms that use traditional cultivation methods. Farmers are noticing the results of the research and are eager to apply the better management practices. With the help of the Department of Agriculture and Rural Development (DARD), the project team is promoting the improved practices across four rice-shrimp farming provinces.

Improving shrimp production has been more challenging, because of the high water temperatures, low dissolved oxygen concentrations and low natural food production for the shrimp. The team is investigating alternative practices to be tested in the next phase of the project, including levels of intensification, species selection and modifying the design of the farm. The project's success lies with its research partnership model, capacity building activities and regular engagement with stakeholders.

7.3 AQUACULTURE SCENARIO IN INDIA

Aquaculture has emerged as one of the fastest growing food farming systems at global level with enormous potential for further development. Although India with a production of approximately 2.3 million tonnes per annum is the second largest aquaculture producer, its contribution is hardly 5 per cent of the global production. However, the country is endowed with a long coastline of 8129 km, 1.2 million hectares of potential brackish water area and 0.5 million sq.km of continental shelf with diverse ecosystems, offering vast scope for development and diversification of coastal aquaculture (Sathidas *et.al.* 2009). According to the information available with MPEDA 13990.53 ha is under shrimp culture during 2000- 2001. This comes around to only 22 per cent of the potential area. Both Public and private sectors are engaged with shrimp farming in Kerala. 7327 metric tons of shrimps are produced at the rate of 530 kg/ha/crop largely through a low input extensive system of farming (Harikumar and Rajendran, 2007).

In India, the traditional system of coastal aquaculture is practiced in approximately 50,000 hectares in the low lying brackish water areas of West Bengal, Kerala, Karnataka and Goa without disturbing the ecological equilibrium. Besides the brackish water area, the shallow coastal region up to a depth of 30 meters are suitable for sea farming of approximately 73 species of various marine organisms of potential drug/pharmaceutical importance including fin fishes (food fishes and ornamental fishes), crustaceans, mollusks, seaweeds and sea cucumber (Devaraj and Appukkuttan 2000).

In India, the traditional system of coastal aquaculture is practiced in approximately 50,000 hectares in the low lying brackish water areas of West Bengal, Kerala, Karnataka and Goa without disturbing the ecological equilibrium. Besides the brackish water area, the shallow coastal region up to a depth of 30 meters are suitable for sea farming of approximately 73 species of various marine organisms of potential drug/ pharmaceutical importance including fin fishes (food fishes and ornamental fishes), crustaceans, mollusks, seaweeds and sea Cucumber (Devaraj and Appukkuttan 2000). This period, a special variety of paddy called '*Pokkali*', which is tolerant to 6-8 ppt salinity, is grown in these fields. Different species of shrimps together with fishes such as *Mugil* spp. *Chanos chanos, Etroplus suratensis, Megalops cyprinoides, Lates calcarifer* and *Oreochromis mossambicus* enter into the fields with the inflow of tidal water. The shrimp is the dominant one, but in most of the areas, it is the commercially least important species like *Metapenaeus dobsoni* and *M. Monoceros*, which dominates in comparison to the commercially important species like *Fenneropenaeus indicus* and *Penaeus mondon*. Regular thinning of crop by filtration is carried out in every new moon and full moon period. Final harvesting will be done at the end of the season by operating sluice net and cast net. The prawn yield from the traditional practice usually varies between 500-1000 kg/ha depending upon the location of the field in relation to the bar-mouth, the distance from main water body, the size of the feeder canal, and the position of the sluice (Sathiadhas et al. 2000).

According to Francis, 2010 benefits of Rice-Fish culture include:

- The recycling of nutrients by the fish through feeding and depositing feces in the soil.

This increases the uptake of nutrients such as phosphorus and nitrogen by the rice.

- An increase in rice yields due to the inflow of Organic residues and the faecal deposits of fish and prawn adds the ecosystem nutrient enriched
- An increase in income from the production of both rice and fish
- A reliable source of protein for farmers and their families, countering the decrease in available wild fish in many countries
- A reduction in the insect pests (such as leaf-hoppers, stem-borers and aphids) and weeds, which the fish eat.
- A reduction in use of chemical fertilizers.

Poly-culture and integrated aquaculture are methods of raising diverse organisms within the same farming system, where each species utilizes a distinct niche and distinct resources within the farming complex. This may involve the rearing of several aquatic organisms together or it could involve raising aquatic organisms in conjunction with terrestrial plants and / or animals. In either case, the wastes from one organism are used as inputs to another, resulting in the optimal use of resources and less pollution overall. Poly-culture systems can provide mutual benefits to the organisms reared by creating symbiotic relationships while allowing for a balanced use of the available aquatic resources, whereas intensive monoculture systems extract resources from the system and place more stress on the surrounding environment.

In addition, integrated systems can increase the economic efficiency of fish farms through improved conversion rates of input materials.

Poly-culture and integrated aquaculture have the potential to address some of the problems that arise from the intensive rearing of single finfish species. For example, the integration of fish culture with the culture of algal and/or shellfish species shows potential for reducing the risks of eutrophication and also for exploitation of the large amounts of wastes produced by fish farms. Further research is needed however, to determine the effectiveness of such systems, especially in open marine environments. Poly-culture systems are not a new concept; on the contrary, they have been used for centuries. For over one thousand years fish farmers in China have produced four of the most widely cultivated fish species together in the same pond: silver carp (a phytoplankton filter feeder), grass carp (a herbivorous plant feeder), common carp (an omnivorous feeder), and bighead carp (a zooplankton filter feeder). This type of system utilizes available food and water resources, with the effect of reducing costs and increasing efficiency and production. Although still experimental, other systems, such as the integration of seaweed, fish, and abalone culture, and the poly-culture of shrimp and tilapia, have proved to be ecologically efficient methods for growing a variety of organisms and may increase profits at fish farms.42, 43 It should also be noted that although poly-culture systems based on net/pens may prove beneficial for waste reduction, they fail to eliminate other problems associated with net pen aquaculture, specifically escape of fish, disease transfer, and discharge of chemicals (White et.al., 2004).

7.4 EVOLUTION OF RICE - FISH CULTURE SYSTEM IN INDIA

Farmers invented the method of culturing rice and fish either concurrently or alternately with rice crops and this method has been in practice in several regions of Asia for a very long time. In India rice-fish culture has been in vogue for centuries in some of the Eastern States of the country such as West Bengal and Southern States like Kerala, Karnataka and Goa. The systems of rice–fish culture that have evolved are known to be ecologically friendly and economically viable. In the pokkali paddy fields of Kerala, even today farmers derive high economic benefits by culturing shrimp and fish. In other areas, rice fields once provided a large amount of fish and other aquatic organisms for consumption but with the adoption of improved varieties of rice for cultivation, pesticide usage became more common and the importance of fish cultivation with rice did not receive the required attention. However, now there is greater emphasis on the use integrated pest management (IPM) practices to reduce or eliminate the usage of pesticides and the new approach is gaining increasing acceptance by the scientific and farming community (Nandeesha, 2004).

7.4.1 Traditional Pokkali Farming Practices in Kerala

The Pokkali fields, a unique eco system cover an area of 1, 25,000 ha, where the age old shrimp filtration practice known as 'Chemmeen Kettu'is carried out after the harvest of paddy. The Pokkali fields are concentrated in Ernakulam, Alappuzha, parts of Kottayam and Thrissur districts. In the traditional system of culture shrimp and fish seeds brought in through tidal water are trapped in the Pokkali fields and are allowed to grow for 4 to 5 months. In this traditional system no selective stocking and supplementary feeding are done. By adopting improved traditional farming with selective stocking and supplementary feeding, the production of shrimp from these traditional fields is increased greatly (Harikumar and Rajendran, 2007). According to Bhattacharya and Ninan, 2009 characteristics of a traditional farming practice include fully tide fed , salinity varies according to monsoon regime, seed of mixed species from the adjoining, creeks and canals by auto stocking, additional stocking of natural seeds, dependence on natural food, water intake and drainage managed through sluice gates, depending on the tidal effects and periodic harvesting during full and new moon periods, collection at sluice gates by traps and bag nets. Out of a total of 2414 farms in private sector more than 58 per cent belongs to the category of small farms with an area of less than 2 ha. In Kollam district almost 90 per cent of the farms belong to this category. Around 44 per cent of farms belong to 2.50-5.0 ha category. Bigger farms of more than 10 ha area are mostly prawn filtration fields belonging to Ernakulam, Thrissur and Alappuzha districts.

7.4.2 Status of Aquaculture in Tamil Nadu and Andhra Pradesh

Most of the aquaculture farmers in Tamil Nadu and Andhra Pradesh have taken up shrimp culture mainly for export. Shrimps cultured in these farms are sold to agencies and exported. Aquaculture is a financially lucrative venture, yielding profits ranging from 20 to 50 per cent per annum, much beyond the returns that can be expected from any other activity. Unemployed coastal folks and agriculturists are offered jobs in the aqua farms for wages that far exceed those in other farming activities. The increased profits and wages are responsible for the overall improved economic activities and better standard of living of the people in the region. However, the advancement of shrimp farming is reported to have led to pollution of drinking water, imposing additional costs on society. These costs incurred are external costs, which are not included in the cost of shrimp production. As per NEERI's estimate, environmental costs exceed the value of shrimps produced and exported. For Andhra Pradesh, the report estimated annual earnings at Rs.1, 500 crores, against annual value of the damage caused by aquaculture at Rs.6, 300 crores. Thus, for each rupee of gain, there would be four rupees of damage. For Tamil Nadu, the estimated annual earnings of Rs.280 crores are outweighed by the damage cost estimated at Rs.430 crores. However, considering the negative externalities, one should not de-activate the spirit of aquaculture industry. Instead, the authorities should come forward to redress the grievances that arose due to negative externalities and an optimum solution through good policy advocacy should be made and protect the aquaculture industry.

7.5 IMPACT OF AQUACULTURE PRACTICES

Aquaculture not only generates positive impacts but also negative impacts too. Sometimes, the negative impacts outweigh the positive impact.

According to Emerson and Rajalakshmi (2002), the positive and negative impacts and the conflicts that arose between different stakeholders were analyzed and documented in their studies. I think it is important to highlight some of the impacts arrived by the scientists through their paper entitled "Environmental and Social Conflicts of Aquaculture in Tamil Nadu and Andhra Pradesh". Under the head impacts, the following are important.

- Impact on Coastal Ecosystem
- Impact created on Surface Water and Ground Water
- Impact on Common Property Resources
- Socio Cultural Impact
- Socio Economic Impact

Among the five impacts, the first three are found to create some negative impacts and the remaining two will be capable of creating positive impacts. However, there arises some conflicts too between different stakeholders. These are also delineated briefly.

- The Owner of Aqua Farms and the Fishermen in the Coastal Belt
- Conflict in the Land Use Pattern
- Conflicts between Aquaculture and Agriculture Practices
- Conflicts between the Residents of Coastal areas regarding their health and the Fish Farms

7.5.1 Impact on Coastal Ecosystem

Aquaculture effluents discharged into the marine environment are reported to threaten the survival of coastal resources. The effects on coastal habitats are mainly due to improper saline water released as pond effluents, which have a pronounced effect on both marine and freshwater organisms. However, this appears to be incorrect since the effluents of aqua-farms consist of only biological wastes like feed remains made up of soya-meal and other fishery products. Further, salinity remains constant since the water is pumped in from the natural source of the same salinity, which is maintained for successful growth of the culture (Emerson and Rajalakshmi, 2002). The effluents released from the brackish water pond may be stored in a pool and the same may be properly aerated and processed without affecting the coastal resources. While giving the approval for the establishment of aqua farms, the authorities can verify whether they have any treatment unit to avoid the hazardous discharge into the natural ecosystem. If this is let into the natural ecosystem, then it will be the failure of monitoring system within the Government.

7.5.2 Impact on Surface Water and Groundwater

The salinization of surface water is another impact from shrimp farms to the receiving waters. It is reported that shrimp farm construction can potentially alter surface water flow patterns and water quality. Water flows were found to be affected by the temporary blockage of permanent division of canals. Discharge of untreated pond effluents into the natural system were known to hinder the quality of water. Further, seepage of saline water from the ponds into the ambient areas also leads to salinization of groundwater. There are cases of stratification occurring on the surface of fresh water zone as it mixes with the deeper saline water that will affect agricultural land when farmers unknowingly pump salinized water into farm land. The shrimp farmers directly discard untreated waste water into the public canal or coastal area or river and inshore areas, causing general pollution and public health hazards. Effective pumping of saline water, in the case of bore water-dependent farms, and fresh water, in the case of coastal farms, lowers the groundwater table. The salinized groundwater also affects the drinking water sources of coastal villages. Shrimp farming causes dispersion of salt into land around the shrimp farms, thereby affecting paddy fields and other plantations. The soil salinization further devalues marginal agricultural land. Abstraction of fresh water from underground aquifers for intensive shrimp farming has also resulted in salt-water intrusion and salinization of freshwater aquifers. While freshwater aquaculture has little negative impact on groundwater and aquifers, brackish water aquaculture is detrimental to subterranean water resources (Emerson and Rajalakshmi (2002). In this circumstance, the zonal delineation for encouraging brackish water fish culture is important. Hence, in the approval of plan, the same should be carefully examined and approved by the authorities of municipality or panchayat.

7.5.3 Impact on Common Property Resources and Rights

Creeks and coastlines have been used to set up shrimp farms along the east coast as they are common property resources. These resources as well as the sea have been, since time immemorial, the means of livelihood of the local fishermen, who are now being denied access to the sea. This has resulted in conflicts and clashes, leading to protests and threats to destroy the shrimp farms in Tamil Nadu.

7.5.4 Impact on Human Health

The social movements against shrimp culture in Sirhazhi, Tamil Nadu, have alleged that shrimp culture activities have resulted in various diseases, which pose a threat to the local population. However, only one or two cases have been reported in Tamil Nadu, and none in the shrimp culture areas of Andhra Pradesh. Further, the public has maintained that shrimp culture industry is the only viable alternative to agriculture, particularly in the context of the present water scarcity, which has been a disaster to agriculturists. In such circumstances, aquaculture has come to their rescue by providing better employment opportunities to the local labor including womenfolk. The public has also pointed out that they are comfortable with the development of the industry, and that any move to destroy it will make them homeless (Emerson and Rajalakshmi (2002).

7.5.6 Socio-Economic Impact from Aquaculture Practices

Shrimp culture is essentially a human activity with varying socio-cultural and economic impacts on human communities involved. An appreciable number of rural people have found employment as owner operators or wage earners in shrimp farms, and their economic standard has risen. The State Government is leasing out sites to co-operatives or to the poor, with a view to helping them earn cash benefits. Since shrimp commands a high price in the market, the overall socio-cultural and economic standards of marginal and small-scale farmers will certainly improve. It is to be noted that only fallow lands and cultivable waste areas have been converted into prawn farms in Tamil

Nadu. Besides, on account of the persisting Cauvery water issue, farmers harvesting only one crop in Poompuhar area have given up agriculture, and prawn farm companies / investors purchased fallow lands from them. In fact, these companies came to the rescue of the farmers (Emerson and Rajalakshmi, 2002).

Although intensive aquaculture practices either in Brackish water environ or Mari culture environment provide foreign exchange from exports, one may strongly question whether it benefits the country and its indigenous population in terms of sustained socio-economic benefits (Meltzoff and Lipuma, 1986; Goodland and Ledec 1987; Bailey 1988; Primavera 1991). The progressive exploitation of the ecosystem to expand the economy has made it necessary to redevelop the ecosystem in order to maintain the economy (Regier and Baskerville 1986). Hence aquaculture practices in both brackish water and marine environment becomes inevitable. The options of clear and clean production system and the ways and means of achieving the clean mechanism alone should be explored through appropriate policy efforts.

7.6 THE POSITIVE ASPECTS OF AQUACULTURE

Aquaculture has a positive impact on the environment, and biodegradable wastes provide nutrition for agricultural production; otherwise it is mineralized by bacteria. The environmental effects of aquaculture include change in resources such as water and land, and competition between users. Aquaculture can contribute positively to the environment in many ways, but this is seldom recognized in environmental forums. Hence, the Fisheries College and Research Institute, Tamil Nadu Fisheries University should come forward to initiate such socio economic studies in detail to solve the issues that arose from environmentalists and the positive aspects of aquaculture should be communicated to the entrepreneurs who are willing to take up this venture through appropriate capacity building programs. The impediments if anyone could come across from the findings of such study may be solved by taking appropriate policy decisions from the Ministerial and Secretarial level by inviting the stakeholders under one forum and a consensus should be developed and implemented to benefit the inland or marine or brackish water fish farmers.

7.7 THE EXISTENCE OF DIFFERENT CONFLICTS IN AQUACULTURE PRACTICES

A number of shrimp farms have been set up in violation of many rules, with no proper systems for their supervision and monitoring. This has resulted in conflicts in the use of common property resources. The shrimp farms fenced their area without prior notice and did not allow the public, especially the fishermen who were using the beaches for years together because of lack of proper rights to use the common property resource. The shrimp farms have been set up without considering the hardships of the local fishermen from open access to the seashore. Open access system has given rights to the fishermen to use the seafront for their livelihood. Likewise, the owners of shrimp farms purchased the cultivable lands from agriculturists and converted them into aqua ponds. This has resulted in conflicts with the users of land for agricultural activities. Some of the farms have not shown any concern when the groundwater sources turned brackish. Very few farms have made arrangements for supply of potable water to the local people. Such conflicts have triggered severe protests from the local people against the shrimp farms (Emerson and Rajalakshmi (2002).

7.7.1 Conflicts between Fishermen and Aqua Farms

A major issue observed by this study was the conflict between the shrimp farmers and the fishermen over accessibility of fishing areas. The shrimp farms do not provide access to the beach for traditional fishermen who have to reach the sea from their villages. As aqua farms are located on the seafront and entry is restricted, the fishermen have to take a longer route of 4–5 kms to sea. The traditional, local fishermen, who have been fishing for years, are subjected to such hardships because of the newly emerging aqua farms. The accessibility of fishing areas / beaches through larger aqua farms is a major source of conflicts between aqua farmers and the local people. It can be resolved through negotiations among themselves on provision of suitable approach roads. While such large farms do help by providing advanced aquaculture know-how, and supply of aqua-feed, hatchery-produced seed, etc., to small farmers, coastal communities in general seldom benefit directly from such Aqua cultural development. The conflict described above is restricted to a few farms near Nagapattinam district and in Kandaleru creek in Andhra Pradesh. It is not a major issue in many shrimp farm areas since the big farms, of more than 20 ha, especially those in the corporate sector, do not hinder accessibility of the fishing areas (Emerson and Rajalakshmi (2002).

7.7.2 Conflicts in the Land Use Pattern

Indiscriminate conversion of fertile agricultural lands into aqua farms in the initial stages of aquaculture development was found to have led to many conflicts prevailing till today. This practice aggravated the landlessness among farmers. Absentee landlords sold their lands to aqua enterprises for a high price. Those who were working or cultivating under tenancy farming lost their livelihood, and crop production was also affected.

7.7.3 Conflicts between Aqua-culturists and Agriculturists

In the early nineties, when aquaculture began, entrepreneurs acquired lands from agriculturists who were incurring tragic losses in agriculture. As this continued, conflicts arose with the support of NGOs against aquaculture. With a view to avoiding such conflicts, the commercial farms that developed subsequently were constructed on fallow wastelands, unsuitable for agriculture. These lands are registered in the revenue records in the agricultural category, and have therefore led to criticism that aqua farms are constructed on agricultural lands. Depending on the soil texture, soil salinization may take place on neighboring agricultural land where coastal aqua farms are set up. It is a well-established fact that the aqua farms, having higher percentage of sand particles, are causing salinization of agricultural lands, the major conflict that led the coastal aquaculture to legal disputes. Similarly, where coastal aqua farms exist on clayey soils, soil salinity could not exceed the permissible limits in adjacent agricultural farms (Emerson and Rajalakshmi (2002).

In respect of Inland Fisheries, the conflict between agriculturists and Aqua-culturists becomes popular and sometimes it went to physical fight and it reached the court elsewhere to solve the disputes. It should be executed by pure monitoring of agricultural crops which need quite amount of water for enhancing agricultural productivity in the tank-fed environment. At the same time, the aqua culturists involved in raising fish culture in the tanks by getting the cultural rights by leasing the water body from Panchayat or the Public Works Department. The interest of Aqua culturists should also be protected by maintaining the minimum level of water in the water bodies and the interests of both agriculturists and the aqua culturists should be balanced.

7.7.4 Conflict between Social Forestry Officials and Fish Culturists

Under the social forestry, the Department of Forests is promoting the "Trees outside the Forests". One of their area for developing the plantation is tank foreshore areas. The tank foreshore areas are planted with ***Acacia nilotica*** species and performs well and the same is capable of meeting the fodder to the livestock and small ruminants and at the same time it is much useful in meeting the small timber requirements of the farm households besides fulfilling the environmental requirements of Carbon assimilation to the greater extent. It also provides appropriate shading mechanisms to the sad loving aquatic forms when the water is flooded with the tanks. The droppings of leaves, pods and the small twigs on decay it promotes the growth of Phytoplankton and Zooplankton which formed the food to the aquatic forms and hence the trees are indirectly promoting the productivity of fish or other crustaceans and mollusks in the fresh water environment. But the fishermen or the aqua culturists involved in promoting the fish culture, has raised objections that the trees are found to be impediment in capturing the fishes using the nets and hence the trees should be uprooted and not adviced to plant in the tank foreshore areas. It is merely an unawareness on the benefits of growing trees in the aquatic environment and hence proper training and capacity building alone could alleviate these misunderstandings.

The debris or silt formed due to the trees are acting as a nutrient and on irrigating the pond water, the agricultural crops and its productivity could also be enhanced. From that the plantation establishment in the tank foreshore areas, the aqua culturists and the agriculturists are mutually benefitted and no other harm is visible to any stakeholders and hence everyone should engaged in planting trees, in the aquatic environment must be promoted. In China, the Silvi-pisciculture initiatives are quite common to get higher productivity under one roof both fish and food materials.

7.8 PRECISION TECHNOLOGIES FOR INLAND FISH CULTURE OR COMPOSITE FISH CULTURE

The technology developed for fish culture in which compatible and non-competing fishes are cultured simultaneously through the utilization of different feeding zones (all the natural niches) from a pond so as to increase the total production from unit area of water is known as Composite Fish Culture and it is the most popular culture technique in the country.

7.8.1 Steps to be followed in Executing Composite Fish Culture

7.8.1.1 *Site Selection:*

Factors to be considered are

A) ***Ecological factors***- depends on location, topography, soil suitability, water quality and quantity, hydrological and meteorological factors.

B) ***Biological factors*** includes species selection, predator and disease control.

C) ***Economical and Social factors*** includes type of facility, marketing, safety and security, social considerations etc.

- Pond Area: It can range between 0.5-1 ha, but even 0.02 ha ponds can also be used based on the convenience of the user.
- Ideal depth to be maintained is 1.5-2 Meter
- pH level to be maintained and monitored is : 7.5-8.5
- Dissolved Oxygen level should be greater than 5ppm
- Salinity should be less than 2 ppt
- Clayey loam soil is ideal for as it has low permeability and high fertility. Clayey loam contains textural components like sand: 20- 45 per cent silt: 15-23 per cent and clay: 27- 40 per cent.
- De-silting of existing ponds
- Deepening of shallow ponds.
- Excavation of new ponds
- Impoundment of marginal areas of water bodies.
- Construction / repairs of embankments.
- Construction of Inlets / Outlets.
- Any other item like civil structures, watchmen sheds, pump sets water supply arrangements, electricity supply arrangements etc. depending on the requirements of the project based on its size one has to choose.

7.8.1.2 *Pond Management*

Pond Management plays a very important role in fish farming before and after the stocking of fish seed.

- Prestocking: New ponds, pre stocking operations starts with liming and filling of the pond with water. The first step for existing pond requiring development deals with clearing the unwanted weeds and fishes either by manual, mechanical or chemical means from the pond. Different methods are employed for this.

- Removal of weeds by Manual/Mechanical, Chemical or biological means. Chemical means- herbicides like 2,4-D (2,4-dichlorophenoxy acetic acid), Aqueous ammonia and Diuron/Karmex,
- Biological means- introduction of Grass carp, Tilapia, Common carp, Pearl spot, Gaint gouramy
- Removal of unwanted and predatory fishes and other animals by repeated netting or using mahua oil cake @ 2500 kg/ha or by sun drying the pond bed. Other toxicants includes tea seed cake with a dose of 15ppm in salinity less than 15ppt, Tamarind seed powder with a dose of 175- 200ppm
- Liming - The soils / tanks which are acidic in nature are less productive than alkaline ponds. Lime is used to bring the pH to the desired level. In addition lime also has the following effects - a) Increases the pH. b) Acts as buffer and avoids fluctuations of pH. c) It increases the resistance of soil to parasites. d) Its toxic effect kills the parasites; and it hastens organic decomposition.

The normal doses of the lime desired ranges from 200 to 250 Kg/ha. However, the actual dose has to be calculated based on pH of the soil and water as follows (Table 7.1)

Table 7.1 Quantity of Lime Applied based on Soil pH for Composite Fish Culture

Sl. No	Nature of the Soil	Soil pH	Lime to be Applied in Kgs per Ha	Liming Materials
01	Highly Acidic	4.0 – 4.5	900 - 1000	
02	Medium Acidic	4.5 – 5.5	700 - 750	Calcium Carbonate
03	Slightly Acidic Nature	5.5 – 6.5	450 - 500	Calcium Oxide
04	Near Acidic	6.5 – 7.5	200 - 250	Calcium Hydroxide
05	Alkaline	7.5 – 8.5	Not Needed	

7.8.1.3 *Fertilization/ Manuring*

Fertilization of the pond is an important means for intensifying fish culture by increasing the natural productivity of the pond. The fertilization schedule has to be prepared after studying the quality of the pond soil. A combination of both Organic and Inorganic fertilizers also be used. The fertiliser program has to be suitably modified depending on the growth of the fish, available food reserve in the pond, physic-chemical conditions of the pond and climatic conditions.

Organic Manure (FYM)- Cowdung @ 5000 kg /ha, Poultry, sheep manure, Crop byproducts like cotton seed meal, mustard oil cake etc can be incorporated into the pond.

Inorganic fertilization to be undertaken after 15 days of organic manuring. Requirement of nitrogenous and phosphate fertilizers would vary as per the nature of the soil fertility indicated below. However any one of the nitrogen and phosphate fertilizers could be used as per the given rate.

Table 7.2 Inorganic Fertilizer Schedule for Incorporation (kg/ha/month)

Sl. No	Soil Fertility Status	Soil Nitrogen Mg per 100 g.	Urea to be Applied in Kgs per Ha	Ammonium Sulphate (Kg per ha)
01	High Fertility	51 -75	30.000	70.000
02	Medium Fertility	26 - 50	40.000	90.000
03	Low Fertility	Up to 25	60.000	140.000

7.8.1.4 *Stocking of Fish Species*

The pond will be ready for stocking after 15 days of application of fertilizers. Fish fingerlings of 50- 100 gm size should be used for stocking @ 7000-8000 nos. per hectare. However, if fingerlings of smaller size are used, suitable allowance may be made accounting for mortality. The present model envisages stocking of advanced fingerlings and rearing for 10-12 months. Depending on availability of seed and market condition, stocking can be of 3, 4 or 6 species combination in the following ratio may be adopted.

7.8.1.5 *Fish Species involved in Composite Fish Culture*

Depending on the compatibility and type of feeding habits of the fishes, Indian as well as Exotic varieties have been identified and recommended for culture in the composite fish culture technology.

Table 7.3 Indian Major Carps, their Nature of Feeding and Feeding Zone

Sl. No	Name of the Species	Common Name	Nature of Feeding	Feeding Zone or Nich
01	Catla catla	Catla	Zooplankton Feeder	Surface Feeder
02	Labeo rohita	Rohu	Omnivorous	Column Feeder
03	Cirrhinus mrigala	Mrigal	Detritivorous	Bottom Feeder
Exotic Carps				
04	Hypophthalmichthys molitrix	Silver Carp	Phytoplankton Feeder	Surface Feeder
05	Ctenopharyngodon idella	Grass Carp	Herbivorous Feeding	Surface, Column and Marginal Grasses
06	Cyprinus carpio	Common Carp	Detritivorous / Omnivorous	Bottom Feeder

Based on the feeding habits of Indian and Exotic carps, the species combination ratio should be decided by the entrepreneur involved in promotion of Composite Fish Culture practices on his own. The species combination ratio is furnished as follows (Table 7.4).

Table 7.4 Species Combination Ratio in Composite Fish Culture

Sl. No	Name of the Species	Three Species Combination	Four Species Combination	Six Species Combination
01	Catla catla	4.0	3.0	1.5
02	Labeo rohita	3.0	3.0	2.0
03	Cirrhinus mrigala	3.0	2.0	1.5
Exotic Carps				
04	Hypophthalmichthys molitrix	--	--	1.5
05	Ctenopharyngodon idella	--	--	1.5
06	Cyprinus carpio	--	2.0	2.0

Table 7.5 Favorable Water Quality Parameters for Composite Fish Culture

Sl. No	Parameters	Safe Limit
01	Turbidity	30 – 45 CM
02	Salinity	< 0.5 PPT
03	Dissolved Oxygen	5.0 PPM
04	Un Ionized Ammonia	< 0.05 PPM
05	Nitrite	< 0.1 PPM
06	Nitrate	50 – 150 PPM
07	Carbon-di-oxide	< 8.0 PPM
08	Iron	< 0.5 PPM
09	Total Alkalinity	20 – 150 PPM
10	Total Hardness	20 – 200 PPM
11	Hydrogen Sulphide	< 0.002 PPM

7.8.1.6 *Post Stocking*

1) Supplementary feeding Fishes need much more food than what is available naturally in the pond. Fishes can be fed with a mixture of rice bran and oilcakes in the ratio 4:1. Due to the high cost of Ground nut Oil Cake (GOC) alternate sources like Cotton seed oil cake which is comparatively cheaper than GOC. The GOC and cotton seed oil cake can be mixed in equal proportions and fed to the fish and is reported to give almost the same growth rate as that of GOC as per the experiences gained by Dr. A. Pounraj, Professor in Fisheries Extension is actively involved in raising composite fish culture. The feed should be placed on a feeding tray or in feeding bags and lowered to the pond bottom or it can be dispersed at the corners of the pond. After some time the fishes will get used to this type of feeding and aggregate at the same place at particular time for regular feeding thereby reducing the feed losses. The recommended feeding rate is 5 - 6 per cent of the body weight upto 500gm size of fish and then reduce to 3.5 per cent of body weight from 500- 1000gm size. The feeding is supplementary in nature.

2) Manuring i) Organic manuring may be done in monthly instalments @ 1000 kg/ha.

ii) Inorganic fertilization may be done at monthly intervals alternating with organic manuring. However, the monthly rate of fertilization will depend on pond productivity and the growth of the fishes. It should be ensured that excess fertilization does not take place which may result in eutrophication.

7.8.1.7 *Harvesting*

Harvesting is generally done at the end of first year, when the fishes attain average weight of 800 gm to 1.30 kg. With Proper management, a production of 4.5 to 6 tonnes per ha can be obtained in a year. Harvesting is done by partial dewatering and repeated netting. In some cases complete dewatering of ponds is resorted to and the water can be fed to the coconut, banana plantations or any other agricultural crops raised in the land nearby fish pond. Some farmers resort to partial harvesting also depending on the season and demand for fish. The Silver carp, Grass carp and the Common Carp are capable of fetching more than one kg weight within ten months period. This is practiced in the ponds of ***Nammazhvar College of Agriculture and Technology, Peraiyur village of Ramanathapuram District*** where the four species combination was practiced to ensure higher productivity per unit of land when the author was working as Dean of that Institution.

7.8.2 Marketing of Fishes

In the urban and Peri-urban markets, the inland fishes which are harvested afresh can be sold live to the consumers at the rate of Rs 100 – 150 per kg of fish. Good demand is visible among the consumers if we started selling the fish alive. Consumer preference to the inland fishes are increasing day by day in the non coastal zones of Tamil Nadu and the ready market is visible in Kerala too. Hence any entrepreneur interested in establishing composite fish culture using precision methods can approach the author or the fisheries department personnel available nearby.

7.8.3 Economics of Composite Fish Culture and Brackish Water Prawn Culture

Productivity in Aquaculture depends on the extent of use of inputs like seed, fertilizer, feed, labor and management. As feed, fertilizer and seed assume a major share of the variable costs, the level of their application is suboptimal (Kund et.al. 2010). Hence, to achieve a particular level of Production, the precision methods of composite fish culture should be followed which are delineated elsewhere for the stakeholders and users of the information. But how exactly, the composite fish culture contributes to the entrepreneur is in question. Various authors had expressed their own economics and delineated in seminars and conferences but the executors are not much exposed on the costs and returns involved in precision composite fish culture and hence an effort has been taken to interview the aqua culturists in the City of Madurai have been interviewed and the Economics is presented as a case study approach.

In general, the costs and returns that are involved in establishing fish culture activity as a separate entity should have the details of Fixed Expenses, Variable or Operating Expenses, the Yield or Return generated from the fish culture activity. On these lines, the costs and returns are carefully examined and are furnished below for the perusal of stakeholders or users of these information.

The costs and returns arrived in the Table 1 for composite fish culture and the brackish water prawn culture is totally a different one. The cost of seed and feeding in respect of Brackish water prawn farms are consuming at the higher end which varies from 3 to 3.20 lakhs per ha. Whereas, in the study, it is arrived at Rs 3 lakhs only. In respect of feeding of prawns, it is worked out to Rs 1.50 lakhs per ha. The total cost of establishing the prawn culture farm per annum is arrived at Rs. 7.24 lakhs per ha. In that the operational cost alone is arrived at 86 per cent to the total cost of establishment of the enterprise. In the study conducted by Rahul during the year 2012, the total cost of establishing a prawn farm is arrived at Rs 10 lakhs. In this study, the average yield of prawn per ha drawn is arrived at 5890 kgs and it could generate a gross return of Rs 17.67 lakhs per annum with the cost of production of Rs 123 per kg. However, the prawn is sold at Rs 300 per kg on the door step and one need not take the produce for sale at the retail centres. This brackish water prawn culture unit could generate an output – input ratio of 2.85 revealed that for a rupee of investment, the entrepreneur could generate a net return of Rs. 1.85. Hence, this enterprise is also highly profitable. One can take up this venture if they have the required facilities with them. But the personnel who have taken up the prawn farming in general is found to be non-aqua culturists and they have gained a knowledge after availing the services of the consultants.

Table 7.6 Comparative Costs and Returns from Inland Fish Culture and Brackish Water Prawn Culture

Sl. No	Details of Costs and Returns from Fish Culture	Composite Fish Culture Farm (Rs per Ha)	Brackish Water Prawn Farm (Rs per Ha)
01	**Capital Investment**		
1.1	The Lease value of one Ha of Land per annum	86956.00	74230.00
1.2	Excavation of Pond of 0.4 Ha	46750.00	42450.00
1.3	Screening the pond	36000.00	39750.00
1.4	Nets and Sieves	19200.00	28500.00
1.5	Electrical Installation and others	49000.00	58700.00
1.6	Pump shed and Pump costs	19800.00	28500.00
	Total Capital Investment	257706.00	272130.00
02	Fixed Expenses		
2.1	Lease value of Land per ha	86956.00	74230.00
2.2	Maintenance of Pond @ 15% of Pond Cost	7012.00	6368.00
2.3	Nets and Sieves @ 10%	1920.00	2850.00
2.4	Land Revenue, Cess and Taxes	4150.00	4620.00
2.5	Miscellaneous Charges under Fixed Expenses	6200.00	14500.00
	Total Fixed Expenses (2)	106238.00	102568.00
3.0	**Operating Expenses**		
3.1	Preparation of Tank / Pond which includes chemicals and manures	3344.00	24500.00
3.2	Cost of Fish / Prawn Seed Fingerlings @ Rs 2 per number Prawn seed @ Rs 0.80 per number for 3.75 Lakhs	8969.00	300000.00
3.3	Manuring the Pond with FYM	2800.00	12000.00

Sl. No	Details of Costs and Returns from Fish Culture	Composite Fish Culture Farm (Rs per Ha)	Brackish Water Prawn Farm (Rs per Ha)
3.4	Fertilizer Application	3385.00	6200.00
3.5	Feeding Materials and Feeding	24258.00	150000.00
3.6	Fuel and Electricity Charges	40000.00	60000.00
3.7	Repairs and Maintenance of Tanks / Ponds	6400.00	27000.00
3.8	Harvesting of Fish / Prawn	7500.00	23000.00
3.9	Miscellaneous Charges	9297.00	18400.00
	Total Operational Costs (3)	105953.00	621100.00
	Total Costs (2)+(3)	**212191.00**	**723668.00**
4.0	Yield of Fish / Prawn (kgs)	3900.000	5890.000
4.1	Sale Price of Fish / Prawn per Kg in Rupees	130.00	300.00
4.2	Gross Return from Fish / Prawn in Rs per Ha	507000.00	1767000.00
4.3	**Cost of Production of Fish / Prawn per Kg**	**54.41**	**122.86**
4.4	Output-Input Ratio	02.39	02.85

In respect of Composite fish culture, the electricity charges per annum is found to be the highest because of pumping of ground water whenever needed which is followed by the feeding cost of fishes which consumed Rs 24258 per ha ponds of 0.4 ha each. The composite fish culture has consumed a total cost of Rs. 2.12 lakhs per ha of 0.4 ha ponds each. The cost of production per kg of fish is arrived at Rs 55/. But the selling price is arrived at Rs 130 per kg of fish and hence the farmer or the entrepreneur could generate a net return of Rs 75 per kg of inland fishes. The output input ratio is arrived at 2.39 which is revealed that for a rupee of investment, the composite fish culture could generate a net return of Rs 1.39. The required water facilities if available with the entrepreneur, he can take up the composite fish culture venture very boldly and confidently. With proper care in feeding and maintenance could generate the return further.

7.9 BEST AQUACULTURE PRACTICES (BAP) AND ITS CERTIFICATION

From the inception of Best Aquacultural Practices (BAP), Global Aquacultural Alliance (GAA) has made the decision to work with all stakeholders whom are involved in fish culture, processing, marketing and quality control and distribution, to bring the ethos of responsible aquaculture development to as many producers as possible. To do this, the GAA has created Best Aquaculture Practices (BAP) facility certification standards that are rigorous and comprehensive. And GAA has created a system to certify compliance using independent, experienced and fully credential ISO 65-accredited certification bodies. (*https://www.aquaculturealliance.org/what-we-do/bap-certification/)*

BAP certification defines the most important elements of responsible aquaculture and provides quantitative guidelines by which to evaluate adherence to those practices for processing plants, farms, hatcheries and feed mills. The number of BAP-certified facilities are increasing globally and promotes precision aquaculture practices.

The Best Aquaculture Practices Certified (BAP Certified) mark on retail packaging tells consumers that seafood came from BAP certified aquaculture facilities.

The Best Aquaculture Practices standards, developed by the Global Aquaculture Alliance, form the basis for BAP certification. The standards specifically protect biodiversity and worker rights within a program that addresses environmental, social, food safety and traceability issues throughout producers' operations.

BAP certification is implemented through the Aquaculture Certification Council, an independent certifying agency that employs an international team of accredited evaluators to inspect facilities to the BAP standards through site inspections, sampling and record reviews. BAP certification is currently available for shrimp farms and hatcheries, and seafood processing plants. (*http://www.ecolabelindex.com/ecolabel/best-aquaculture-practices*).

7.10 SUMMARY

The economics of aquaculture comprising composite fish culture and the brackish water prawn culture was taken up on case study basis and the details were compared and delineated that the brackish water prawn culture is investment oriented and commands higher profit per unit of investment. However the enterprises chosen are capable of contributing to the practitioners on the higher side revealing that the enterprises are most viable and hence one can take up. Little conflicts are visible in running the enterprise. These are manageable by the policy makers in Tamil Nadu.

References

1. Bailey, C. (1988). "The Social Consequences of Tropical Shrimp" *Mariculture Development, Ocean and Shoreline Management*, 11: pp. 31–44.
2. Goodland, R., and G. Ledec (1987). "Neoclassical Economics and Principles of Sustainable Development", *Ecological Modelling*, 38:19–46.
3. Emerson, K and N. Rajalakshmi (2002) "Environmental and Social Conflicts of Aquaculture in Tamil Nadu and Andhra Pradesh", Journal of Social and Economic Development, 4(1): pp. 13-26
4. Kund, G.C., R. Mishra and P.K. Sethi (2010) "An Economic Analysis of Composite Fish Culture Ponds in Sundergarh District, Orissa", *The Asian Journal of Animal Sciences*, 5(2): 139-141
5. Meltzoff, S. K., and E. Lipuma (1986). "The Social and Political Economy of Coastal Zone Management: Shrimp Culture in Equador", *Coastal Zone Management Journal*, 14: 349–80.
6. Primavera, J.H. (1991). "Intensive Prawn Farming in the Philippines: Ecological, Social and Economic Implications", *Ambio* , 20: 28–33.
7. Rahul, M and K.T. Thomson (2012) "Cost and Earnings of Brackish Water Farm in Pallipuram Panchayat of Cherai Poyil" (*Unpublished Thesis Submitted to School of Industrial Fisheries, Cochin University of Science and Technology: Kochi*) pp. 1-60.
8. Regier, H. A., and G. L. Baskerville (1986). Sustainable Redevelopment of Regional Ecosystems Degraded by Exploitative Development. In Clark, W.C., and R.E. Munn (eds.). Sustainable Development of the Biosphere, (Cambridge University Press: Cambridge) 75–101.
9. Sammut, J. (2017). "Rice and Shrimp Farming in the Mekong Delta", In: Partners in Research and Development, Australian Centre for International Agricultural Research: Government of Australia, pp. 26-27.

Web References

- http://www.nmfs.noaa.gov/aquaculture/what_is_aquaculture.html
- https://www.aquaculturealliance.org/what-we-do/bap-certification/
- http://www.ecolabelindex.com/ecolabel/best-aquaculture-practices

CHAPTER VIII

Precision Silvicultural Practices in India

Dr. K. K. Suresh and Dr. C. Sekhar

Objectives

- To learn the in-and-outs of Precision Silviculture and the parameters to be considered for executing the Precision Silviculture for Maximum output realization
- To illustrate the role of GIS and GPS in implementation of Precision Silviculture Practices
- To learn the Environmental impact that arose due to the implementation of Precision Silvicultural practices

Organization of the Chapter

- Introduction to Precision Forestry
- Study of Locality Factors
- Choice of Tree Species
- Quality Planting Stocks
- Site Specific Silvicultural Operations
- Irrigation Management for Tree Crops
- Weed Management of Tree Crops
- Nutrient Management of Tree Crops
- Role of GIS and GPS in Precision Silviculture
- Optimal Rotation Age

- Reduced Impact Logging and Conversion Technologies
- Environmental Impact and Sustainable Yield
- Summary

8.1 INTRODUCTION TO PRECISION FORESTRY

In this era of plantation forestry and under the 1980 Forest Conservation Act of India, a lot has to be done in establishing forest plantation outside the natural forest to fetch high quality and quantity wood for meeting the ever increasing demand from various wood based industries and the domestic needs. Precision Silviculture is one, which aims at developing a Silvicultural package to improve productivity, obtain sustainability in yield and fetch maximum profit from such laid out plantation. In Precision Silviculture the foresters work towards developing a site specific prescription for different tree species grown or a particular site with a set of management practices.

Figure 8.1 One year old Neolamarckia cadamba under precision silviculture

Figure 8.2 One year old Acrocarpus fraxinifolius under precision silviculture

According to Taylor *et al.* (2002) Precision Silviculture is a planning and conducting of site specific forest management activities and operations to improve wood product, quality and utilization, reduce waste and increase profit and maintain the quality of the environment.

Most important part of precision forestry is new and modern technologies. Hence according to Schmoldt and Thomsan (2003), precision technologies are instrumentation, mechanization and information technologies that measure, record, process, analyze, manage or actuate multi source data and high spatial or temporal resolution to enable information based management practice or to support - scientific discovery.

The Precision forestry is a high technology sensory and analytical tools to support site specific, economic environmental and sustainable decision making for the forestry sector supporting the forestry value chain for bare land to the customer buying a sheet of paper or board (Ziesak, 2006).

Most important part of Precision Forestry are putting together new and modern technologies at the appropriate site at the appropriate time with the site specific tree species to fetch maximum yield within a short rotation period.

India being a developing Country, forest plays a vital role in the sustenance and survival of large population. Being a renewable resource, forest can contribute much for the economic development of a country. In India forest contribute a 2 % share of the Gross Domestic Product.

Increasing population growth, higher demand for forest products, higher rate of economic growth and trade liberalization are putting pressure on all resources including forest. India occupies 2.5 % of the world geographic area supporting 17 % of human population and 18% of livestock of the world (Oberai, 2000). Recorded forest area of India is 76.5 million ha (23 % of total land area). However the actual forest/tree cover is just about 19 % of total geographical area the per capita forest area in the country is 0.08 ha as compared to the world average of 0.64 ha. The demand and supply of various kinds of wood in India is presented in Table 8.1.

Table 8.1 Demand and Supply of Various Kinds of Wood in India (million cubic metre)

Industry	**Year**							**Percentage Change during 2015 over 2000**
	1998	**1999**	**2000**	**2005**	**2010**	**2015**	**2020**	
Paper & Paper board	4.48	4.48	4.48	8.96	15.4	26.24	35.84	485.71
Newsprint	1.46	1.63	1.78	2.56	3.42	4.63	6.22	160.11
Rayon Graded Pulp	2.1	2.3	2.5	2.8	3.1	3.4	3.8	36.00
Construction Industry	13.6	14.6	15.9	19.4	22.1	26.3	28.5	65.41
Packaging	4.36	4.49	4.62	5.54	6.4	7.55	9	63.42
Furniture	2.25	2.38	2.52	3.36	4.62	5.9	7.53	134.13
Automobile Industry	0.17	0.18	0.19	0.28	0.41	0.6	0.87	215.79
Agricultural Implements	2	2.06	2.12	2.33	2.5	2.5	2.5	17.92

Industry	Year							Percentage Change during 2015 over 2000
	1998	1999	2000	2005	2010	2015	2020	
Railway Sleepers	0.04	0.04	0.03	0.02	0.02	0.02	0.02	(-) 33.33
Sports Goods	0.27	0.31	0.35	0.58	0.98	1.67	2.73	377.14
Handicrafts	0.4	0.42	0.45	0.54	0.65	0.78	0.95	73.33
Plywood	10.1	10.5	11	14	17.96	22.9	29.2	108.18
Veneer	0.25	0.26	0.27	0.34	0.43	0.54	0.7	100.00
Particle Board	0.13	0.13	0.14	0.18	0.22	0.28	0.35	100.00
MDF Board	0.13	0.14	0.14	0.17	0.21	0.24	0.28	71.43
Match Box	2.1	2.2	2.3	2.6	3	3.4	4	47.83
Mining	2.9	3	3.2	3.5	4	4.5	5	40.63
Catamaran	0.03	0.03	0.03	0.05	0.07	0.11	0.16	266.67
Miscellaneous	5.14	5.4	5.7	6.7	9.4	11.2	15.15	96.49
Total	**51.9**	**54.55**	**57.72**	**73.9**	**94.89**	**122.8**	**152.8**	112.75

Source: www.indiastat.com

The productivity of Indian forest is low as 0.7cum/ha as against the world average of 2.1cum/ha. The average growing stock is 32cum/ha as compared to the world average of 110cum/ha. Bridging this gap will go a long way in meeting the raw material demand for the various woods based industries and also fetch higher economic returns for forest tree growers in India.

Precision Silviculture aiming on wood production to supply raw materials to any wood based industries would have two key aspects.

1. Sustainable and increased supply of wood from a fixed production area
2. Understanding the locality factors, improving site properties and adopting site specific silviculture for achieving the required yield both in quantity and quality.

Forest plantations in recent time are actually managed for the commercial production of forest products. Industrial plantations are usually of large scale. The following parameters have to be considered while formulating any Precision Silvicultural package

a) Study of locality factors
b) Choice of species
c) Quality planting stock
d) Site specific silvicultural operation
e) Optimum rotation age
f) Reduced impact logging and conversion technologies
g) Environmental impact and sustained yield

Wood production from a tree plantation is generally higher than that of natural forest. While forest managed for wood production commonly yield 1 to 3 m^3/ha/year, plantation of fast growing species commonly yield 20 - 30 m^3/ha/year. Hence to fetch high output from plantation forestry, precision silviculture can play a vital role.

8.2 STUDY OF LOCALITY FACTORS

Locality factors play a vital role in deciding the quality and quantity of wood produced for various end users (Table 8.2). Climatic and the Edaphic factors are the major factors that act upon the forest vegetation and could bring about any changes in the biomass production. The actual agency or factors of the locality depends on the nature of soil and the climate. The work of a silviculturist should be well acquainted with the manner in which soil and climate act on forest vegetation, inorder to decide in each case which species and method of treatment are best adopted under a given set of condition, to yield the most favourable results.

Table 8.2 Locality Factors Deciding Quality and Quantity of Wood

Climatic factors	Edaphic factors	Topographic factors	Biotic factors
a) Temperature	a) Soil depth	a) Configuration of land surface	a) Weeds
b) Light	b) Physical properties	b) Altitude	b) Pest and disease
c) Rainfall	c)Chemical properties	c) Slope	c) Domestic and wild animals

Climatic factors	Edaphic factors	Topographic factors	Biotic factors
d) Wind	d)Biological properties	d)Aspect and exposure	d) Human being
e) Relative humidity			
f) Snow			
g) Forest			

8.2.1 Climatic Factors

8.2.1.1 *Temperature*

The solar radiation warms up the atmosphere and the soil, thereby raising their temperatures. The temperature of the atmosphere affects the activities those of their roots. Factors affecting temperature includes latitude, altitude, distance from the sea, wind, mountains, cloudiness and presence or absence of forest vegetation. The mean annual temperatures under the major groups of forest area as follows (Table 8.3)

Table 8.3 Major Forest areas and their Mean Annual Temperature

Zone	Mean Annual Temp
Tropical	over 24°C
Sub tropical	17°C to 24°C
Temperate	7°C to 17°C
Alpine	under 7°C

8.2.1.2 *Light*

All trees require light to enable them to live and grow, as soon as the available reserve materials have been consumed in a general way, it may be said that trees with a dense crown are shade bearing, and those with a thin crown light demanding. The light requirement does not always stand exactly in inverse proportion to the density of the crown. From the incoming radiation about 42 % is reflected, by clouds (33 %), and by the molecules and suspended matter in the atmosphere (9 %). The albedo or reflecting power of the earth is therefore taken as 0.42 %.

Light is an important factor, because it influences chlorophyll formation, functioning of stomata, photosynthesis, growth, form and quality of trees, species stratification, and size and structure of leaves. The requirement of light various with species

For Eg: Tropical moist deciduous forest

Light demanders - *Tectona grandis, Dalbergia sissoo,* and *Bombax ceiba*

Shade bearers - *Dalbergia latifolia* and *Pterocarpus marsupium*

Shade demanders - *Schleichera oleosa, Mallotus philippinesis*, and *syzigium cumini*

8.2.1.3 *Rainfall*

In India rainfall occurs under the influence of

(a) Advancing south west monsoon

(b) Retreating north east monsoon

(c) Cyclonic depressions

(d) Westerly depression

(e) Northwesters and other thunder storms

In most of India, it is the moisture bearing south west monsoon winds which has the most influence on the vegetation. This besides the North East monsoon bring rain to the Southern part of India

In addition to the quantity of rain and its seasonal distribution, rainfall intensity (i.e., the rate at which the rain falls) is of vital importance to the foresters particularly in taking up artificial regeneration and planning for water conservation, and erecting anti - erosion works. The rainfall received based on various major forest types are as follows (Table 8.4)

Table 8.4 Major Forest areas and their Mean Annual Rainfall

Type of Forest	Rainfall
Tropical forest	800 - 2500mm
Montane sub tropical	1000 - 3000mm
Montane temperate	1500 - 5000mm
Sub alpine	2900 - 3500mm

8.2.1.4 *Wind*

Wind has a great influence not only on the form of trees but also on their distribution. In a forest stand the calming influences is also reinforced by the length and width of the stand and the reduction in velocity may be 20 to 60 %. The greater the air velocity in the open, the larger the reduction. Shelterbelts and wind breaks work against the prevailing wind direction and prevent any damage.

The favourable effects of winds includes fresh supplies of CO_2 to the foliage of trees thus help in photosynthesis and it helps in pollination of anemophilous flowers and dispersal of seed in many forest trees, Eg. *Holoptelia intergrifolia, Bombax ceiba.*

8.2.1.5 *Relative Humidity*

In a forest stand, the humidity is high as compared to the open. it is at a high level from the ground to above the crowns before sunrise but later on as the crown warm up a sharper gradient is established and reduction in humidity gradually takes place in a downward direction, until a mid day minimum is reached. Relative humidity plays a vital role in pest and disease incidence and irrigation.

8.2.1.6 *Snow*

The amount of snow fall and the period during which it remains on the ground, depends upon temperature and the amount of winter precipitation. Snow occasionally falls down to 1200mm in north western and Central Himalayas but it stays only above 2000meters. Snow fall affects plantation growth and at times during melting of snow supply water to trees in the temperate zone.

8.2.1.7 *Frost*

Because of its moderating influence on both air and ground temperature fluctuations, the forest canopy has a marked influence on the frost incidence as the cold air is prevented from settling on the ground surface which therefore does not get chilled to the extent that it does in the open. Frost affects young seedlings and proper protection has to be taken for frost tender species.

8.2.2 Edaphic Factors

8.2.2.1 *Soil Depth*

Depth of soil is an important factor so far as the trees are concerned because it affects developments of roots and the amount of moisture that can be stored in it to be available for plant growth. Depth of soil decides the height growth of trees.

8.2.2.2 *Physical Properties of Soil*

- Soil texture - Relative proportion of the various size groups of individual soil particles.
- Soil texture - Arrangement of individual's soil particles into aggregates of definite size and shape.
- Soil porosity - Extent to which the gross volume of the soil is unoccupied by solid particles

8.2.2.3 *Chemicals Properties of Soil*

- Cation exchange capacity
- Soil acidity
- Silica - sesqiuoxide ratio
- Major nutrients - N, P2O5, K20,
- Minor nutrients - Fe, Zn, Bo, Mn, Ca, Mg, etc

8.2.2.4 *Biological Properties of Soil*

A variety of organism inhabits the soil. They decompose organic matter, fix atmospheric nitrogen, and cause de-nitrification and plant disease. Cultivated soil harbour bacterial, actinomycets, fungi, algae, protozoa, nematodes, worms, insects, and rodents. Soil depth, organic matter and nutrients, microclimate, and physical and chemical soil environment influence the structure and function of soil biota.

8.2.2.5 *Litter and Organic Matter*

The upper most layer of organic debris (dead vegetation matter) on a forest floor, freshly fallen or only slightly decomposed and consisting chiefly of leaves but also including bark fragments, twigs etc., play a vital role in nutrient cycling and thereby can support tree growth.

8.2.3 Topographic Factors

8.2.3.1 *Configuration of the Land Surface*

It influence vegetation through its effect on temperature, wind movement etc., Due to proximity of high hills on either side, sunlight reaches the valley late in the morning and disappears early in the afternoon. The flowing of strong noisy wind up and down the valley is locally called 'Dadu' in mother India. This land configuration also affects plantation productivity.

8.2.3.2 *Altitude*

Altitude affects solar radiation, temperature, rainfall, which in turn have a great effect on vegetation.

(i) Solar Radiation

Altitude in meters 100 500 1500 4000

Radiation Cal/cm^2/min 0.8 1.2 1.4 1.6

(ii) Temperature

Generally upto 1500m, there is a fall of 1°C in the mean temperature with a rise of 270m in the hills but after 1500m, fall in temperature is more rapid.

8.2.3.3 *Slope*

Slope modifies the intensity of the insolation and thus affects the temperature and moisture of the surface soil. Slope also affects erosion as well as depth of soil. The depth of soil in the hills varies inversely with increasing slope but in the plains or in the valleys in the hills have deep soil.

8.2.3.4 *Aspect and Exposure*

Aspect is the direction towards which a slope faces and exposure is the relation of a site to weather conditions especially sun and wind. This has a profound effect on the type of vegetation and also on the growth of trees.

8.2.4 Biotic factors

8.2.4.1 *Weeds*

Any unwanted plants in the main plantation are called as weed. There are certain invasive weeds and one of the most important in that is *Lantana camera*. It colonizes new areas by the seed dispersed by birds. It also coppice well and is difficult to eradicate. It becomes a serious obstacle in natural regeneration of important species like Sal. Many invasive weed species like *Dodonea viscosa, Eupatorium, Prosopis* etc causes great hindrance in plantation establishment and growth.

8.2.4.2 *Influence of Wild Animals*

The animal helps in pollination of flowers and dispersal of seed as well as it causes considerable damage to forest trees.

Eg: Indian gaur repeatedly browses seedlings and saplings of *Artocarpus hirsuta, Dalbergia sissoo,* Elephant feed on *Ficus* sp., *Artocarpus chaplasha*

8.2.4.3 *Influence of Human Beings and Domesticated Animals*

Through man has been managing forests on scientific basis for the last 100 years or more, yet he continues to be the most powerful agent in destroying or otherwise damaging forests in many parts of this country. The following are some of the most conspicuous effects of man and his animals on forest vegetation.

(i) Deforestation and denudation

(ii) Fire

(iii) Lopping

(iv) Removal of dry leaves

(v) Collection of flowers & fruit

(vi) Grazing

8.2.4.4 *Impact of fire on the growth of trees*

Through controlled use of fire many advantages are there. Uncontrolled fire results in enormous damage to forests. The following are some of the advantages of controlled use of fire in forest area.

(a) Improvement in soil and undergrowth conditions for including natural vegetation

(b) Clearance of site for artificial regeneration

(c) Hazard reduction

8.3 CHOICE OF TREE SPECIES

Matching tree species to specific site for various end users of wood is the most important for factor success of precision silviculture. Based on the locality factors and silvics of tree species the best tree for that locality has to be chosen. Preference should always be given to native species. If it is exotic the naturalized exotic species in that area can be chosen. The tree suitability map has to be referred to in the choice of any species to a locality. The following Table 8.5 provides the name of the plantation species grown in the tropical region.

Table 8.5 Major Tree Species Grown in Tropical Region

S. No	Name of the species	Natural/ Exotic	Rainfall required	Tempe rature °C	Soil	End use
1.	*Eucalyptus tereticornis*	E	400-1200	13-38°C	Deep rich well drained	Paper, plywood, fuelwood
2.	*Eucalyptus camaldulensis*	E	400-1400	13-38°C	Deep rich well drained	Paper, plywood, fuelwood
3.	*Casuarina equisetifolia*	E	800-1500	18° C- 26° C	Fine coastal sand and loamy soil	Poles, scaffolding, fencing, paper making ,fire wood
4.	*Casuarina junghuhniana*	E	800-1500	18° C- 26° C	Fine coastal sand and loamy soil	Poles, Scaffolding, fencing, paper making ,fire wood
5.	*Leucena leucocephala*	E	600-2800	25°C-30°C	Red soil, alluvial	Fire wood , fodder

S. No	Name of the species	Natural/ Exotic	Rainfall required	Tempe rature °C	Soil	End use
6.	*Tectona grandis*	N	1100-2000	15 °C-45 °C	Sandy alluvium, red soil	Ship building, bridges, cabinet work, carpentry
7.	*Dalbergia latifolia*	N	750-5500	18°c to 45°c	Sandy, alluvial	Furniture, fencing, timber, fire wood
8.	*Dalbergia sissoo*	N	740-6500	6° C- 43° C	Sandy, well drained alluvial soil	Furniture, fencing, timber, fire wood
9.	*Melia dubia*	N	700-2600	21°C -36°C	Red soil, sandy loam	Pulp wood , ply wood
10.	*Santalum album*	N	750-2000	15°C - 38°C	Well drained alluvial, red loam	Carving and fancy, oil distilled, is used perfumery and medicines
11.	*Pterocarpus santalinus*	N	700-2800	21°C - 34°C	Red soil, alluvial	Musical instruments ,ship building, bridges, cabinet work, carpentry
12.	*Pterocarpus marsupium*	N	800-2000	22°C - 34°C	Red soil, well dined soil alluvial ,	Poles, Scaffolding, fencing, paper making ,fire wood
13.	*Azardiracta indica*	N	400–1,200	21° C -32 °C	Red soil ,clay and well-drained soil	Neem oil, timber, gum. Furniture,
14.	*Pongamia pinnata*	N	500-2 500	19°C-42°C	well-drained sandy loams	Oil seed, fire wood ,
15.	*Simarouba glauca*	E	500 - 4000	10°C - 40 °C	well-drained 16.soil	Oil seed, fire wood ,ply wood

S. No	Name of the species	Natural/ Exotic	Rainfall required	Tempe rature °C	Soil	End use
16.	*Bassia longifodia*	N	600 - 3000	15°C- 35°C		Oil seed,Timber
17.	*Gmelina arborea*	N	750- 4500	20°C-38°C	Moist, fertile, freely drained soils	Pencil wood, pulp , ply wood ,firewood
18.	*Acacia nilotica*	N	250- 1500	15°C- 40°C	Red soil , clay soil, sandy loam soil	Fire wood, furniture.
19.	*Acacia leucophloea*	N	250- 1600	15°C- 40°C	Red soil , clay soil, sandy loam soil	Fire wood, furniture , gum
20.	*Acacia planifrons*	N	350- 1500	15°C- 40°C	Red soil , clay soil, sandy loam soil	Fire wood, furniture , gum Arabic
21.	*Acacia auriculiformis*	E	700 -2000	13°C-32°C	Sand, clay soil	Pulp wood, ply wood, furniture.
22.	*Acacia mangium*	E	600- 2500	13°C T-21°C	Red soil , alluvial	Pulp wood, ply wood,furniture.
23.	*Bambusa bamboo*	N	750- 1900	17°C-42°C	Alluvium, loam and clay loam	Pulp wood , basket making , edible ,poles
24.	*Bambusa balcooa*	N	1200 to 4000	8°C to 36°C	Well drained soil, alluvial soil	Poles , pulp wood , carpet making,scaffolding
25.	*Bambusa vulgaris*	E	800- 4200	8°C to 38°C	sandy loam and clay loam	Poles , pulp wood , carpet making, scaffolding
26.	*Dendrocalamus strictus*		1000 - 3000	20°C -30°C	sandy loam soil	Poles , pulp wood , carpet making, scaffolding

S. No	Name of the species	Natural/ Exotic	Rainfall required	Tempe rature °C	Soil	End use
27.	*Neolamarkia cadamba*	E	600-2200	18°C-33°C	Well drained soil alluvial , red soil	Pencil wood, pulp and ply wood, firewood
28.	*Acrocarpus fraxinifolius*	N	1500-2000	19°C-28°C	Clayey loam soils	Poles, Scaffolding, fencing, paper making ,fire wood

8.4 QUALITY PLANTING STOCK

Quality seedlings play a vital role in establishing healthy plantation which could yield higher utilizable biomass. Targeting seedling production to the anticipated field site is an important step in producing strong healthy seedlings.

To produce quality seedlings for outplanting under precision silviculture the following practices have to be followed

1) Genetic quality of seedlings should be assured by sowing seeds of good quality in the nursery. To achieve this seeds should be collected from seed orchards.
2) Collected seeds should be size graded and should also be free for pest and disease.
3) Depending upon the tree species. The seeds can also be given seed scarification using hot water or commercial grade H_2SO_4 (*Acacia* seeds, *Albizia* seeds) and then used for sowing after repeated washing.
4) Treated seeds could be directly sown into poly bags Ex. *Acacia, Albizia* etc or raised in mother bed and transplanted after considerable growth Ex. *Eucalyptus, Casuarina*, or raised for 8-12 months for stump preparation. Ex. *Teak*
5) An per the species grown and for the duration kept in the nursery container size will vary 20 x 10cm, 20 x 15 cm, 40 x 30cm etc of 250 - 300 gauge thickness.
6) Enriched container mixture using native, soil, sand, FYM, Vermi compost Biofertilizers, Pseudomonas & Trichoderma should be used for getting

healthy seedlings. Irrigated plantation seedlings may be enriched with chemical fertilize like DAP, urea of Super phosphate, Muriate of potash etc.

7) Seedlings should be given optimum irrigation during its growth. Hardening of seedlings in the nursery will help the seedlings to withstand drought in the main field.

8) Seedlings raised should be free from weeds, pests and diseases with constant weeding and proper pest and disease managed.

9) The seedlings should have a straight stem without much branches and with tap root and sufficient lateral roots. Macro propagated seedlings and tissue culture seedlings should ered with good root system. Nursery hygiene and good nursery practices is very much essential to supply healthy tree seedlings for future higher biomass forest plantation.

10) Optimimum aged (cloves - 3-4 months, Seedlings > 6 months) and seedlings with minimum growth (Height >30 cm, Collar Diameter > 4 mm) should be chose for outplanting.

The main factors that decide seedling quality are

(i) Nursery site (Soil, fertility, moisture, weather)

(ii) Genetic make-up of the stock, seed (size, variability, germination treatments)

(iii) Method of production from seeds or macro and micro propagation (bare-rooted, rooted - cuttings, container seedlings)

(iv) The space occupied by the seedlings in the nursery bed.

(v) Time of growing and length of growing period

(vi) Age of seedlings

(vii) Weeding

(viii) Method of seedling conditioning

(ix) Pest management

(x) Disease management

The nursery growth seedlings quality could be judged by its physiological and morphological, characteristics (Top length and Collar diameter).

Figure 8.3 Quality seedlings of Tectona grandis

Figure 8.4 Quality seedlings of Pterocarpus santalinus

8.5 SITE SPECIFIC SILVICULTURAL OPERATION

Site preparation plays on important role in making the physical environmental suitable for early establishment and faster initial growth of a tree seedling after outplanting. Site preparation involves the following steps:

(i) Cutting and removal or incorporation of competing vegetation present on the site.

(ii) Depending on the topography of the land on steeper land or undulating land go in for spot preparation of the planting area by digging pits of suitable size. On plain lands open up the soil using disk ploughing and the inverted soil could be broken of its clods and leveled by drawing a cultivator. Site preparation should always be done well in advance of planting for proper curing.

(iii) As forest plantation are laid out on a larger land area, depending upon the slope of the land large size (1 meter) contour bunds may be raised to arrest run off and erosion. This will conserve rain water for better growth of plantation. On heavy rainfall sites sufficient drainage pits could be dug and the water can be taken into farm ponds to be used during water scarce period.

(iv) Staking and pitting has to be taken depending upon the species and the locality. On drier sites larger pits of 45-60 cm^3 and on wetter site $30 cm^3$ pits can be dug.

(v) Enrichment of pits based on soil test values with organic manure (FYM, Compost, Vermi compost, Decomposed leaf litter etc) Chemical fertilizers (urea, Super phosphate, Muriate of Potash) Biofertilizers (VAM, Rhizobium, Azospirillum, Phosphobacteria, Azatobacter) could be used.

(vi) Time of planting is very much needed for rainfed plantation. In case of those region receving rain from South west monsoon June - July months are ideal. Those region which are benefited from North east monsoon September - October are the ideal season. In case of assured irrigation, Better to avoid hot summer and seedlings can be planted at any time of the year (sivere winter) period for planting.

(vii) Best quality seedlings as per certification standards have to be used for out planting

(viii) Age of seedlings is vital for survival and establishment of tree seedlings in the main field. Colonel seedlings of age 3 - 4 months, tissue cultured seedlings of 4 - 6 months and seed origin graded seedlings of 4 - 6 months are ideal for planting. On drier site 6 - 8 months old seedlings have to be planted.

(ix) Disinfecting pits with insecticide like Chloropyripos 25 EC - 2ml/lit/pit could ward of termites and Bvistine 2gm/lit/pit could be used to control any fungal disease.

(x) Proper staking has to be given for tall seedlings for early anchorage and establishment of seedlings and for its erect position.

8.6 IRRIGATION MANAGEMENT FOR TREES

The main objective of irrigating forest plantation is to avoid unwanted moisture stress and its negative impact on tree growth. Soil water potential, best measured by tensiometer, decreases as soil water content drops. This relationship changes with soil texture. The secret of effective irrigation is to keep soil pores filled with the proper balance of water and air to minimize moisture stress. Irrigation to a plantation varies with the species grown, environment and soil. Hence plantation managers should have a through knowledge of these factors to implement the most effective irrigation monitoring and application programme for judicious use of water for maximum utilizable biomass yield. Determining when to irrigate is important as knowing how much water to apply. Water consumption by trees varies with species age and the environment from 5 lit to 100 lit/day.

Factors involved in scheduling of irrigation for forest plantation

(i) Plant water use

(ii) Soil water holding capacity

(ii) Water infiltration rate

(iv) Plant rooting depth

(v) Irrigation system output

The about factors have to be studied and worked out for effective scheduling of irrigation. To boost growth of trees drip irrigation can be followed to save water and improve growth. Irrigation is a must during summer and can be avoided during winter and raining months.

Figure 8.5 Drip irrigation under Dalbergia sissoo

Figure 8.6 Drip irrigation under Swietenia macrophylla

8.7 WEED MANAGEMENT FOR TREES

Weed growth and unwanted ground vegetation affect plantation growth to a layer extent during its early growth stage. Precision Silviculture aims at providing a weed free environment during the initial and at early growth stage of any plantation. It also aims at faster growth rate of trees during the 1and 2nd year of the stand to cut of light to the soil surface and prevent weed growth. The number and intensity of weeding various according to the locality.

Ex. Teak - 3 weeding in the first year, 2 in the second and the one in the third year are required.

Weeds		Type of Locality
Eupatorium odoratum	-	Moist locality
Dodonia viscosa	-	Moist and dry locality
Lantana camera	-	Dry locality
Parthenuim hysterophorus	-	Moist locality

The general concept of weeding under precision silviculture is to provide a weed free environment during the early establishing period of the tree (1 - 2 years after planting).

Figure 8.7 Weed free environment under Precision Silviculture

8.8 NUTRIENT MANAGEMENT FOR TREE PLANTATIONS

Soil nutrient status and the added manures (FYM, Vermi compost), fertilizers (N, P_2o, K_2o), Biofertilizers (Azospirillum, Phosphobacteria, VAM) play a vital role in early growth and establishment of a tree. High exchangeable calcium favours teak growth (Totey ***et.al*** 1991), By use of organic bio-stimulants as stress vitamin can stimulate tree growth and can provide optimum yield with up to 50% reduction in fertilizer use (Russo and Berlyn 1992).Based on soil nutrient status judicious combination of the above nutrients based on the tree species have to be derived and applied in split doses to achieve best result .In case of liquid fertilizers they may be combined along with drip irrigation as drip fertigation to sane application cost and for effective application and usage by the trees with minimal loss.

Figure 8.8 INM under Swietenia macrophylla

Figure 8.9 Drip fertigation under Melia dubia

Table 8.6 Management of Major Pest of Tree Crops

Name of the Tree Species	Pest	Insect name	Management
Teak	Teak defoliator	*Hyblaea Puera*	Clip badly affected leaves with larva Spraying of Neem Seed Kernel Extract (NSKE) 5% Setting up of light trap to trap adult moths Mixed plantations
	Teak Skeletonizer	*Eutectona machaearails*	Clip severely affected leaves Spraying of Neem Seed Kernal Extract (NSKE) 5% Spraying of Quinalphos @ 2 ml/lit
	Teak beehcle borer	*Xyleutes ceramicus*	Weeding control Thinning and inter croffy Coffee borer – Zauyera Coffeea 0.5% dichlorvos 75 EC 5ml/lit

Name of the Tree Species	Pest	Insect name	Management
Alianthus	Aliathus defoliator	*Eligma nacrcissus*	Spraying of Malathion @ 1ml/lit of water. Alianthus Web Worm – Atteva fabriciella Spraying of Malathion @ 1ml/lit of water.
Sal	Sal Heartwood borer	*Hoplocerambyx spinicornis*	Felling in winter Debarking before march Conversion of infested trees into Sleepers Cotton soaked in 0.5% dichlorvos 76 EC (5ml/lit.) can be kept inside the bore hole and plugged with mud
	Mango Stem Borer	*Batocera rufomaculata*	Unhealthy trees should be felled and converted before the emergence of adults Cotton soaked in 0.5% dichlorvos 76EC (5ml/lit.) can be kept inside the bore hole and plugged with mud
Eucalyptus	Eucalyptus gall wasp	*Leptocybe invasa*	Providing nutritional supplements like salicylic acid @ 500 ppm and vermiwash @ 50 ml/root trainer or DAP @ 10gm/lit of water Use of yellow and green colored sticky traps Spraying of imidacloprid @ 1ml/lit or dimethoate @ 2ml/lit

	Termite	*Odontotermes sp*	Avoid water stress conditions Soil drenching with chlorpyriphos 25 EC – 2ml/lit
Casurina	Root and shoot borer	*Celosterna scaprator*	Grubs can be removed from attacked roots by cutting down from the side of root at 40 cm below the ground Pushing petrol or dichlorvos through the exposed hole and ceiling with moist soil
Gmelina	Phassus borer	*Sahydrassus malabaricus*	Avoid alternatehost Avoid close planting Destruction of severely damaged plants
Mahua (Illupai)	Bark caterpillar	*Indarbela quadrinotata*	Cotton soaked in 0.5% dichlorvos 76 EC (5ml/lit.) can be kept in the borechole and plastered with mud paste
Silk cotton	Manage Stem Borer	*Ulata Batocera rufomac*	Cotton soaked in 0.5% dichlorvos 76 EC (5ml/lit) can be kept inside the bore hole and plastered with mud
Cassia	Longhorn beetle	*Aristobia approximator*	Cotton soaked in 0.5% dichlorvos 76 EC (5ml/lit) can be kept inside the bore hole and plugged with mud
Neem	Tea Mosquito Bug	*Helopeltis antonii*	Spraying of phosalone @ 2 ml/lit of water during evening hours

			Clipping of affected leaves
Pungam	Leaf gall mite	*Aceria pongamiae*	Application of acaricides like Fenazaquin or propargite @ 1 ml/lit of water during new flush formation in summer
Jatropha	Papaya mealy bug	*Paracoccus marginatus*	Release of parasitoid, Acerophagus papaya Spraying of chlorpyriphos, 20 EC @ 5 ml/it of water (or) Profenofos 50 EC @ 2 ml/lt of water.

Table 8.7 Diseases of forest crops and Management

Tree Species	Name of the diseases	Causal organism	Management
Azadirachta indica	Damping off	*Pythium, Phytophthora, Fusarium and Rhizoctonia*	Formalin is applied as soil fumigant and the fungicides can be applied either as soil drench or soil mix. Drenching with Bavistin @ 2g/litre of water. Potting media should be properly sterilized by solar heating before use.
	Leaf Web Blight	*Rhizoctonia solani.*	Drenching with Bavistin @ 2g/litre of water.
	Colletotrichum Leaf spot and blight	*Colletotrichum gloeosporioides*	Blitox fungicide (0.2% a.i.) twice at weekly intervals
	Alternaria Leaf spot and blight	*Alternaria alternata*	Blitox fungicide (0.2%) at fortnightly intervals

Tree Species	Name of the diseases	Causal organism	Management
***Albizia* spp**	*Pseudocercospora* Leaf spot:	*Pseudocercospora subsessilis*	Mancozeb (0.2%) in combination with Brestan (0.1%)
	Leaf spot and blight	*Cercospora albizziae, Colletotrichum sp., Alternaria alternata, Camptomeris albizzia, Pleiochaeta setosa and Epicoccum sp*	Captaf (0.2%) fungicide solution as foliar spray
	Seedling wilt	*Fusarium oxysporum*	Captaf (0.2%) fungicide solution as foliar spray
	Rhizoctonia Leaf Web Blight	*Rhizoctonia solani*	Bayleton (0.1% a.i.) as foliar spray
	Leaf Rust	*Ravenalia clemensiae*	0.2% Dithane M-45 or Sulfax fungicide.
	Little leaf disease	*Pytoplasma like organisms*	foliar spraying of systemic insecticide
Gmelina arborea	Foot –rot	*Fusarium oxysporum*	Soil drenching with 0.2% Bavistin or Dithane M-45 at monthly intervals
	Poria root-rot	*Poria rhizomorpha*	
	Root-rot and Collar rot	*Sclerotium rolfsii*	Soil drenching with 0.2% Bavistin or Dithane M-45 at monthly intervals
	Leaf spot	*Pseudocercospora ranjita, Deptoshaeria gmelinae, Phoma tropica, Alternaria laternata and Macrophomina phaseolina*	Bavistin (0.1%) and Dithane M-45 (0.1%)
	Leaf and shoot blight	*Glomerella cingulata in association with Fusarium solani*	two applications of Bavistin at weekly interval

Tree Species	Name of the diseases	Causal organism	Management
	Powdery Mildew	*Phyllactinia suffulta var. gmelina*	Application of sulphur fungicide 0.2%
	Phoma stem rot	*Phoma nebulosa*	applying 2-3 foliar sprays of Dithane M-45 (0.05% a.i.) at weekly intervals
	Canker disease	*Thyronectria pseudotricha and Hendersonula toruloidea*	
Pongamia pinnata	Leaf spot and blight	*Fusicladium pongamiae*	Foliar spray of Bavistin fungicidal solution (0.1%)
	Leaf Rust	*Ravenelia hobsoni*	Sulphur based fungicide (0.05%)
	Powdery mildew	*Oidium sp.*	Bavistin fungicidal solution (0.01%)
Tectona grandis	Leaf blight	*Rhizoctonia solani*	Application of Dithane M-45 (0.1%) is found effective
	Leaf rust	*Olivea tectonae*	sulphur based fungicide (Sulfax)
	Leaf spots		Immediate removal of infected plants
	Powdery mildew	*Phyllactinia corylea Uncinula tectonae*	Sulphur dust was most effective
	Heart rot	*Fistulina hepatica*	The prevention of heart rot can be a very difficult task but, there are effective measures to minimize damage. These methods include facilitating healthy growth, minimizing wounds and proper pruning of branches.

Tree Species	Name of the diseases	Causal organism	Management
	Basal Root Rot	*Phellinus noxius*	Trenches are a sound cultural Strategy in minimizing spread of the disease to other planting rows. Construction of isolation trenches is the most common and popular method of root disease control in plantations to prevent disease spread
Dalbergia sissoo	Leaf spot	*Cercospora sissoo, Colletogloeum sissoo, Phyllachora dalbergiae, Phyllachora spissa, Phyllosticta sissoo, Mycosphaerella dalbergiae, Myrothesicum roridum and Alternaria alternata*	
	Leaf Blight	*Rhizoctonia solani*	Proper sanitation, weeding and foliar application of fungicide solution (Bayleton – 0.1% at fortnightly intervals).
	Powdery Mildew	*Phyllactinia dalbergiae*	sulphur based fungicide was found most effective followed by Baycor, Mortesan
	Rust	*Maravalia achroa*	foliar application of 0.08% Bayleton at fortnightly intervals

Tree Species	Name of the diseases	Causal organism	Management
Ailanthus excelsa	plant parasite	*Helicanthus elastica*	Severely infested trees should be uprooted and kept in isolation or even burnt in order to prevent further spread of the plant parasite to other trees. Prevent flowering and fruits in *H. elastica* and thus seed production.
***Atrocarpus* spp**	Alternaria leaf blight	*Alternaria* sp	foliar spray with Dithane M-45 (0.01% a.i.), Dithane M-45 (0.1% a.i.) or Fytolan (0.2% a.i.) at weekly intervals
Terminalia chebula	**Leaf Rust**	*Uredo terminaliae*	fungicide (Sulfax @ 0.05% a.i.) as foliar spray and soil drenching at fortnightly intervals
Eucalyptus	root, collar, foot or crown rot	*Phytophthora*	drenching of copper oxychloride or mancozeb + metalaxyl-M
	leaf spot	*Mycosphaerella*	Bavistin 0.2% as foliar *spray*
	Powdery mildew		sulphur based fungicide was found most effective followed by Baycor, Mortesan and Calixin
	canker	*Botyrosphaeria*	copper *sprays* monthly during the growing season
	Damping Off	*Cylin-drocladium, Pythium, Fusarium and Rhizoc-toma.*	copper oxide and Benlate (Benomyl 50%

Tree Species	Name of the diseases	Causal organism	Management
Casuarina	Damping off	*rhizoctonia, fusarium* and *phytophthora*	Providing proper drainage Seed treatment with captan or thiram @ 4g/kg.Soil drenching with carbendazim (Bavistin) @ 0.1 %
	Stem canker	*Phomopsis casuarinae*	Spray with mancozeb @ 0.25 %
	Die back	*Botryosphaeria ribis*	Removal of infected plant parts and spray with mancozeb @ 0.25 % or copper oxy chloride @ 0.25 %.
	Pink disease	*Corticium salmonicolor*	Removal of severely affected plant parts or scrap the Infected portions and apply with Bordeaux Paste.
	Wilt disease	*Pseudomonas solanacearum* [*Ralstonia solanacearum*],	Remove the infected trees immediately. Digging the trenches around the infected tree. Scrap the infected portion and spray with copper oxy chloride @ 0.25 %.
Leucaena	gummosis, Canker	*Camptomeris leucaenae, Phytophthora drechsleri*	spraying of Fosetyl-Al or Copper
	seedling disease	*Colletotrichum capsici*	Seed treatment withThiram 0.2 % or 2 *sprays* of Bavistin (300g/ha)
	root and collar rots	*Ganoderma species, Fusarium species*	Application of foliar *spray*with 0.2% Bavistin, Dithane M - 45, Fytolan

Tree Species	Name of the diseases	Causal organism	Management
	Fan-leaf di sease	In absence of any fungal or bacterial Association with the disease it is suspected to be viral in nature.	*Management* of Thrips is the ultimate control

Table 8.8 Nematodes of Forest crops and their management

Tree species	Nematode genera	Management
Casuarina	*M. incognita* *M. javanica* *M. arenaria* *Rotylenchulus reniformis*	Solarization of nursery beds Nursery beds can be treated with carbofuran 3G @ 3g a.i /m2. Organic amendments with neem/mustard/ castor cake @ 1 kg/ sq. m. Seedling root dip with carbofuran @ 500 ppm for 1 hr. In main field, carbofuran or phorate as spot treatment @ 1.5 kg ai/ha
Pongamia pinnata	*M. incognita,* *M. javanica* *Rotylenchulus reniformis*	Soil application of *Pseudomonas flourescens* @ 2.5 kg/ha Application of carbofuran @ 1.5 kg/ha.
Bixa orenella	*Helicotylenchus* *Pratylenchus* *Tylenchorhynchus* *Longidorus*	Seedling root dip with carbofuran @ 500 ppm for 1 hr. In main field, carbofuran or phorate as spot treatment @ 1.5 kg ai/ha

Tree species	Nematode genera	Management
Dalbergia sissoo	*M. incognita* *Helicotylenchus*	Nursery beds can be treated with carbofuran 3G @ 3g a.i /m2. Organic amendments with neem/mustard/ castor cake @ 1 kg/ sq. m. Seedling root dip with carbofuran @ 500 ppm for 1 hr. In main field, carbofuran or phorate as spot treatment @ 1.5 kg ai/ha
Ailanthus excelsa	*Tylenchorhynchus* *Longidorus*	Seedling root dip with carbofuran @ 500 ppm for 1 hr. In main field, carbofuran or phorate as spot treatment @ 1.5 kg ai/ha
Terminalia chebula	*Helicotylenchus* *Tylenchorhynchus*	Soil application of *Pseudomonas flourescens* @ 2.5 kg/ha Application of carbofuran @ 1.5 kg/ha.
Melia dubia	*Helicotylenchus* *Tylenchorhynchus* *Longidorus* *Rotylenchulus reniformis*	Soil application of *Pseudomonas flourescens* @ 2.5 kg/ha Application of carbofuran @ 1.5 kg/ha.
Pterocarpus santilinus	*Longidorus* *Xiphinema* *Helicotylenchus*	Nursery beds can be treated with carbofuran 3G @ 3g a.i /m2. Organic amendments with neem/mustard/ castor cake @ 1 kg/ sq. m. Seedling root dip with carbofuran @ 500 ppm for 1 hr. In main field, carbofuran or phorate as spot treatment @ 1.5 kg ai/ha

8.9 ROLE OF GIS AND GPS IN PRECISION SILVICULTURE

Geographical information system and global possitioning system are used in identifying soil types and growth of stand and is widely used under precision silviculture. The fundamental link between soil, geology, climate and species is required to make precision silviculture a reality. Climatic factors like Solar Radiation, Temperature, Rainfall are used along with species specific ecophysiological parameters (Photosynthesis, Respiration, Carbon allocation) to produce site specific estimate system that integrate geology, topography, and soil data. Leaf area estimate obtained from Landsal satellite imagery are put into 3PG an ecophysiological based process model of forest growth, to predict potential productivity, identify the climate and soil factors that limit growth and determine the potential response to silvicultural treatments designed to ameliorate these limitary factors. This has been tried in Pinvs teada in southern United States. (Mervy et al 2008). Precision silviculture in the 21st century links GIS and Remote Sensing to develop site specific silviculture in southern pine plantation (Thomas R. Fox ***et al.***)

8.10 OPTIMAL ROTATION AGE

Precision silviculture should always aim at optimal rotation age to fetch higher biomass and early returns. Optimal rotation age is the growth period required to derive maximum value from a stand of timber. The calculation of this period is specific to each species and to the economic and sustainability goals of the harvester. Precision silviculture aims at economically optimum rotation age to generate the maximum revenue or economic yield by calculating the maximum Net Present value. Through precision silviculture locality factors are studied, choice of species is made and with quality seedlings combined with high-tech silvicultural operation like land preparation, enrichment of pit, planting, weeding, soil working, irrigation and nutrient management rotation age has been brought down in many plantation species.

Table 8.9 Tree Species and their Rotation Age under Normal Cultivation and Precision Silviculture

S.No	Name of Species	Normal Rotation Age	Rotation Age is Under Precision Silviculture
1.	*Eucalyptus hybrid*	8 - 10 years	4 - 6 years
2.	*Casuarina equisetifolia*	6 - 8 years	3 - 5 years
3.	*Casuarina junghuhniana*	5 - 6 years	3 - 4 years
4.	*Melia dubia*	8 - 10 years	5 - 6 years
5.	*Ailanthus excelsa*	10 - 12 years	6 - 8 years
6.	*Tectona grandis*	60 - 80 years	20 - 30 years
7.	*Gmelina arborea*	30 -40 years	20 - 25 years
8.	*Acrocarpus fraxinifolius*	20 - 30 years	10 - 15 years
9.	*Swietenia macrophylla*	20 - 30 years	15 - 20 years
10.	*Grewia tiliaefolia*	20 - 30 years	15 - 20 years

8.11 REDUCED IMPACT LOGGING (RIL) AND CONVERSION TECHNOLOGIES

Reduced impact logging (RIL) is defined as the intensively planned and carefully controlled implementation of timber harvesting operation to minimize the environmental impact on the forest stand and soil. Reduced impact logging is a felling operation carried out in a matured tree stand without much damage to the nearby standing trees and disturbing the soil much during the impact created by the fall of the felled tree. While RIL technique are often disregarded for their initial cost, studies often find them beneficial in the long term due to reduced wood west and less damage to future growing stock. RIC technique can save labour and machinery operating time. RIL is also able to reduce carbon emission from forestry activity Precision silviculture produces trees of evenaged and with even spacing and this facilitates mechanization for easy felling. Application of RIL is such plantation will be easy without much damage to the unfelled trees and the soil.

Planning optimum logging operation is done through precision forest approach: The important precision forestry tools are data collection and data analysis. The data collection is done by ground survey, remote sensing and real time season approach. The data analsyis is done through decision support system and geographical informtion system. (Taylor ***et al.,*** 2002)

The felled wood has to be converted into billets or logs so as to be utilized for various end uses like pulp, plywood, match wood, timber etc. This process of converting the felled wood to such end use materials is called conversion. Of late on site sawmilling and timber conversion techniques, on site billeting using chain saw or machines billeters and on site chipping mechanism are available. Precision silviculture helps in providing logs of uniform size and shape and this facilitate easy conversion operation.

8.12 ENVIRONMENTAL IMPACT AND SUSTAINED YIELD

Precision forestry practices aims at achieving higher economic returns with adequate environmental protection to achieve sustainable yield. The site specific silviculture should always aims at sustaining the site quality by adequate soil, moisture conservation method and enrichment of site for sustained yield. Runoff control measures like contour bunding alley cropping will go a long way in reducing runoff and conserving moisture through better infiltration and reducing soil erosion. Precision silviculture aims at higher production of wood at the shortest rotation period. In this process for sustained yield the soil heath has to be property refinished with adequate nutrients. Enrichment of soil could be done by the following ways.

(i) Powdering and ploughing insitue the left over brush wood after felling and conversion.

(ii) Raising of green manure crops or using green leaf manure and ploughing institute during raining season when there is adequate moisture in the soil.

(iii) Based on soil test value enrichment of pits with organic (FYM, Compost - Vermi compost etc., inorganic fertilizer Nitrogen (Urea, DAP), Phosphorus (Super Phosphate, DAP) Potash (Muriate of Potash) could be applied in pits at planting

(iv) Intercropping the inter space of the plantation with leguminous crops and after harvesting the stubbles could be ploughed into the soil.

(v) Rotation of non nitrogen fixing trees with nitrogen fixing trees after the rotation age.

Ex. Irrigated *Eucalyptus* rotated with *Casuarina* or *Leucaena*

8.13 SUMMARY

Thus Precision Silviculture, analyses the locality factors prevailing in a site and through judicious choice of species and site specific silvilculture technologies will help in generating high utilizable biomass within the shortened rotation period without affecting the environment on a sustainable basis. Hence adequate planning keeping in mind all resources have be to taken up for a successful Precision Silviculture operation.

"NATIONS HEALTH IS FOREST WEALTH - SOIL HEALTH IS PLANTATIONS WEALTH"

References

1. Chin 2007. How much water do trees take? 22 Tech. News Africa.
2. Dhruv Sood 2014. Wood and wood products in India. GAIN Report No. In 409 dt. 24.06.2014.
3. Mervy K, Fei S, Drake N, Nibbling and J. Hepenstall. 2008. Southern Forestry and Natural Resources GIS Conference (2008) University of Georgia, Athens, GA.
4. Network for Certification and Conservation of Forest. 2017. Draft Policy Paper on Promoting Sustainable Trade of Wood and Wood based Products in India.P 1-22
5. Russo, R.O and G.P Berlyn (1992). Vitamin - humic acid - algal bio-stimulant increases bean yield. Hort. Sci. 27:847.
6. Robert Neil Pallett (2005). Precision forestry for pulpwood re - establishment silviculture. South African Forestry Journal vol 23 (1).
7. Oberai, C. P (2000). Forest for New Millennium. Employment News, Ministry of Information and Broadcasting, Govt. of India. New Delhi 29 (51): 1 - 3.
8. Sheikh Ibrahim bin Sheik Ali, (1991). A manual on grading of Nursery seedlings. Forest Department. Peninsular Malaysia and ITTO PD185 Rev. 2 (F) - Phase II.
9. Schmoldt D.L and Thomsan A.J. 2003. Precision Technologies: Data availability Past and Future Proceeding of the Second International Precision Forestry Symposium - Washington
10. Totey, N. G, Arun Prasad, Bhaumik, A.K, Khatri, P.K (1991). Soil productivity of teak forests. J. Indian Soc. Soil Sci. New Delhi.

11. Thomsan R. Fox, Lee Allen, Randolph H. Wynne and Christine E. Blinn 2008. Precision Silviculture is to 21st century, linking GIS and Remote sensing to develop site specific silvicultural Regimes in Southern Pine Plantation. In the proceedings of 6th Southern Forestry and Natural Resources GIS Conference (2008). University of Georgia, Athens. GA.

12. Taylor S.E, MW veal, TE grift, T.P. Mc Donald and F. Conley. 2002. Precision Forestry operated Tactics for today and tomorrow, Aubrun University.

13. Yadav, M. Basera, K. 2013. Status of Forest Products Production and Trade. Indian Institute of Forest management, Bhopal.

14. Ziesk, M. (2006). Precision Forestry An overview on the current status of precision forestry. IUFRO Precession Forestry Conference. 20006. Technical University Munich 5-10 March 2006.

CHAPTER IX

Precision Forestry Practices for Carbon Neutrality

Dr.A.Balasubramanian and C.N. Hariprasath

Objectives

- To introduce the students on the concept and significance of Carbon Neutrality
- To give an overview on the Role of Trees and Forests in performing Carbon Sequestration
- To highlight the needed strategies to reduce the Carbon Emissions from different sources and
- To delineate the success stories and their methodologies for implementing the Carbon Sequestration.

Organization of the Chapter

- Introduction to Climate Change
- The Precision Concept of Carbon Neutrality
- The Concept of Carbon Sequestration
- Role of Trees and Forests for Carbon Sequestration
- Carbon Neutrality Through Afforestation Efforts
- Estimation of Carbon Sequestration
- Strategies for Reducing Carbon Emissions
- Carbon Neutral Certification

- Benefits of Carbon Neutrality
- Success Stories in Carbon Neutrality Practices
- Summary

9.1 INTRODUCTION TO CLIMATE CHANGE AND GREENHOUSE GAS EMISSION

Climate change has been acknowledged by most of the world community as the greatest threat to mankind. If we look at the reason for such a threat to occur, we would clearly find our own activities, which have led to global warming and ultimately climate change. Rising concentration of greenhouse gases (GHGs) in the atmosphere could lead to a change in solar energy balance and consequently the world's climate. Climate change or global warming is largely dictated by carbon dioxide (CO_2). The CO_2 is the most abundant and responsible for half of the radioactive rays forcing associated with the greenhouse effect that leads to global warming (Reddy, 2010). An average atmospheric CO_2 concentration during 2017 is 402.16 ppm and 400.65 ppm in the year 2016. During past decade (2008-2017), the average annual increase in CO_2 level is 4.1 ppm and the average for the envious decade (1993-2002) is 1.7 ppm (NOAA, 2017). The largest growth in CO_2 emissions has come from energy supply, transport and industry, while residential and commercial buildings, forestry (including deforestation) and agriculture sectors have been growing at a lower rate (IPCC, 2007). Cars, trucks and other road transport vehicles, while essential for commercial and humanitarian work, exert a considerable adverse impact on the environment. The vehicles are a significant source of air pollution which can adversely affect human health and the environment. The Kyoto protocol and its implementation from 2008 have been a turnaround for the attitude of several developed nations regarding their various emission sources.

Intergovernmental Panel on Climate Change (IPCC) ascertained that the pre industrial level of carbon in the atmosphere rose from 285μ mol l^{-1} (600 gigatonnes (Gt)) to the current level of 340 mol l^{-1} (800 gigatonnes (Gt)) and the future predicted rise in the atmospheric CO_2 would approach 1000 Gt by the year 2050 with an increased CO_2 concentration level of 480 ppm. Climate scientists revealed that we have years, not decades to stabilize CO_2 and other greenhouse gases. So, they developed an efficient strategy for reducing the CO_2 level in the atmosphere. Carbon sequestration is ought to be a promising solution for reducing the carbon dioxide in the atmosphere by capture and storage of carbon in trees as an effective and efficient strategy. Thus the concept of Carbon Auditing or Carbon foot printing of an organization or an individual is gaining momentum. One of the biggest challenges in combating climate change is lack of awareness about climate change and the role every individual plays to enhance it.

Hence sequestering carbon on earth surface as a carbon naturalistic strategy will enable us to promote environmentally safe. In order to promote carbon neutral situation, a multi stakeholder's participation is required to make industries/ institutions/ public to achieve environmentally friendly living.

9.2 CARBON NEUTRAL – AN INNOVATIVE CONCEPT

Carbon neutrality refers to absorbing of carbon equivalent to the amount of carbon emission by an organization /school/industry. It means absorbing an equivalent amount of carbon emitted by various activities like transport, using electricity for lighting / industries / air conditioning etc., where we burn fossil fuels, wood, etc. in turn emitting a large amount of CO_2 into the atmosphere. Carbon neutral mechanism help to reduce greenhouse gas emissions where ever and where ever possible, using offset units to compensate for the remainder and achieving net zero emissions.

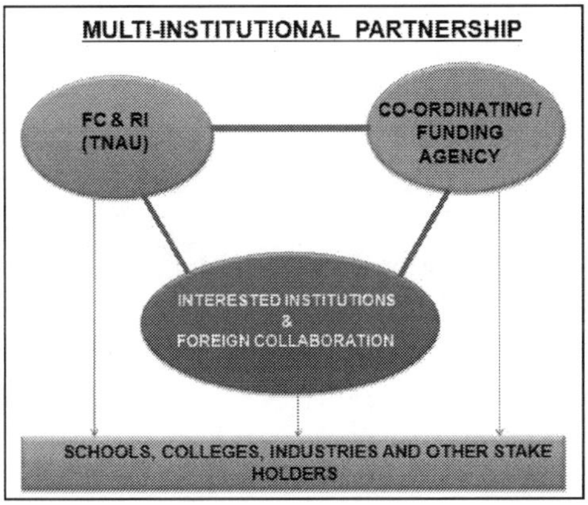

Figure 9.1 Multi – Institutional Partnership for attaining Carbon Neutrality

By becoming carbon neutral, leading businesses and organisations are positioning themselves for growth and competitiveness in a lower emissions future. In working towards carbon neutrality, organisations are building their capacity for emissions management and responding to growing expectations of corporate social responsibility. Beyond this, businesses are looking for carbon smart procurement choices and are making new demands on their supply chain.

9.2.1 Steps for Achieving Carbon Neutral

1. Identify the area/subject - A clear description of the area/subject
2. Measuring the CO_2 emissions - Provide a complete and accurate account of the GHG emissions and CO_2 emissions of the area.
3. Achieving carbon neutrality through afforestation - Achieve the target through a combination of internal reductions and environmental instruments (Afforestation).
4. Certification - Provide accurate and transparent information on how carbon neutral certification is achieved.

9.2.1.1 *Identify the Area / Subject*

The area/subject to which the carbon neutral protocol is being applied must be clearly mentioned by name and description of the relevant legal and/ or physical boundaries. The duration and starting date of the carbon neutral certification must be defined. The major area need to be certified is industries, institutions, villages and major CO_2 source areas. Schools play a major role in shaping the lifestyle of any human beings. Though environment education is now a part of regular course curricula it leaves a gap in the individual and people oriented climate education. Schools are also an enterprise which involves various activities that emit CO_2 during their functioning. Several International schools also want to be certified as green schools which would improve their marketing their brand name and also result in lower energy and resources bill.

9.2.1.2 *Measuring the CO_2 Emissions*

The second step in going carbon neutral by calculating the amount of greenhouse gases emitted within a defined boundary. The boundary will depend on whether it is seeking to become carbon neutral for an organisation, product or service, and in some cases what level of control it is having over those emissions. Within this boundary, all activities or sources that lead to greenhouse gas emissions must be identified. These include sources that you have direct or indirect control over, such as transport fuels, purchased electricity, waste generated in operations, business travel and other purchased goods and services.

9. 2.1.2.1 *Carbon Footprint*

Carbon Footprint is a methodology to estimate the total emission of greenhouse gases (GHG) in carbon equivalents from a product across its life cycle from the production of raw material used in its manufacture, to disposal of the finished product (excluding in-use emissions).

For eg. BASIC COMPONENTS OF CARBON FOOTPRINT ESTIMATION IN SCHOOLS

9. 2.1.2.2. *Different Forms of Carbon Emissions from Schools*

- Direct CO_2 emissions from schools comprise all the emissions released from schools. They arise from fuel use of school-owned buildings and equipment. CO_2 account, the direct CO_2 emissions from electricity use are not part of the direct emissions from schools as they are emitted from electricity generating plants elsewhere in the economy. Instead they are accounted for as part of the embodied emissions.
- Transport CO_2 emissions comprises all the emissions from operating transport equipment for the commuting to schools. These include emissions from private vehicles as well as from public transport.
- Embodied CO_2 emissions comprise all emissions from the procurement of goods and services consumed in schools arising in the industrial supply chain.

9. 2.1.2.3. *Data Collection*

A questionnaire was formed by an understanding of the available data within schools to calculate a carbon footprint. This questionnaire was tailored towards the needs of the hybrid model. The questionnaire was divided into five consumption categories:

- Energy – Total Grid electricity and Fuel used.
- Transport – Includes everyday travel to school, field trips and site visits.
- Food – Breakfast, lunch and soft drinks.
- Resources – Paper, cleaning materials
- Water – Water pumped or used
- Waste – Waste generated/month

The date collected was transferred to excel sheet and analysed levels of percentage among different grades of students by calculating the carbon footprint using the emission factor for different consumption categories.

Figure 9.2 Carbon Footprint Training to Environmentalist and Students

9. 2.1.2.4 *Emission Factor*

An emission factor is defined as the average emission rate of a given GHG for a given source, relative to units of activity. The emission factor is the ratio between the amount of pollution generated and the amount of a given raw material processed.

9.2.1.3 Achieving Carbon Neutrality through Afforestation in Urban Areas

9.2.1.3.1 URBAN FORESTRY

An urban forest is a forest or a collection of trees that grow within a city, town or a suburb. In a wider sense, it may include any kind of woody plant vegetation growing in and around human settlements. In a narrower sense, it describes areas whose ecosystems are inherited from wilderness leftovers or remnants. Care and management of urban forests is called **urban forestry**. The concept of planting trees in urban areas was first introduced at University of Toranto in 1965 and the term "**Urban Forestry**" was first used by Jorgensen in 1970. Urban forestry is the care and management of single trees and tree population in urban settings for the purpose of improving the urban environment. Urban forestry is a practical discipline, which includes tree planting, care, protection and the overall management of trees as a collective resource. The main objective of urban forestry is plan and manages the tree resource to create a healthy and sustainable forest to enhance the benefits and safety of urban dwellery.

Benefits of Urban Forestry

a. Urban forests intercept and absorb air pollution from the atmosphere. The major air pollutants are sulfur dioxide (SO_2), nitrogen oxides (NO_2), ozone (O_3), and fine particulate matter such as soot, dust, pollen and emissions from diesel vehicles.

b. Trees also reduce air pollutants (**i.e.** Particulate matter, volatile organic compounds and nitrogen oxides) that are detrimental to human health (**E.g.** Asthma and related illnesses) and contribute to more complex air quality problems such as ground-level ozone or smog (EPA).

c. Trees lower air temperatures making urban areas more inhabitable for humans and wildlife.

d. Urban forests help reduce global warming by absorbing (sequestering) and storing carbon dioxide CO_2 from the atmosphere.

e. The increased heat in cities can cause health problems including general discomfort, respiratory difficulties, heat cramps, heat exhaustion, heat stroke and even death. To combat the high temperatures in urban climates, the use of air conditioning/cooling systems in urban heat islands requires energy production, which not only increase utility costs, but also releases dangerous greenhouse gasses into the atmosphere, exacerbating the rate ozone is formed. Green spaces help offset the formation of urban heat islands by cooling the regional micro-climate through shading and evapo-transpiration, thusly reducing the energy needed to cool buildings during hot periods. In addition, trees planted as wind shields can significantly reduce heating costs for buildings during the winter. Planting trees (more than other kinds of vegetation) is the best way to combat urban heat islands because trees have a higher potential to cool the climate and reduce the urban heat island effect and carbon dioxide (CO_2).

f. In urban settings, trees can reduce the amount of runoff and pollutant loading. In fact, 100 mature trees catch about 250,000 gallons of rainwater per year decreasing flood risk as well as soil and hill erosion.

- **Trees Purify Water**: Tree leaves and roots naturally filter water and remove pollutants.
- **Trees help store water**: Trees help water soak into the ground and recharge water tables.
- **Trees promote cleaner waterways**: Trees absorb sediment, nutrients and chemicals before they enter streams and other bodies of water.
- **Trees fight floods**: Trees reduce soil erosion and flooding by slowing the movement of storm water.

g. Trees and other plants help remediate soils at landfills and other contaminated sites by absorbing, transforming, and containing a number of contaminants.

h. Urban residents living in greener areas build strong feelings of belonging and attachment to their neighborhood, and the trees that comprise their neighborhood. Active participation in urban forestry provides: a sense of pride in themselves and the community, improved health (mental and physical) due to increased time in natural environments, and feelings of trust and connectedness towards the community as a result of social interaction with residents

i. Trees can not only help make money, but save money, too. During the summer, trees shade houses reducing building temperatures and the need for air conditioning, thereby reducing energy bills. By reducing building electricity needs in the summer, trees (urban forests) indirectly reduce carbon dioxide emissions in energy power plants.

j. Urban forests are a great way of offering refuge to flora and fauna alike, offering increased plant and animal biodiversity. Urban forests contribute to the long-term health of urban ecosystems by providing wildlife habitat (i.e. Shelter, food and water) for birds, mammals, insects, amphibians, and reptiles. Those birds and insects can then help further pollinate our natural environment, giving us a more verdant and robust natural environment.

9.3 CARBON SEQUESTRATION

Carbon sequestration refers to the capture and long-term storage of carbon in forests and soils or in the oceans, so that the build-up of carbon dioxide (one of the principal greenhouse gases) in the atmosphere will reduce or slow. Managing land and vegetation to increase carbon storage can buy valuable time to address the ultimate challenge of reducing greenhouse gas emissions.

Article 3 of the Kyoto Protocol allows for the offset of emissions by investing in activities that increase carbon sequestration. This would generally involve an investor or buyer being issued with "carbon credits" corresponding to the amount of carbon sequestered by these activities. These credits could then be used to offset part of the buyer's net greenhouse gas emissions (for example from their electricity or steel manufacturing plant). This provides a relatively low-cost opportunity for the private sector to reduce emissions while promoting a comprehensive and environmentally responsible approach to climate change.

There are three main types of carbon sequestration

- Carbon sequestration in terrestrial ecosystems - Increasing the amount of carbon stored in vegetation (Forest and Trees) and soils
- Carbon Sequestration in the Oceans - Enhancing the net uptake of carbon from the atmosphere by the oceans, through fertilisation of phytoplankton with nutrients and injecting carbon dioxide to ocean depths
- The subsurface sequestration of carbon dioxide in underground geological repositories.

All of these options are commonly known as carbon "sinks". The prime most increasing carbon storage is terrestrial ecosystems, which is currently focusing with almost attention and it is the easiest and most immediate option at the present time.

Terrestrially, carbon is stored in vegetation and in the soil. Plants store carbon for as long as they live, in terms of live biomass. Once they die, the biomass becomes a part of the food chain and eventually enters the soil as soil carbon. If the biomass is incinerated, the carbon is reemitted into the atmosphere and is free to move in the carbon cycle.

9.4 TREES AND FOREST FOR CARBON SEQUESTRATION

The role of forests in carbon sequestration is probably best understood and appears to offer the greatest near-term potential for human management as a sink. Unlike many plants and most crops, which have short lives or release much of their carbon at the end of each season, forest biomass accumulates carbon over decades and centuries. Furthermore, carbon accumulation potential in forests and trees is large enough that forests offer the possibility of sequestering significant amounts of additional carbon in relatively short periods. Fortuitously, forests managed for timber, wildlife or recreation sequesters carbon as a by-product. Forests may also be managed strictly to sequester carbon. There are four components of carbon storage in a forest ecosystem.
These are trees, plants growing on the forest floor (under-storey material), detritus such as leaf litter and other decaying matter on the forest floor, and forest soils. Carbon is sequestered in the process of plant growth as carbon is captured in plant cell formation and oxygen is released.

Forest transitions from one ecological condition to another will produce substantial carbon flows – forests can be a carbon source or a sink. It is important to carefully assess exactly what is happening to the carbon as the forest changes to determine the forest's sink/source contribution.

9.5 CARBON NEUTRALITY THROUGH AFFORESTATION IN URBAN AREAS

Implanting the concept of carbon neutral through tree planting in young minds of helps for permanent sequestration of carbon from the atmosphere. Scientific estimation of carbon and carbon emission for synergizing emission equivalent to sequestration. The procedure for tree planting in urban area is

9.5.1 Preparation of Planting Site

It is ideal if the area to be planted is first leveled with the help of earth movers if needed, so that it is clear of obstructions or depressions or tree stumps that could interfere with proper establishment and facilitate easy access to the entire site for men and machinery. This step also ensures that weeds are also removed and delays their reestablishment compared to an unprepared site. It is preferred that no large trees are present except those that come outside the planting area. If trees are to be retained then it would be a good idea to leave an area adjacent to the canopy around it free. Once the land is prepared for plantation after leveling and weeding, the next step is to align and place stakes to mark spots for taking the pits.

Depending on the spacing chosen and modifications made to avoid large trees and earth formations if any, wooden or bamboo stakes are driven into the ground to mark the centre of the square or round pits. This will help avoid misalignment of pits when unskilled labour or machinery is used for pitting, particularly when planting is to be done along contours on undulating sites in hilly areas. A North –South orientation is ideal for aligning the rows to get the best benefit of sunlight.

Figure 9.3 Planting Trees with High Carbon Sequestration Values

9.5.2 Local Factors to be considered before Raising Tree Plantations

Since the purpose of planting, site conditions, social and economic factors vary widely across the country, providing a common protocol. However the following major aspects are to be considered when planning for a plantation.

- Preparation of planting site (Leveling, Weeding, Pitting) - Manual or mechanical
- Cost of planting material (Seedling, vegetative propagation, tissue culture)
- Mortality replacement (3- 30 %)
- Fertilizer and mulching (Chemical fertilizer /Manure/compost)

- Protection (Perimeter Fencing, individual guard, Watch and ward)
- Irrigation (Rain fed, drip irrigation, piped, fertigation)
- Plantation Management (Soil mounting, harvesting)
- Labour costs

9.5.3 Tree Species selected for Urban Planting

The commonly grown urban trees with high carbon sequestration are listed in Table 9.1.

Table 9.1 Commonly Grown Urban Trees with High Carbon Sequestration

Botanical Name	Common Name	Tamil Name	Family
Dalbergia sissoo	Indian rosewood	Sissoo	Fabaceae
Dalbergia latifolia	Black rosewood	Nukkam	Fabaceae
Terminalia arjuna	Arjuna	Neermaruthu	Combertaceae
Ficus retusa	Cuban laurel	Kallithi	Moraceae
Gmelina arborea	Gamhar	Kumil	Verbanaceae
Pongamia pinnata	Indian beech tree	Pungan	Fabaceae
Azadirachta indica	Neem	Vembu	Meliaceae
Bixa orellana	Lipstick tree	Sappiravirai	Bixaceae
Thespesia populnea	Indian Tulip Tree	Poovarasu	Malvaceae
Aleurites fordii	Tung oil tree	Nattakkarottu	Euphorbiaceae
Neolamarckia cadamba	Kadamba	Cadamba	Rubiaceae
Mitragyna parvifolia	Kaim tree	Neerkkatampu	Rubiaceae
Acrocarpus fraxinifolius	Pink Cedar	Malankonnai	Fabaceae
Terminalia bellerica	Belliric Myrobalan	Tanri	Combertaceae
Pterocarpus marsupium	Indian Kino Tree	Vengai	Fabaceae
Pterocerpus santalinus	Red Sanders	Sen santhanam	Fabaceae
Mimusops elengi	Mimusops	Mahilamaram	Sapotaceae
Swietenia macrophylla	Honduras mahogany	Mahogany	Meliaceae
Butea monosperma	Flame of the Forest	Ilai Purasu	Fabaceae
Syzygium cuminii	Jamun	Naaval Pazham	Myrtaceae
Tamarindus indica	Tamarind	Pulia maram	Caesalpiniaceae

Figure 9.4 Tree Seedlings for Afforestation

9.5.4 Planting

Best time for planting is during the pre-monsoon showers, so that establishment during the monsoon is successful and requires less watering. The seedling/cutting is placed in the pit and the polybag is removed with a sharp blade taking care not to disturb the roots and rhizome of the plant. Soil around the plants should be consolidated tightly by pressing with feet. The surface of the pit, after filling should be slightly sloped to one side so as to promote moisture retention. Generally, three weedings in the first year, two in the second and one in the third year has to be carried out.

Planting should be carried out between October and March. Dry areas are best planted in early autumn and wet areas in February and March. No planting should be done in very wet, frosty or snowy conditions.

9.5.5 Planting Methods

During planting, it is important to keep the plants in a plastic bag to prevent drying out and exposure to light. It is essential that the roots are kept moist and dark as even a short period of exposure could damage the plants.

9.5.5.1 *Pit Planting*

This is generally used when planting larger stock or container trees. The trees should be planted by digging out a pit which must be deep and wide enough to take the roots. The base of the pit should be forked over to improve drainage and the roots should be evenly spread around the pit. The pit can then be back-filled with the soil being firmed in every 10cm to prevent air pockets. Trees should be planted no deeper than the soil mark on the stem.

Normally tree species planted in 45 cm x 45 cm x 45 cm pits (or) with larger pits of 60 X 60 X 60 cm or even larger (1 X 1 X 1 m) can be considered if the soil is of poor quality or if the area is rocky or hard laterite so that the filled soil permits easy establishment of plantations. Bigger pits will benefit the tree species for establishing well under harsh climate and condition. A trench is dug and the soil used to make a bund of about 1 m in width and 0.5 m in height along the length of the plot.

Figure 9.5 Field Spacing and Pitting Experiments

9.5.6. Plant Protection

Newly planted trees will require protection from grazing animals. The major plant protection types are

- Tree shelters
- Protective fencing
- Individual tree enclosures
- Cattle / horse proof enclosure
- Sheep proof enclosure

9.5.7 Plantation Management – After care

Tree plantations do not require much care except sufficient moisture to prevent drying up in summer. Mortality of plants can be expected in extremely difficult sites and replacements should be considered then and then in sites where drip irrigation is provided. However the following activities are recommended for improving the performance of plantations and ensuring profits for the highest carbon sequestration.

9.5.7.1 *Weeding*

Regular weeding in the initial two to three years is very important for quicker establishment and faster growth of trees. Soon after the rainy season clear knife weeding will have to be carried out depending on the site condition and amount of weed present in the plot. At least an area of 60 cm around the trees should be kept clear of weeds particularly the climbers which can smother the young plants.

9.5.7.2 *Soil Working and Moisture Conservation Methods*

The soil around the plant should be loosened twice every year for better growth and production. Soil working also improves the water retention capacity and fertility of the soil; adjusts temperature and air permeability, improves soil physical and chemical composition and decreases weed competition for water and nutrients. Spreading the leaf branches and other mulching materials like leaf litter in plantations will help in increasing the moisture retention.

9.5.7.3 *Mulching*

Mulching is providing for a cover over the soil around the tree plant which reduces loss of moisture through evaporation and checks weed growth around the trees. Degradation of the mulch by the natural decay process eventually releases nutrients in to the soil and improves the texture through addition of organic carbon.

9.5.7.4 *Excessive Pruning to be avoided*

Practice of excessive pruning of crops should be avoided. Pruning of plants In a well nurtured garden end pruning of roadside plants should be differentiated. Excessive pruning may lead to upsetting the root: shoot ratio.

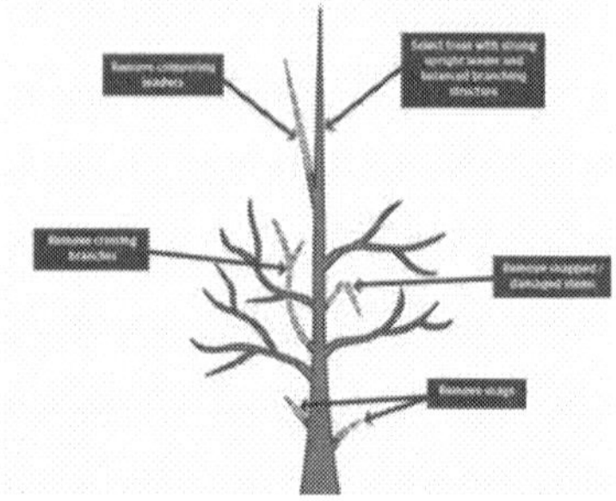

Figure 9.6 Proper Trees Pruning Techniques in Urban Area

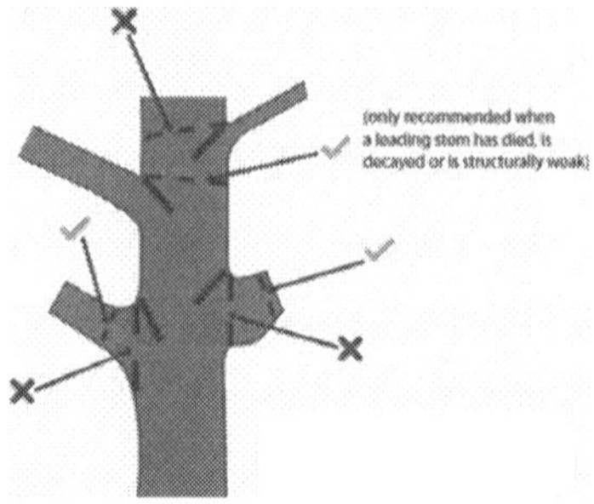

Figure 9.7 Position of Tree Pruning in Urban Area

9.5.7.5 *Pollarding*

Pollarding is when the main stem or stems of a tree are deliberately severed to stimulate new growth, which is then periodically removed to be used for construction purposes or animal fodder. Pollarding is started when a tree is young and repeated periodically so that the wood adapts to wounding by developing strengthened tissues close to where the cuts are made. Some urban trees are managed as pollards, with the pollarding regime beginning in the nursery.

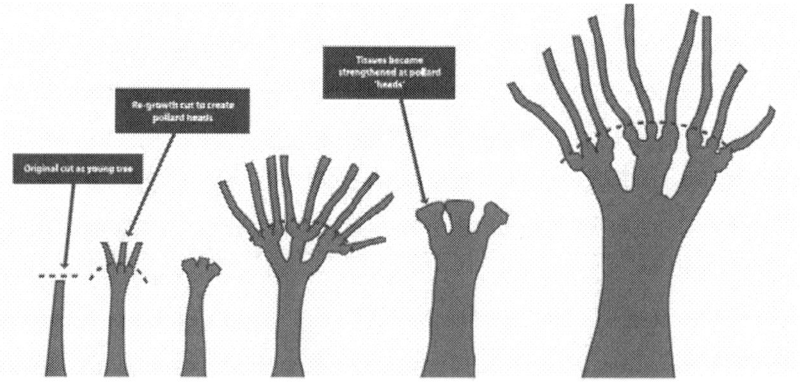

Figure 9.8 Tree Pollarding Techniques in Urban Area

9.5.7.6 *Protection from Pests and Grazing*

For initial two years, the planted area should be protected from grazing. Fencing of the plantation with protection against wild boar and grazing animals may therefore be needed in some areas. The tree species will be highly affected by pest and diseases, and hence the pest and diseases management should be carried out regularly.

Figure 9.9 Pest and Disease Attack in Plantations

9.5.7.7 *Irrigation*

During the first year of planting watering helps obtain higher survival rates especially in areas where the dry period is longer than two months. Irrigation generally helps to increase the productivity at least by three times. Natural water conservation methods like ditches or crescent shaped trenches and use of mulch help in moisture conservation. However the requirement of irrigation will vary with the local climatic conditions and microenvironment. In water scarce areas, drip irrigation has been found cost-effective, but that requires technology and investment during site preparation.

Figure 9.10 Drip Irrigation and Mulching

9.5.7.8 *Fertilizer Application*

Intensively managed plantations require application of fertilizers as fast growing trees is a heavy feeder and respond well to fertilizers. Fertilizer application will also ensure higher yield and overall profitability from plantations. In general trees respond well to phosphorous, potassium and especially nitrogen. It is a good idea, however to carry out soil analysis prior to establishment of the plantation and at annual intervals thereafter to fix and confirm the fertilizer dosages. Fertilizers should be applied in ditches around the seedlings when the soil has enough moisture and then covered with soil. Organic fertilizers are best applied during the onset of monsoon. Organic manure and compost are established sources of nutrients, and simple to prepare and use. They encourage microbial activity, which allows the fertilizer to provide nutrients over a longer period of time. Unlike chemical fertilizers, the nutrients in organic fertilizers are less likely to leach away in rainwater.

Figure 9.11 Organic Fertilizer Application

9.6 ESTIMATION OF CARBON SEQUESTRATION

9.6.1 Non Destructive Method

The biomass of the planted trees was estimated using non destructive sampling method. The biomass of the planted trees was classified into

9.6.1.1 *Above Ground Biomass (AGB)*

The above ground biomass of the planted tree species was estimated using non destructive sampling method. In order to estimate the above ground biomass of the planted trees, the volume and wood density are required (Pandya *et al.*, 2013). The volume calculated and the value of wood density for required tree species was obtained from the website www.worldagroforestrycentre.org/sea/products/AFDbases/WD. The above ground biomass was calculated by using the formula given below

$$\text{AGB (kg/tree)} = \text{Volume of tree (m}^3\text{) x Wood density (kg/m}^3\text{)}$$

9.6.1.2 *Below Ground Biomass (BGB)*

The below ground biomass was calculated by multiplying above ground biomass by 0.26 factors as the root: shoot ratio (Ravindranath and Ostwald, 2008). The formula for calculating the below ground biomass was given below

$$\text{Below Ground Biomass (kg. /tree)} = \text{AGB (kg/tree) x 0.26}$$

9.6.1.3 *Total Biomass*

Total biomass of trees includes both above ground and below ground biomass of the tree (Ravindranath and Ostwald, 2008) was calculated by following method

Total Biomass (kg. /tree) = AGB + BGB

9.6.1.4 *Biomass Carbon Estimation*

The biomass carbon content of planted tree species was calculated by utilizing the arithmetic value of biomass (Pearson *et al.*, 2005). The formula for calculating biomass carbon given below

Carbon Storage = Biomass x 50 per cent

Figure 9.12 Carbon Sequestration in Trees by Non Destructive Method

9.6.2 Destructive Method

The amount of carbon present in the afforestation site was assessed for estimating the carbon capturing potential of planted tree species. This estimation was done as per the procedure detailed below.

(i) Estimation of biomass of trees in plantation

(ii) Estimation of biomass carbon in plantation

9.6.2.1 Biomass Estimation in Plantation

9.6.2.1.1 *Above Ground Biomass (AGB)*

The above ground biomass estimation was estimated by destructive sampling method. The trees were felled at ground level using a mechanical chain saw (Poulan / Pro, USA). After recording the total height and GBH of the felled trees, the above ground portions were separated into stem, branches and leaves. Fresh weights of all the above ground tree components were recorded immediately after felling using appropriate spring scales. A small sample (500 g) of stem, branches and leaves were immediately transported to the laboratory in double sealed polythene bags. The collected samples were dried at 80°C till constant weight was obtained. The oven dry weight of the whole sample was calculated using the formula given below by Lasco *et al.* (2005). From the oven dried weight, carbon content in the tree biomass was analysed through the appropriate laboratory technique.

$$ODW(t) = \frac{TFW - \{TFW * (SFW - SODW)\}}{SFW}$$

Where, ODW = Total oven dry weight

TFW = Total fresh weight

SFW = Sample fresh weight

SODW = Sample oven dry weight

9.6.2.1.2 *Below Ground Biomass (BGB)*

The below ground biomass was also estimated through destructive analysis. Pits were excavated by removing soils around the tree and complete recovery of roots were done upto depth of roots. The roots were carefully separated and the root weight was recorded immediately using spring scales. Representative sample (500 g) of root was immediately transferred to the laboratory in double sealed polythene bags. The collected samples were dried in hot air oven at 80°C till constant weight was obtained. The oven dry weight of the whole sample was calculated using the formula given below (Lasco *et al.*, 2005).

$$ODW(t) = \frac{TFW - \{TFW * (SFW - SODW)\}}{SFW}$$

Where,

ODW = Total oven dry weight

TFW = Total fresh weight

SFW = Sample fresh weight

SODW = Sample oven dry weight

Figure 9.13 Carbon Sequestration in Trees by Destructive Method

9.6.2.2 *Carbon Estimation of Biomass in Afforested Plantation*

The samples of various components of trees *viz.*, stem, branches, leaves and roots of the targeted trees species were collected separately, air dried and oven dried. Oven dried biomass samples were grounded in Willey Mill and carbon concentration in different tree components were determined based on ash per cent and determined by the procedure given by Allen *et al.* (1986). To estimate the ash per cent and biomass carbon the following methods are used in tree species.

9.6.2.2.1 *Ash Content Estimation*

Porcelain crucibles were washed with 6 N hydrochloric acid and distilled water and dried in an oven at 65°C for one hour. One gram of powdered sample was taken in pre weighted crucibles. The crucibles were taken inside the cool furnace. After adjusting the furnace at 550°C, heating was increased slowly and after reaching 550°C, ignition was continued for 1 hour. Then, the crucibles were cooled slowly keeping them inside the furnace. After complete cooling, the crucible with ash was weighed and the percentage of ash was calculated as per the procedure given by Allen *et al.* (1986) with the following formula.

$$\text{Ash\%} = \frac{(W_3 - W_1)}{(W_2 - W_1)} \times 100$$

Where,

W1 = Weight of crucibles

W2 = Weight of oven dried powdered samples + crucibles

W3 = Weight of ash + crucibles.

Figure 9.14 Carbon Estimation by Ash Content Estimation

9.6.2.2.2 *Carbon Per cent Estimation in Biomass*

Carbon per cent in above ground biomass, below ground biomass, litter and dead organic matter was estimated followed by Negi *et al.*, (2003) and Dhruw *et al.* (2009), using the follwing formula given below.

Carbon % = 100 % - {Ash % + Molecular weight of O_2 (53.3 %) in $C_6H_{12}O_6$}

The carbon stock in the above ground biomass, below ground biomass, litter and dead organic matter was computed by using the following formula given below.

Carbon (MT) = Biomass (MT) x Carbon per cent

Percentage of organic carbon using the above procedure was estimated for all samples of fractinated biomass *viz.*, leaf, stem, branch and root. Using the carbon per cent value, the above ground organic biomass carbon (t ha^{-1}), below ground organic biomass carbon (t ha^{-1}) and total organic biomass carbon (t ha^{-1}) were calculated on individual species and also per hectare basis. The total biomass carbon was calculated by using the following formula.

i) AGB carbon (t C ha^{-1}) = Components of above ground biomass (t ha^{-1}) x Carbon (%)

ii) BGB carbon (t C ha^{-1}) = Components of below ground biomass (t ha^{-1}) x Carbon (%)

iii) Total biomass carbon stock (t C ha^{-1}) = AGB carbon + BGB carbon

9.6.3 Carbon Estimation in Decomposed Leaf Litters

The decomposed leaf litter samples were collected and washed using the distilled water and oven dried. The oven dried samples were grounded separately in Willey Mill. The carbon concentration in leaf litter sample was estimated using the procedure given by Allen *et al.* (1986). As per the technique the following estimations were carried

- Ash content estimation
- Carbon reduction estimation in the decomposed leaf litter biomass

9.6.4 Soil Organic Carbon (SOC)

Soil sample from sample plots were collected according to procedure given in the IPCC guidelines (IPCC, 1997). The soil samples were collected randomly from all sample plots by digging a fresh rectangular pit in the native vegetation and by clearing the top layer of grass, litter and humus in an area of 50 cm x 50 cm.

The soil samples were collected from 0-15 cm, 15-30 cm and 30-45 cm layers separately, dried and powdered using wooden mallet and sieved through 2 mm sieve. The soil organic carbon content in the 0.5 g sieved soil samples were estimated as per Walkley and Black's wet oxidation method (Walkley and Black, 1934). Soil organic carbon was expressed in percentage.

Figure 9.15 Soil Organic Carbon Estimation

9.7 STRATEGIES FOR REDUCING CARBON EMISSION

Another way of reducing carbon emission is utilizing organic wastes generated from industries/organisations. This waste will be utilizing as a resource base for environmental friendly products for human use by the following the scientific techniques.

9.7.1 Organic Mushroom Production

The organic waste will be chopped and used as bedding material for mushroom cultivation. The mushroom spawn will be seeded in the bedding material and kept in the mushroom sheds developed by the institutions. The scientific techniques will be given for industries/organisations to establish semi-permanent shed.

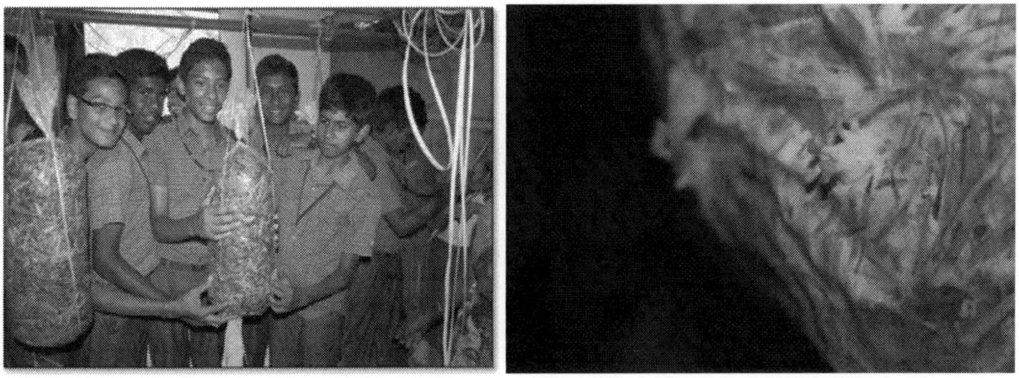

Figure 9.16 Mushroom Cultivation Techniques to Schools/Institutions

9.7.2 Composting

The organic wastes generated from industries/organisations will be converted into useful manure using composting techniques. The industries/organisations will be taught various methodologies like handling and segregation of wastes and practically explained about the process and production of nutrient rich useful compost from wastes

9.7.3 Vermicompost

The secondary organic wastes generated from industries/organisations will be converted into useful manure using Vermicomposting techniques. The organisations will be taught with handling of earthworms and practically explained about the role of earth worms in production of nutrient rich useful compost from wastes.

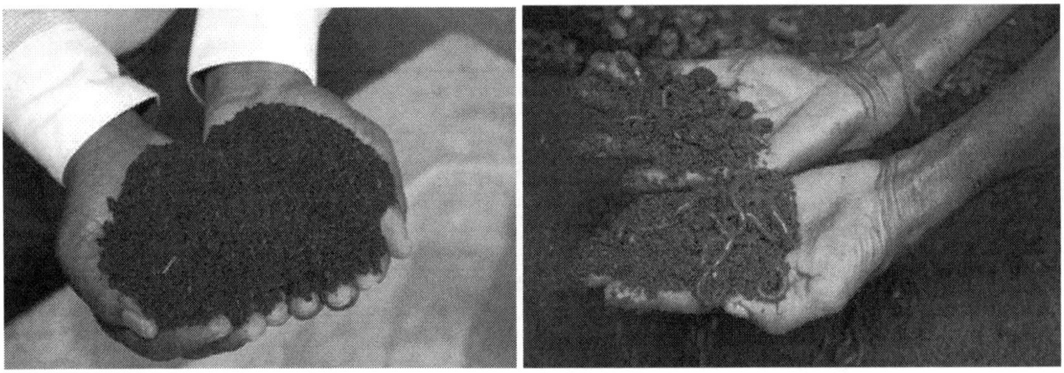

Figure 9.17 Vermicompost Technology

9.8 CARBON NEUTRAL CERTIFICATION

The carbon footprint assessment will be done in different phases in organisation/ industries.

Carbon Neutral = **Carbon Emission (Footprint)** - **Carbon Sequestration** = **Carbon Neutral Certification**

At the end, the organisation/industries successfully implement and made net carbon emission zero through the programme will be certified carbon emission free based on scientific estimation of total carbon sequestered which can be neutralized by total carbon emissions by the organisation/industries.

9.9 BENEFITS OF CARBON NEUTRAL ACTIVITIES

- **Increased Customer Recognition:** Customers recognise and value organisations that lead on sustainability and climate change action.
- **A New Competitive Edge**: Going carbon neutral can create brand differentiation opportunities and connections with new clients and customers.
- **Entry to the Carbon Neutral Supply Chain**: As climate change becomes a mainstream business concern, large organisations are looking beyond their internal carbon management to make carbon-smart procurement choices. Service providers and product manufacturers have the opportunity to meet this demand by supplying carbon neutral products and services.
- **Energy and Cost Savings**: Reducing emissions to become carbon neutral commonly involves making operational changes that reduce energy or fuel use. These efficiency improvements can lead to significant cost savings.
- **Improved Employee Engagement**: Carbon neutral organisations increase their reputation as an employer of choice, resulting in better staff retention and attraction of quality employees.
- **Enhanced Corporate Social Responsibility**: Organisations that go carbon neutral can purchase offset units from projects that align with their organisational values and corporate social responsibility strategy.
- **Internal Carbon Management Expertise**: Becoming carbon neutral helps organisations to build internal carbon accounting competence and management processes to meet compliance and voluntary emission reduction commitments.
- **Community Connection**: Many organisations and businesses enjoy greater recognition and engagement in their community as a result of their carbon neutrality.
- **Environmental Outcomes**: Benefits to the environment can be delivered by reducing emissions and by purchasing offset units from projects that support various environmental co-benefits.

9.10 SUCCESS STORY

A research programme on "**Vision Carbon Neutral - A Joint Venture Initiative for Promoting Carbon Neutral Institution**" was implemented at Forest College and Research Institute, Tamil Nadu Agricultural University, Mettupalayam with the co-ordinating partners of Confederation of Indian Industries – Young Indians from 01.01.2014 in 9 educational institutions at Western Zone (Coimbatore) of Tamil Nadu. The educational institutions are

- K' Sirs International School, Chinnavedampatti, Coimbatore.
- Western Ghats International School, Ettimadai, Coimbatore.
- Delhi Public School, Onaapalayam, Coimbatore.
- Sri Jayendra Saraswathi School, Singanallur, Coimbatore.
- Park College of Engineering and Technology, Kanniyur, Coimbatore.
- KG International School, Annur, Coimbatore.
- Park College of Technology, Karumathampatti, Coimbatore.
- Park Global School, Kanniyur, Coimbatore.
- Sasi Creative School of Business, Othakal Mandapam, Coimbatore.

The Successful activities carried out in the 9 Educational Institutions are

- Totally 10,500 indigenous tree seedlings with the height of 5 feet were planted in 9 educational schools/institutions. The technical input of raising the tree nursery, planting techniques and management of planted trees were given to 1150 school and college students.
- Nearly 20 varieties of tree species were characterized for high carbon sequestration value and planted in the urban environment (Schools/Colleges).
- Carbon footprint trainings were given to 1900 students and 150 teachers of the nine institutions based on the carbon footprint questionnaire. The carbon footprint questionnaire was categorized separately for institutions, teaching/non-teaching staffs, college students and school students (LKG to 5th students, 6th to 10th students and 11th to 12th students). The general awareness was also given to reduce the usage of the electricity and vehicles.

- All the schools and colleges were encouraged to go zero waste technology through complete recycling of organic waste generated in the premises through microbial composting, vermicomposting and mushroom production. The technical inputs towards establishment of vermicomposting and mushroom shed were given to all the organizations. The training on vermicomposting technology to 2700 students and mushroom cultivation to 1800 students were given to institutions.

"Institutes are managing their waste and producing nearly 1 tonne of vermicompost per month"

- Scientific estimation on carbon and carbon emission for synergizing emission equivalent to sequestration was given to 2000 students and teachers from nine institutions.
- Nature awareness programmee on jungle strolling (Importance of forest and its conservation), wildlife education (Identification of wildlife, rescuing of wildlife and safety measures to be carried out), trekking to different forest types and Non-timber forest products museum were given to 5000 students and teachers from all the nine institutions.

SUMMARY

Climate scientists revealed that we have years, not decades to stabilize CO_2 and other greenhouse gases. So, they developed an efficient strategy for reducing the CO_2 level in the atmosphere. Carbon sequestration is ought to be a promising solution for reducing the carbon dioxide in the atmosphere by capture and storage of carbon in trees as an effective and efficient strategy. Carbon neutral mechanism help to reduce greenhouse gas emissions where ever and when ever possible, using offset units to compensate for the remainder and achieving net zero emissions. Besides, the steps and strategies to achieve carbon neutrality is prescribed for follow up to reduce the CO_2 level in the atmosphere.

REFERENCES

1. Allen, S.E., H.M. Grimshaw and A.P. Rowland. 1986. Chemical analysis. In: P.D. Moore and S.B. Champan (Eds.). Methods in plant ecology. Blackwell Scientific Publications, Pp: 285-344.

2. BTSG-KFRI. 2012. Manual for establishment and management of bamboo plantations. National Bamboo Mission, Pp: 7-19.

3. Dhruw, S.K., L. Singh and A.K. Singh. 2009. Storage and sequestration of carbon by leguminous and non-leguminous trees on red-lateritic soil of Chhattisgarh. Indian Forester, 135(4): 531-538.

4. Hari Prasath, C.N., A.Balasubramanian and S.Radhakrishnan.2015. Carbon Capture and Storage – A Specific Study on Ocean Storage Technologies. LAMBERT Academic Publishing (ISBN No: 978-3-659-18558-8).

5. Hari Prasath, C.N. 2013. Studies on afforestation and carbon sequestration of few fast growing tree species. M.Sc. Thesis, Tamil Nadu Agricultural University, Coimbatore.

6. IPCC. 2007. Climate change 2007: Impacts, adaptation and vulnerability. Fourth Assessment Report. Parry M.L., O.F. Canziani, J.P. Palutikof, P.J. van der Linden and C.E. Hanson (Eds.). Cambridge University Press, Cambridge, UK.

7. Lasco, R.D., F.B. Pulhin, R.V.O. Cruz, J.M. Pulhin and S.S.N. Roy. 2005. Carbon budgets of terrestrial ecosystems in the Pantabangan-Carranglan Watershed. In: Assessments of Impacts and Adaptations to Climate Change (AIACC) Working Paper. www.aiaccproject.org.

8. Manoj, K., A.Balasubramanian, S.Radhakrishnan, C.N.Hari Prasath, M.Tilak and P.Durairasu. 2013. Urban Forestry – A Visualization. Shri Garuda Graphic, Coimbatore (ISBN No: 978-9-380-76952-3).

9. Narentheran, G.G. 2012. Carbon Auditing of Nanjing International School, China (Major Project Report). Department of Natural Resources, TERI University, New Delhi.

10. Natural Capital Partners. 2016. The Carbon Neutral Protocol - The global standard for carbon neutral programmes. Pp: 20-30. www.carbonneutral.com

11. Negi, J.D.S., R.K. Manhas and P.S. Chauhan. 2003. Carbon allocation in different components of some tree species of India: a new approach for carbon estimation. Current Science, 85(11): 1528-1531.

12. NOAA. 2017. National Climate Data Centre, State of the Climate: Global Analysis for August 2017, published online September 2013, retrieved on August 18, 2013 from http://www.ncdc.noaa.gov/sotc/global/.

13. Pandya Ishan, Y., H. Salvi, O. Chahar and N. Vaghela. 2013. Quantative analysis on carbon storage of 25 valuable tree species of Gujarat, Incredible India. Indian J. Sci. Res., 4(1): 137-141.

14. Pearson T., S.Walker and S. Brown. 2005. Sourcebook for land use, land use change and forestry projects. Win rock International and the Bio Carbon Fund of the World Bank. 57 p.

15. Ravindranath, N.H. and M. Ostwald. 2008. Carbon inventory methods handbook for greenhouse gas inventory, carbon mitigation and round wood production projects. Springer, 29.

16. Reddy, R., K.R. Girish Kasineni and Agepati S.Raghavendra. 2010. The impact of global elevated CO_2 concentration on photosynthesis and plant productivity. Current Science, 99(1): 46-47.

17. Suresh Kumar, M. 2011. Carbon sequestration potential and baseline carbon studies through afforestation. M.Sc. Thesis, Tamil Nadu Agricultural University, Coimbatore.

18. Walkley, A. and T.A. Black. 1934. An examination of the wet acid method for determining soil organic matter and proposed modification of the chromic acid Titration Method. Soil Science, 37: 29-38.

CHAPTER X

Farmer Producer Companies in India

Dr. C. Sekhar

Objectives

- To document the role and Significance of Farmer Producer Companies
- To infuse ideas to the students on the activities and services provided by the Farmer Producer Companies
- To provide information on the institutional support provided from the Government for successful establishment and run of the companies

Organization of the Chapter

- Introduction to Farmer Producer Companies (FPCs)
- Comprision of Cooperative Society and Producer Companies
- The Activities of Farmer Producer Organizations (FPOs)
- The Services Provided by Farmer Porducer Organizations
- The Farmer and the Producer Organizations
- Role of Central Government Institutions in Supporting FPOs
- Role of State Government Institutions in Supporting FPOs
- FPOs in India
- Factors Governing the Size of Companies
- Success Stories of FPOs
- Summary

10.1 INTRODUCTION TO FARMER PRODUCER COMPANIES

Precision Technology is capable of enhancing the output per unit area in respect of fruits and vegetables, cereals, aquaculture, poultry and livestock production activities. The increase of output aggregated in one village quite normally attract very low price among the wholesalers or traders whom will be periodically visiting the farm gate for their benefits. Since the produce produced from Precision Farming Technique is a quality produce, the pricing mechanism in the market should be high when compared to the normal price for the vegetables and fruits. But this is not happening across different markets. In this circumstances, farmers themselves started establishing their own organization after witnessing the model promoted by Dr. E.Vadivel, the architect of Precision Farming in Tamil Nadu Agricultural University, Coimbatore, Tamil Nadu. Following this practice, the farmer producer organizations in each district has been established and few of the organizations are so successful.

Collectivization of producers, especially small and marginal farmers, into producer organisations has emerged as one of the most effective pathways to address the many challenges of agriculture but most importantly, improved access to investments, technology and inputs and markets. Department of Agriculture and Cooperation, Ministry of Agriculture, Govt. of India has identified farmer producer organisation registered under the special provisions of the Companies Act, 1956 as the most appropriate institutional form around which to mobilize farmers and build their capacity to collectively leverage their production and marketing strength.

10.1.1 The Meaning of Producer Organizations

A Producer Organisation (PO) is a legal entityformed by primary producers, viz. farmers, milk producers, fishermen, weavers, rural artisans, craftsmen. A POcan be a producer company, a cooperative society or any other legal form which provides for sharing of profits/benefitsamong the members. In some forms like producer companies, institutions of primary producers can also become member of PO

10.1.2 The Need of Producer Organizations

The main aim of PO is to ensure better income for the producers through an organization of their own. Small producers do not have the volume individually (both inputs and produce) to get the benefit of economies of scale. Besides, in agricultural marketing, there is a long chain of intermediaries who very often work non-transparently leading to the situation where the producer receives only a small part of the value that the ultimate consumer pays. Through aggregation, the primary producers can avail the benefit of economies of scale. They will also have better bargaining power vis-à-vis the bulk buyers of produce and bulk suppliers of inputs

The Farmer Producer Organization (FPO) is one type of PO where the members are farmers. Small Farmers' Agribusiness Consortium (SFAC) is providing support for promotion of FPOs. PO is a generic name for an organization of producers of any produce, e.g.agricultural, non-farm products, artisan products, etc.

The Essential Features of Producer Organization

- It is formed by a group of producers for either farm or non-farm activities.
- It is a registered body and a legal entity.
- Producers are shareholders in the organization.
- It deals with business activities related to the primary produce/product.
- It works for the benefit of the member producers.
- A part of the profit is shared amongst the producers.
- Rest of the surplus is added to its owned funds for business expansion.

10.1.3 The Ownership of Producer Organization (PO)

The ownership of the PO iswith its members. It is an organization of the producers, by the producers and for the producers. One or more institutions and/or individuals may have promoted the PO by way of assisting in mobilization, registration, business planning and operations. However, ownership control is always with members and management is through the representatives of the members.Any individual or institution can promote a PO. Individualpersons or institutions may promote PO using their own resources out of goodwill or with the noble objective of socio-economic development of producers. If, however, the facilitating agency wishes to seekfinancial and other support, then they have to meet the requirements of the donor/financing agency

10.1.4 The Institutional Support for Producer Organizations

NABARD, SFAC, Government Departments, Corporates and Domestic &International Aid agencies provide financial and/or technical support to the Producer Organisation Promoting Institution (POPI) for promotion and handholding of the PO. Each agency has its own criteria for selecting the project/promoting institution to support.The

NGO can promote PO which will provide better income to the members. Sharing of profit among members is an important objective of the PO.

10.1.5 The Legality Aspects of Producer Organization

Producer Organisation can be registered under any of the following legal provisions:

- Cooperative Societies Act/ Autonomous or Mutually Aided Cooperative Societies Act of the respective State
- Multi-State Cooperative Society Act, 2002
- Producer Company under Section 581(C) of Indian Companies Act, 1956, as amended in 2013
- Section 25 Company of Indian Companies Act, 1956, as amended as Section 8 in 2013
- Societies registered under Society Registration Act, 1860
- Public Trusts registered under Indian Trusts Act, 1882

Institutions registered as cooperative societies and producer companies have legal provisions for sharing of profit earned by the PO by way of dividend. Other legal forms do not explicitly provide for profit sharing. However, the PO can offer better price for the produce it procures from the members, thus, benefiting the latter. Similarly, it can procure inputs/raw material in bulk and sell to members with low margin. Such activities are permissible for POs under all legal forms.

10.2 COMPARISION OF COOPERATIVE SOCIETY AND PRODUCER COMPANY

A comparative chart of important features of Cooperative Society and Producer Company is given below (Table 10.1).

Table 10.1 Comparative Picture of Cooperative Society and the Producer Company

Sl. No	Parameters	Cooperative Society	Producer Company
01	Registration	Cooperative Societies Act	Indian Company Act
02	Objectives	Single	Multiple
03	Area of Operation	Restricted; Discretionary	Entire Union of India

Sl. No	Parameters	Cooperative Society	Producer Company
04	Membership	Individuals	Any Individual, Group, Association, Producer of Goods and Services
05	Share	Non Tradable	Not Tradable but Transferable; lImited to Members at par value
06	Profit Sharing	Limited Dividends on Shares	Commensurate with Volume of Business
07	Voting Rights	One Member-One Vote. But Government and Registrar of Cooperatives hold Veto Power	One Member-One Vote. Members not having transactions with the company cannot vote
08	Government Control	Highly patronized to the extent of interference	Minimal, Limited to statutory requirements
09	Autonomy	Limited in Real World Scenario	Fully Autonomous, self ruled within the provisions of act
10	Reserves	Created if there are profits	Mandatory to create every year
11	Borrowing Powr	Restricted as per bye-law. Any amendment to bye-law needs to be approved by the Registrar and time consuming	Borrowing limit fixed by Special Resolution in general meeting. Companies have more freedom to raise borrowing power
12	Relationships with Other Corporate / Business / NGOs	Transaction Based	Producers and Corporate entity can together float a producer company.

(Source: http://www.nabard.org)

10.3 THE PRINCIPAL ACTIVITIES OF PRODUCER ORGANIZATIONS

The primary producers have skill and expertise in producing. However, theygenerally need support formarketing of what they produce. The PO will basically bridge this gap. The PO will take over the responsibility of any one or more activities in the value chain of the produce right from procurement of raw material to delivery of the final product at the ultimate consumers' doorstep. In brief, the PO could undertake the following activities:

- Procurement of inputs
- Disseminating market information
- Dissemination of technology and innovations
- Facilitating finance for inputs
- Aggregation and storage of produce
- Primary processing like drying, cleaning and grading
- Brand building, Packaging, Labeling and Standardization
- Quality control
- Marketing to institutional buyers
- Participation in commodity exchanges
- Export

10.4 THE SERVICES PROVIDED BY THE PRODUCER ORGANIZATIONS

The Services provided by the farmer producer companies or organization are many which are not available with the individual farmer or a small group. For a better understanding, the usefulness of Farmer Producer Companies and their services are discussed with many of the pioneering farmers and the members of successful farmers' organization and the same are presented in the following Table 10.2.

Table 10.2 The Nature of Services Extended by the Farmer Producer Company

Sl. No	Nature of Services Extended	Type of Services
01	Organizational Services	Organizing the Farmers Catalyzing Collective Action Capacity Building Establishing Monitoring and Control Systems
02	Production Services	Input Supply Facilitation of Production Activities Provision of Packing Materials
03	Marketing Services	Transport and Storage Functions Output Pricing, Processing and sale Market Information and Analysis Branding Certification
04	Financial Services	Savings, Loans, Forms of Credit Financial Accounting and Management
05	Technology Services	Training (T), Research (R), Education (E) and Extension (E) (TREE)
06	Educational Services	Business Development Skills Production Techniques Core Competency Development to Human Resource Selection of Training and Skill Development Models
07	Welfare	Health Management Safety Nets

Sl. No	Nature of Services Extended	Type of Services
08	Management of Resources	Water, Pasture, Fisheries, Forests and Soil Conservation
09	Policy Advocacy	Need Based Consultancy Services for Facilitating the Functioning of FPC

Table 10.2 highlighted the different type of services. Among which the organizational services are much important to the FPC. The individual farmers are not conversant with the organizing services. The organization of services should listen to the time variable which is of prime importrance in business followed by production services. The production services should also match the input requirement and the output quality maintenance to fulfill the needs of the consumers. If consumers are provided with good quality output, the company will have sustenance in its growth. Among the marketing services, the transportation and storage of the materials on time found to be the principal activity and hence the logistics support should be properly planned by the Farmer Producer Company.

10.5 THE FARMER AND THE PRODUCER ORGANIZATIONS

A PO will support the members in getting more income by undertaking any/many/all of the activities listed under point listed above. By aggregating the demand for inputs, the PO can buy in bulk, thus procuring at cheaper price compared to individual purchase. Besides, by transporting in bulk, cost of transportation is reduced. Thus reducingthe overall cost of production. Similarly, the PO may aggregate the produce of all members and market in bulk, thus, fetching better price per unit of produce. The PO can also provide market information tothe producers to enable themhold on to their producetill the market price become favourable.All these interventionswill result inmore income to the primary producers. (http://www.nabard.org)

Table 10.3 The Cost Estimate for Incorporation of a Producer Company

Sl. No	Particulars of Expenses	Item of Expenditure	Amount in Rs
01	Cost of Application for Name of the Producer Company	Fees	500.00
02	Digital Signature	Fees	2600.00
03	Stamp Duty	Memorandum of Association and Articles of Association	1500.00
04	Registration / Filing	MoA, AoA, Form-I, Form-18, Form 32	17200.00
05	Fees of Chartered Accountant or the Company Secretary	Consultancy Charges	10000.00
06	Stamps Cancellation	Fees	300.00
07	Affidavit Expenses	Fees for Notary Public	500.00
08	Share Transfer and Processing	Fees	5000.00
09	Miscellaneous Expenses	Expenses	2000.00
	Total Expenses	Rupees	39600.00

(Source: http://www.nabard.org)

Table 10.3 has highlighted some important services which are beneficial to the farmers in the Farmer Producer Company. How to start a company? What will be the Cost to be incurred for establishing the Farmer Producer Company should be answered well. Table 3 has presented the cost which are to be incurred for establishment of FPC which is approximately arrived at Rs 39600/. Among the different cost, the registration and the fees for chartered accountant alone found to be the three-forth component of total fees paid. Hence before establishing the company, the organizers should plan for the investment to be made once for the establishment of FPC.

10.5.1 The Members of the Producer Company

- In a producer company, only primary producers or producer organisations can become members. That is the Farmers who are actively involved in farming can become members.

- Membership is acquired by purchase of shares in a Farmer Producer Company
- A Producer Company can act only through its Members
- Members Create and Govern the Company
- Members can also Wind up the Company
- Members act through their General Body Meetings

10.5.2 The Minimum Share Capital Needs of Producer Company

- The minimum Authorized Capital of Producer Company is Rs.5 lakh.
- The Authorized Capital of the company can be more than Rs. 5 lakh as indicated in the Memorandum of Association.
- The authorized share capital should be sufficient for carrying out the objects mentioned in the memorandum.
- The authorized share capital should be realistic.
- The minimum paid up capital for Producer Company is Rs. 1 Lakh

10.5.3 The Needed Documents for Registration of Producer Company

Every Producer Company name should end with "Producer Company Limited"which indicates its status as Producer Company. The Producer Company may be named in such a way that it inspires the entire membership and creates a sense of ownership for its members and is to be indicative of the objectives of the company.

- Copy of the letter of RoC confirming the availability of name.
- Memorandum of Association (MoA) and Article of Association (AoA) duly stamped and signed.
- Form INC-22 indicating the Registered Office of the company with full address.
- Form DIR-12 in duplicate with details about the directors of the company.
- Form INC-7 on stamp paper declaring compliance with all and incidental matters regarding formation of companies.
- Consent of each of the Directors along with form DIR -12 for appointment of Directors.
- An affidavit indicating that MoA is fully understood by the subscribers /signees, if they sign in Hindi.
- Power of Attorney to the agent who is dealing with the Registrar of Companies (RoC) to make corrections in MoA and AoA, if necessary, to the satisfaction of the RoC.

The Board of Directors are elected by the Members of the company and any collective decision should be taken by forming the meetings. A producer company can have a minimum of 5 board of directors and not more than 15 members as Board of Directors. The Names of the first Board of Directors will be indicated in the Memorandum of Association.

10.5.4 The Functions of Chief Executive Officer of Producer Company

The Chief Executive Officer of the Producer Company involved in the production of agricultural commodities normally have the practice of recruiting the Agricultural Science graduate with the minimum qualification of B.Sc. (Ag). The desirable qualification for the position is M.Sc. (Ag). Currently, the producer companies dealing with farm produce are engaging or appointing the agricultural science graduate.

- The main functions of a CEO include the following:
- Do administrative acts of routine nature including managing the day-to-day affairs of theCompany;
- Operate bank accounts or authorize any person, subject to the general or special approvalof the Board;
- Make arrangements for safe custody of cash and other assets of the Company;
- Sign business related documents as may be 'authorized by the Board' for and on behalf of the Producer Company;
- Maintain proper books of account, prepare annual accounts, place theaudited accounts before the Board and in the Annual General Meeting of the Members;
- Furnish the members with periodic information to appraise them of the operation and functions of the Company;
- Make appointments to posts in accordance with the powers delegated to him by the Board;
- Assist the Board in the formation of goals, objectives, strategies, plans and policies;
- Advise the Board with respect to legal and regulatory matters concerning the proposed and ongoing activities and take necessary action in respect thereof;
- Exercise the powers as may be necessary in the ordinary course of business;

- Discharge such other functions, and exercise such other powers, as may be delegated by the Board;
- To provide timely information to the Members and Board of Directors for scheduled company meetings or emergency or short notice meetings
- The CEO shall be the Ex-Officio Director of the Producer Company and his qualifications for appointment can be decided by the Board of Directors

10.5.5 The Financial Delegation to Chief Executive Officer of the Producer Company

- The CEO can withdraw cash up to the limit of Rs. 5000/ (Rs. Five thousand) from the Company's bank account;
- The cash payment against any purchase of goods or services in any circumstances shall be limited to Rs. 500/- (Rs. Five Hundred Only).
- All payments above Rs. 500/-(Rs. Five Hundred Only) shall be paid by cheque only. In case of non-acceptance of cheque by any institution or individual, cash payment only with the approval of a committee comprising of 3 directors.
- Purchase of all consumable goods and services for use by the Company for its business operations or managing its affairs up to Rs. 5000/- (Rs.Five Thousand Only) following stipulated purchase procedure.

10.6 ROLE OF CENTRAL GOVERNMENT INSTITUTIONS IN SUPPORTING FPOs

Department of Agriculture, Cooperation and Farmers' Welfare (DACFaW), Ministry of Agriculture, Govt. of India willact as the Nodal Agency for the development and growth of FPOs.

Small Farmers' Agribusiness Consortium (SFAC), a Society under DAC, will be the designated agency of DAC to act as a single-window for technical support, training needs, research and knowledge management and to create linkages to investments, technology and markets. SFAC will provide all- round support to State Governments, FPOs and other entities engaged in promotion and development of FPOs. In particular, SFAC will create sustainable linkages between FPOs and inputs suppliers, technology providers, extension and research agencies and marketing and processing players, both in the public and private sectors.

The mandate of National Cooperative Development Corporation (NCDC) will be expanded to include FPOs in the list of eligible institutions which receive support under the various programmes of the Corporation.

NAFED will take steps to include FPOs in the list of eligible institutions which act on its behalf to undertake price support purchase operations.

DACFaW will work with Food Corporation of India (FCI) and State Governments to encourage them to include FPOs as procurement agencies under the Minimum Support Price (MSP) procurement operations for various crops.

DACFaW and its designated agencies will work with NABARD and other financial institutions to direct short and medium term credit for working capital and infrastructure investment needs of FPOs. DACFaW will also work with all relevant stakeholders to achieve 100% financial inclusion for members of FPOs and link them to Kisan Credit Cards.

DACFaW will work with Ministry of Corporate Affairs and other stakeholders to further clarify and strengthen provisions of the law relating to the registration, management and regulation of FPOs with a view to fostering fast paced growth of FPOs.

10.7 ROLE OF STATE GOVERNMENT INSTITUTIONS IN SUPPORTING FPOs

Besides encouraging State Governments to take up formation of FPOs on a large scale through Centrally-sponsored and State-financed programmes and schemes, DAC suggests the following steps to be taken by State Governments to support and strengthen FPOs:

By declaring FPOs at par with cooperatives registered under the relevant State legislation and self-help groups/federations for all benefits and facilities that are extended to member-owned institutions from time to time.

By making provisions for easy issue of licenses to FPOs to trade in inputs (seed, fertilizer, farm machinery, pesticides etc.) for use of their members as well as routing the supply of agricultural inputs through FPOs at par with cooperatives.

By using FPOs as producers of certified seed, saplings and other planting material and extending production and marketing subsidies on par with cooperatives.

By suitable amendments in the APMC Act to allow direct sale of farm produce by FPOs at the farmgate, through FPO owned procurement and marketing centres and for facilitating contract farming arrangements between FPOs and Bulk Buyers

10.8 FPOs IN INDIA

A new initiative is brewing among farmer producer companies (FPCs) across the country. Several federations of these companies are attempting to enter partnerships with federations of other states. This would not only enable companies to learn from each other, share technical inputs but also market products to different states.

Shriram Gadhave, president, Vegetable Growers Association of India (VGAI) and also president of the Maharashtra Federation of Farmer Producers Organisation (MFFPO), is presently in Gujarat attempting a tie-up with the Gujarat Pro Agri Business Consortium Producer Company. "There is a lot we can learn from the federation of farmer producer companies in Maharashtra," says Kuldeep Solanki, its director. "We have around 25 farmer producer companies in our federation and are attempting to increase these numbers to some 50 companies in a year or two.

The attempt is to be able to bring some 100-120 companies under the umbrella of the federation gradually. The objective is to be able to service these companies well and strengthen their marketing, and help them market both commodities and value-added products to sellers across the country", he said.

"Our FPCs are strong in mango, mango pulp, groundnut and wheat, among others, and our attempt is to be able to sell these in other states as well," Solanki said. His federation is also in talks with similar federations in Madhya Pradesh and Rajasthan. Gadhave says the attempt is to bring all the state federations under a common umbrella and form a national apex body. The country has some 600 FPCs, of which a third are in Maharashtra alone. Madhya Pradesh was earlier leading the pack with 100 FPCs but Maharashtra has overtaken it and has also formed the first apex FPC in the country, the Maharashtra Rajya Shetkari Company Federation. The apex FPC, which has representatives from 10 major FPCs in the state, will help the rest of the FPCs to find better markets for their products.

There are similar smaller markets running across the country. However, these are run entirely by farmers without any form of government support. Gadhave said that the process has been on for a couple of years and after a short lull, is being re-attempted again. VGAI has been in talks with several farmer producer companies in the north and south India to help them form federations and bring them into the proposed apex organisation.

The idea is to prepare a website for the apex FPC that will also contain links to markets in London and Dubai, and will help the FPCs get export licences as well, Gadhave said. VGAI has already been meeting farmer producer companies in Chandigarh, Madhya Pradesh, Punjab, Rajasthan, Gujarat and Kerala so far. Gadhave also visited Dubai recently where he visited Gulfood 2016, one of the biggest annual food and hospitality trade events in the world. Gulfood takes place at a critical time in the world harvest cycle. To keep on top of pricing trends, buyers are present at the event.

At Gulfood, over 90,000 visitors from 160 countries network with thousands of industry peers and people passionate about driving the market forward. Gadhave was in Dubai to obtain an export licences for the Maharashtra federation and also meet up with traders and exhibitors at the show. The licensing process may take several months but the federation has received inquiries from traders at the event, he said.

Recently, some 276 agreements were signed between FPCs and traders, exporters and bulk buyers at the first buyer-seller meet in Pune, organised under the auspices of the Maharashtra Agriculture Department and the Maharashtra Agriculture Competitiveness Project (MACP) through an ambitious World Bank-aided project. These agreements have been signed between FPCs and buyers for a volume of 79,751 tonne of grains, pulses, fruits and vegetables at a value of some R260 crore. Gadhave said his organisation is also part of this effort.

Maharashtra companies under Gadhave's organisation currently have a turnover of Rs 15 crore.

10.9 FACTORS GOVERNING THE SIZE OF PRODUCER COMPANIES

- A Farmers Producer Organisation is to be designed in such a way to cover all the lands that fall in one or two micro watersheds
- The sustainability of the farmers of the micro watersheds is already in existence with various types of productive activities of the farmers to take care of risk factors, likevariations in the market prices of various produce, continuity of income etc. which include sustainable agricultural practices.
- The secret to sustainability of a PO depends on comprehensive engagement of the PO with their membersthroughout the year

- The design variables for a PO are mainly size, scope, technology, ownership of resources, management and purpose. These variables need to be aligned to meet the sustainability requirements indicated in the earlier two points.
- The size of the PO should be small to be able to be managed by the local talent available in the area of the PO.
- The scope of the PO should be defined in such a way that there shall be good number of crops to be grown to maintain the soil health, support for allied activities like dairy, nutritional security of the local people and to mitigate risk.
- The technologyadopted by the PO should be such that majority of the local people or members of the families of the PO should be able to adopt to it and work with it with minimaltraining, effectively.
- The management of the PO should take into account the incubation of the local youth in such a way that in a few years' time, say in 3-7 years, local youth should be able to take over andmanage the PO effectively.
- The purpose of the PO at all times must be to serve the larger needs of the community and the ownership of the PO always should rest with all itsmembers.

10.10 THE SUCCESS STORIES OF FPOS FROM COIMBATORE, TAMIL NADU

Mr. Subburathinam whose educational status is less than the higher secondary, is presently the Director of the Kovai Farmers Producer Company. It has 520 shareholders capable of producing about 50 tonnes of gourd a day. It was formed in August 2013, more than a decade after a dozen farmers in Navakkarai and Vettaikaranpudur decided to band together to get a better deal from traders who short-changed them on various pretexts. They would buy by the sack full, not by weight. Bent gourd would be paid half the price, but sold at the full rate. There would be deductions for transportation.

Figure 10.1 Picture of Snake Gourd Brought from Garden to the Trading Point

Coimbatore is the transit point for vegetables to Kerala. Exports to that state and the city's own consumption add up to about 200 truckloads of vegetables. The produce of the Kovai Company that is about five truckloads a day - is not market moving, but enough for traders to fall in line. Prices are fixed for the next fortnight based on the rates prevailing at various markets. A live and let live policy is followed.

Actually, the trader has also to survive. So the FPC leave a legroom of Rs 1-1.50per kg. If FPC found that the Trader has earned more, then the FPC will pluck it from him the next time. Snake gourd was selling for Rs 19 per kg, ridge gourd for Rs 15 and bitter gourd for Rs 14 on Sunday, 12 June, in Coimbatore. In the government-regulated *mandis* a flat 10 percent commission is charged.

At Thondamuttur near Coimbatore, it was a spiritual leader known as Sadguru, who provided the initial thrust for a producer company of coconut growers in 2012. Though aware of the gains of collective action, the growers were wary of grand intentions crashing into poor outcomes. It took scores of meetings to win them over.

FPC held more than 80 meetings with the help of volunteers from Sadguru's Isha Foundation, says OKR Radhakrishnan, Coordinator of the Velliangiri Farmers' Producer Company. Radhakrishnan is a cane specialist and has worked with many South Indian sugar companies. He had also set up a sugar factory in Cameroon at the initiative of the World Bank, before volunteering his services to the foundation. "With great difficulty FPC registered 220 farmers initially.

Individually, the small holders suffered many indignities. Coconut pluckers would charge more than a rupee per nut or about Rs 80 per tree. The shellers were equally demanding as agricultural labour is in short supply in this part of the country. Traders would claim more than their fair share. The farmers company has stopped these exactions. It has engaged teams of pluckers and shellers. They get regular work and charge lower rates. Traders cannot beat down prices.

Farmer producer companies have been enabled through a 2002 amendment to the Companies Act so that they do not face the political pulls and pressures which cooperatives are subject to. Like cooperatives, they have the one-member-one-vote rule, but, unlike them, there is no provision for veto-wielding government representation on the board. This is supposed to assure their autonomy.

There are 445 FPCs in the country, with Tamil Nadu leading at 51, Madhya Pradesh with 48 and Telangana with 40. The Delhi-based Small Farmer Agribusiness Consortium sponsors them with matching grants of equity and credit guarantees without collateral - up to a certain limit. Incidentally, it was Manmohan Singh as finance minister who proposed SFACs in his 1992 budget that opened up the economy to serve as incubators of rural enterprise.States have also set up their own SFACs as sponsoring institutions and Nabard, the agricultural lending agency, has got into the act.

The thinking behind FPCs is to help relieve rural distress arising from the relentless sub-division of landholdings. Easier leasing laws can help consolidate fragmented holdings but states have not amended tenancy legislation (which regard the tiller as the owner) to allow longer leases (though there is a vibrant informal leasing market). "The idea of the FPO (farmer producer organisation) is to avoid the legal challenges of consolidation but to consolidate farmers as shareholders in one entity," according to Pravesh Shama, managing director of SFAC.

Though farmers tend be sceptical, the promise of higher price realisation through better bargaining power persuades them to sign on. FPCs bring economies of bulk purchases and the assurance of quality in inputs like fertiliser, seeds and chemicals. Since modern agriculture must be run as a business, where technology and good agronomic practices play a key role, they can jointly hire agronomists as well. Quite a few do, because the government pays their salaries for the first three years.

In states it is understood that the number of FPOs has crossed 25 to 30, facilitating the promotion of state-level producer companies, which can arrange finance, deal with input suppliers and look after marketing on behalf of their constituents. One has made a business plan for Rs 100 crore for the coming year.Sharma believes that e-commerce can give FPCs an enormous leg-up. They have breathed life into handicrafts by connecting artisans in far-flung areas with urban consumers.

Dr. E Vadivel, whom SFAC has engaged as a '*Resource Person*' or consultant to hand-hold the FPCs, believes even if half of them fail, stars could emerge from the rest. Dr. E. Vadivel was formerly Director, Extension Education and Professor of horticulture at the Tamil Nadu Agricultural University, and active in agricultural extension initiatives even before the FPC movement gained traction. "Sometimes village wisdom is more sensible than corporate wisdom. A few enterprising farmers from the village will emerge as giants. They will give the lead to the rest.

One cannot discount his optimism. Many of our entrepreneurs have emerged from peasant communities like Gounder, Patidar, Kamma, Kapu and Reddy. It was township and village enterprises that set the pace for China's manufacturing boom and forced state-owned enterprises to reform. But many of the consultants guiding the FPCs have an NGO background. They tend to be development - not business - oriented. Farmer members themselves might not want to let go and entrust operations to professional managers. It is hard to see FPCs, with a few exceptions, evolve differently from cooperatives.

10.11 SCHEMES AND SUPPORT MEASURES FOR FPOS

Many State Governments have schemes for preferential procurement of produce from POs. For example, procurement of certified seeds through POs has been implemented by the Government of Chhattisgarh. The facilitating agency should be able to get the relevant information from the respective Governments.

10.12 MARKET LINKAGE WITH CORPORATES AND OTHER AGENCIES

The corporates need continuous supply of desired quality produce for processing and value addition. Therefore, they prefer to enter into contract with few producer organisations who will meet their requirement. Usually the following mechanisms are adopted while performing procurement from producer organizations. They are

- Retail chains tie up with POs for procurement.
- Corporates extend dealership for farm machinery and inputs to POs.
- Corporates provide primary processing machinery to PO with buy-back arrangement for the produce.

In the city like Salem, Madurai and Coimbatore, the precision products are distributed even to the college hostels, hotels etc. Since, the hoteling business depends on the low cost resource for profit maximization, the second grade vegetable products are being supplied to them by keeping the first grade for the retail outlets manned by the corporate.

10.13 SUMMARY

The Farmer Producer Organization is a recent concept to coordinate the producers under one roof to collectivize their output, process the same, grading the produce and distribute the produce after branding the same by establishing market linkages with big corporates, hotels and hostels. Few of the farmer producer organizations are successful in different parts of the country. They do input business particularly in distributon of quality seed to the producer organizations, fertilizer both chemical and biofertilizers, consultancy and have mechanisms to buy back the produce for processing and distribution. The success stories of producer organizations in different parts of the country is also discussed and their achievements are highlighted for reference and follow up.

References

1. Anika Trebbin, Markus Hassler (2012) "Farmer Producer Companies in India- A New Concept for Collective Action"***Environment and Planning,*** 44: pp.411-427
2. Government of India (2015). "Farmer Producer Organizations – Frequently Asked Questions", (National Bank for Agriculture and Rural Development: Mumbai) pp. 01-47

Web References

- http://www.nabard.org/auth/writereaddata/File/FARMER%20PRODUCER%20ORGANISATIONS.pdf
- http://www.firstpost.com/business/agricultures-big-hope-farmer-producer-companies-learn-to-tame-middlemen-2417802.html

CHAPTER XI

Organic Farming and its Certification in India

Dr. C. Sekhar and Dr. K. Uma

Objectives

- To highlight the advantages of organic farming over conventional farming
- To infuse ideas on the importance of organic production and certification
- To share the organic experiences in different regions of the state
- To give an orientation to the Organic standards and regulations

Organization of the Chapter

- Introduction to Organic Farming
- Perspectives of Organic and Conventional Farming
- Characteristics of Organic Agricuture
- Certified Organic Farming in the World
- Standards and Regulations
- Organic Farming Experiences in India
- Organic Certification
- Participatory Guarantee System (PGS)
- Institutional Support for PGS Initiatives
- National Accreditation Body for Certification
- Summary

11.1 INTRODUCTION TO ORGANIC FARMING

The Precision Technology is for enhancing highest productivity per unit of land by applying the required input for achieving the target productivity without wasting any resources. But currently, awareness among the consumers of both urban and rural consumers got increased with respect to organically produced fruits and vegetables and it commands highest price over the precision vegetables. In this respect, the progressive farmers and the farmer producer organizations are tiring themselves in realization of organic vegetables through precision agricultural or horticultural practices. For that few of the progressive farmers are relying themselves in identifying and developing their own bio inputs for maximizing their output. However, the output of organically produced fruits and vegetables are to be certified whether these are the real output of organic source or origin. For that state level agencies are available.

Tamil Nadu Organic Certification Department

Tamil Nadu Organic Certification Department (TNOCD) was established in the year 2007-2008 to carryout inspection and certification of organic production system in accordance with NPOP (National Programme for Organic Production), which was launched by Government of India in the year 2000 and notified in October 2001 under the Foreign Trade and Development Act (FTDR Act).

Tamil Nadu Organic Certification Department is accredited by APEDA (Agricultural and Processed Food Products Exports Development Authority), New Delhi, Ministry of Commerce and Industry, Government of India. The accreditation number allotted to Tamil Nadu Organic Certification Department is NPOP / NAB / 0019. Organic Certification carried out by this Department is on par with standards of European Union.

Tamil Nadu Organic Certification Department also imparts free training to registered organic farmers on National Standards for Organic Production, and Tamil Nadu Organic Certification Department Standards. During the year 2017-18, 30910 acres of land have been registered under Organic Certification by 3672 farmers in Tamil Nadu. The average size of holding of Organic farms is arrived at 8.50 acres (03.40 Ha). The farmers who are interested in entering into organic agriculture can register the same by using the following format. Several formats are available in the website of Tamil Nadu Organic Certification Department. One can make use of the same for registration purpose.

TAMIL NADU ORGANIC CERTIFICATION DEPARTMENT (TNOCD)

ANNUAL CROPPING PROGRAMME

Registration No :

Name of the Farmer :

Address :

Area in (ha)

S.NO	Perennial Crops		Annual Crops					
			Kharif		Rabi		Summer	
	Crop	Area	Crop	Area	Crop	Area	Crop	Area

Signature of the Farmer

Organic Farming is very much native to India. While writing the history of organic farming one cannot ignore the countries like India and China whom are the pioneers in performing Organic Agriculture. The farmers of these two countries are farmers of 40 centuries and the organic farming has sustained them. The Concept of organic farming is based on the following principles

- Nature is the best role model to farming, since it does not use any inputs which are unreasonable
- The entire system is based on intimate understanding of nature's ways. The system does not believe in mining of the soil of its nutrients and do not degrade it in any way for today's needs.
- The soil in the system is a living entity
- The soil's living population of microbes and other organisms are significant contributors to its fertility on a sustained basis and must be protected and nurtured at all cost
- The total environment of the soil, from soil structure to soil cover is more important

Organic farming is a method of farming system which primarily aims at cultivating the land and raising crops in such a way to keep the soil alive and in good health by use of organic wastes and other biological materials along with beneficial microbes to release nutrients to crops for increased production and productivity in an eco-friendly manner accompanied with pollution free environment.

11.1.1 Origin of Organic Concept

The term "Organic" was first used in relation to farming by Northbourne during the year 1940 in his book Look to the Land. Organic Agriculture is a holistic production management system which promotes and enhances agro-ecosystem health, including biodiversity, biological cycles, and soil biological activity. It emphasizes the use of management practices in preference to the use of off-farm inputs, taking into account that regional conditions require locally adapted systems. This is accomplished by using, where possible, agronomic, biological and mechanical methods, as opposed to using synthetic materials, to fulfil any specific function within the system (FAO, 1999)

11.1.2 The Principles of Organic Agriculture

To understand the motivation of organic farmers, the practices they use and what they want to achieve is important to understand the guiding principles of organic agriculture. These principles encompass the fundamental goals and caveats that are considered important for producing high quality food, fibre and other goods in an environmentally sustainable way. The principles of organic agriculture have changed with the evolution of the movement and are now codified. The principles apply to agriculture in the broadest sense, including the way people tend soils, water, plants and animals in order to produce, prepare and distribute food and other goods. They concern the way people interact with living landscapes related to one another and shape the legacy of future generations. The principles of organic agriculture serve to inspire the organic movement in its full diversity. They are the roots from which organic agriculture grows and develops. They express the contribution that organic agriculture can make to the world and a vision to improve all agriculture in a global context. The principles of Organic Agriculture serve to inspire the organic movement in its full diversity.

11.1.3 The International Federation for Organic Agriculture Movements (IFOAM) definition of Organic Agriculture is based on the following concepts

- The Principle of Health of the Human Being
- The Principle of Preserving the Ecology
- The Principle of Being Fairness in all Aspects and
- The Principle of Caring the People and the Environment

Each Principle is articulated through a statement followed by an explanation. The principles are to be used as a whole. They are composed as ethical principles to inspire action.

11.1.3.1 *The Principle of Health of the Human Being*

Organic Agriculture should sustain and enhance the health of the soil, plant, animal, human and planet as one and indivisible

Here, the role of organic agriculture, whether in farming, processing, distribution or consumption, it is to sustain and enhance the health of the ecosystems and organisms from the smallest in the soil to human beings. In particular, the organic agriculture is intended to produce high quality, nutritious food that contributes to preventive health care and wellbeing. In view of this it should avoid the use of fertilizers, pesticides, animal drugs and food additives that may have adverse health effects.

11.1.3.2 *The Principles of Preserving the Ecology*

Organic Agriculture should be based on the living ecological systems and cycles work with them, emulate them and help sustain them

This principle roots organic agriculture within living ecological systems. It states that production is to be based on the ecological processes and recycling. Nourishment and well-being are achieved through the ecology of the specific production environment. Organic agriculture should attain ecological balance through the design of farming systems, establishment of habitats and maintenance of genetic and agricultural diversity. Those who produce, process, trade or consume organic products should protect and benefit the common environment including landscapes, climate, habitats, biodiversity, air and water.

11.1.3.3 *The Principle of Being Fairness*

Organic Agriculture should build on relationships that ensure fairness with regard to the common environment and life supporting opportunities.

This principle emphasizes that those involved in organic agriculture should conduct human relationships in a manner that ensures fairness at all levels and to all parties like farmers, workers, processors, distributors, traders and consumers. Natural and environmental resources that are used for production and consumption should be managed in a way that is socially and ecologically just and should be held in trust for future generations. Fairness requires systems of production, distribution and trade that are open and equitable and account for real environmental and social costs.

11.1.3.4 *The Principle of Caring the People and the Environment*

Organic Agriculture should be managed in a precautionary and responsible manner to protect the health and well-being of Current and Future Generations and the Environment in Toto.

The practitioners of organic agriculture can enhance the efficiency and increase the productivity. But it should not be at the risk of jeopardizing the health and well-being of the human. Consequently, new technologies need to be assessed and existing methods reviewed. This principle states that precaution and responsibility are the key concerns in management, development technology choices in organic agriculture. Organic agriculture should prevent significant risks by adopting appropriate technologies and rejecting unpredictable ones, such as genetic engineering. Decisions should reflect the values and needs of all who might be affected, through the transparent and participatory processes.

11.2 COMPARATIVE PERSPECTIVES OF ORGANIC AND CONVENTIONAL FARMING

Over the years, the researchers have been busy in working out the comparisons between the two systems with respect to yield, economics, resource use status and its efficiency, environmental impacts and social impacts, diverse range of farm types such as diaries, orchards and mixed farming systems. Several studies have confirmed that the organic agriculture is productive and sustainable. Some key findings from the research on yields suggests that;

- Yield is equivalent to or better than the Conventional Farming under organic systems
- Yields decrease during the conversion period and thereafter it improves upwards
- Organic farms have higher soil biological activity and biodiversity
- Weeds can have major impact on yields and pests on specific crops, are problematic
- Organic agriculture causes no or very little pesticide contamination in food, people and the environment
- The beneficial aspects of organic agriculture in food quality are appealing but are yet to be authenticated
- The fruits drawn from organic farming gets good colour and taste when compared to other type of farming output

11.3 CHARACTERISTICS OF ORGANIC AGRICULTURE

- Organic Agriculture relies primarily on local renewable resources
- It makes efficient use of solar energy and the production potential of biological systems

- It maintains the fertility status of the soil
- It maximises recycling of plant nutrients and organic matter
- It does not use organisms or substances foreign to nature
- It maintains diversity in the production systems as well as the agricultural landscape

11.4 THE CERTIFIED ORGANIC FARMING IN THE WORLD

- Around 35 million ha of agricultural land are managed organically by almost 1.4 million producers.
- The regions with the largest areas of organically managed agricultural land are Oceania with 12.10 Million ha; Europe with 8.20 Million Ha and the Latin America had 8.10 million ha. The countries with the most organic agricultural land are Australia, Argentina and China.
- The highest shares of organically managed agricultural land are in the Falkland Islands (36.90 per cent), Liechtenstein (29.80 per cent) and Austria (15.90 per cent)
- The countries with highest number of producers of organic produce are India (3.40 Lakhs of Producers), Uganda (1.80 Lakh producers) and Mexico (1.30 Lakh Producers). More than one third of organic producers are in Africa.
- On a global level, the organic agricultural land area increased in all regions, in total by almost three million ha or nine per cent, compared to the data from the year 2007
- Twenty-Six per cent more land is brought under organic management with Latin America. It was mainly due to the strong growth in the Argentina. In Europe, the organic land increased by more than half a million ha in Asia by 0.4 million
- About one-third of the World's organically managed agricultural land (12 Million Ha) is located in developing countries. The countries with the largest area under organic management of land are Argentina, China and Brazil

11.5 STANDARDS AND REGULATIONS

- The Year 2009 has witnessed several developments in the field of standards and regulations with respect to Organic Agriculture
- The new EU regulation on Organic production came into force as well as the Canadian Organic Standards

- Australia has also developed the organic standards in their own nation
- Canada and the US concluded the world's first fully reciprocal agreement between the regulated organic systems and the EU introduced the Procedures for approving the certification bodies from outside the EU.
- The number of countries who have drafted the organic standards rose to the level of 73 and there are another 16 are drafting their standards
- In 2009, FAO, IFOAM and UNCTAD started the Global Organic Market Access (GOMA) project. The aim of GOMA is to facilitate equivalence, harmonization and other type of cooperation in order to simplify the process for trade flow of products among the various organic guarantee system
- The growing number of organic producers are certified through Participatory Guarantee Systems (PGS) across the world. PGS are locally focused quality assurance systems.
- It is estimated that around 10000 small operators are involved in PGS world-wide.

11.6 ORGANIC FARMING EXPERIENCES IN INDIA

The growth of organic agriculture in India has three dimensions and is being adopted by the farmers for different reasons. They are

First Category of Organic farmers are those which are situated in no-input use zones, for them organic is a way of life and they are doing it as a tradition.

Second Category of farmers are those which have recently adopted the organic farming in the wake of ill effects of conventional farming.

The Third Category comprised of farmers and enterprises which have systematically adopted the commercial organic agriculture to capture emerging market opportunities and premium prices

During the year 2014, India has brought more than 4.48 million ha area under organic certification process. Out of this cultivated area accounts for 1.08 million ha while remaining 3.4 million ha is from wild forest zone either by way of collection or by harvesting the produce (Table 11.1).

Table 11.1 Area under Organic Management in India

Sl. No	Year	Area Under Organic Management (Ha)	Percentage Increase
01	2003-2004	42000	00.00
02	2004-2005	76000	80.95
03	2005-2006	173000	127.63
04	2006-2007	538000	210.98
05	2007-2008	865000	60.78
06	2008-2009	1085648	25.51
07	2009-2010	1207000	11.18
08	2010-2011	1812000	50.12
09	2013-2014	4408000	143.27

11.6.1 Organic Farming Experience from a Farmer

A Village called "Thali", Udumalpet Taluk of Coimbatore District in which a farmer named Mr. Sathyanathan, who has left the agriculture because of several constraints in doing vegetable farming. Based on the motivation article appeared in "Pasumai Vikatan", Mr. Sathyanathan has acquired 12 acres of land on lease and entered into Floriculture instead of Vegetable Farming in his village after sharing the expertise from a farmer named "Mr. Vijayaragavan", from Shrivilliputhur, Virudhunagar District of Tamil Nadu. Let us share the Experiences of Floriculture under organic farming approach in one acre of farm.

- Add 8 tonnes of Sheep and Goat Manure and plough the land well
- Prepare Channels and Funds to the width of 3 feet and height 20 CM
- Quantity of Sambangi Seeds Required 600 Kg
- Prepare pits and add one kg Farm Yard Manure + 50 Grams of Neem Cake and sow one Sambangi Rhizome per pit
- Sow continuously leaving a feet as gap between the rhizome
- Provide Drip Irrigation System or Flood Irrigation based on the water availability
- Weeding once or twice in a month

- Add Azospirillum 500 Gram + Phospho bacteria 500 gram in a Mix of 15 kg Groundnut oil cake combined with 30 kg of Cotton seed cake are mixed with 200 litres of water and add the mixture in every root zone of Sambangi immediately after planting of 30 days.
- During the second month, add 500 kg of Vermicompost in the root zone of every pit.
- During the Third Month, add 10 litres of Jeevamirtham in 150 litres of water and apply through a sprayer
- During the end of Fourth Month, 5 tonnes of compost has to be powdered and applied in the root zone of Sambangi
- During the Fifth month, 10 litres of Amuthakkaraisal have to be added with 100 litres of water and the same should be mixed well and for each and every plant around one litre of the mix may be added.
- Add organic fertilizer which contains NPK to the tune of 250 kg every month and apply the fertilizer in the root zone of Sambangi or one can even rotate the application.
- On examining the drying of leaf tip in Sambangi, add 100 gram Psuedomonas mixed with 200 ml butter milk added with 10 litres of water and prepare and spray through the hand sprayer.
- If the flowers are infected with Sucking insects or pests, then take one litre of pungam oil, 4 litres of neem oil and mix the same with 200 litre water mixed with 50 gram soap brought from khadi kraft and mix thoroughly and apply through the hand sprayer

Table 11.2 The Yield and Income from Sambangi Flowers from 6 Acres

Sl. No	Month	Yield of Flower in Kgs	Income in Rs
1	January 2017	2945	191000
2	February	1400	230000
3	March	2852	370760
4	April	3420	523260
5	May	4557	546840
6	June	4890	361860
7	July	2914	241862

Sl. No	Month	Yield of Flower in Kgs	Income in Rs
8	August	1240	179800
9	September	1620	196000
10	October	1519	119000
11	November	2940	306000
12	December 2017	1953	250000
	Total	32250	3516382

(Source: Pasumai Vikatan, 10th February, 2018)

Table 11.2 revealed that through organic farming approach the farmer could earn a gross income of Rs 35.16 lakhs from six acres of Sambangi cultivation. After deducting the expenditure towards, land preparation, fertilizer application and manuring and plant care, the farmer could earn a net profit of Rs 23.75 lakhs. Though this 'Thali' Village is not a flower cultivation track, the farmer becomes a pioneer in cultivation of flowers and many of the farmers have been motivated and in future this track will be enriched with the flowering plants which are having good market demand. The farmer himself has started cultivating different varieity of flowers in his holdings. The cropping pattern in vogue are summarized as under.

Name of the Flower Crop	Area under Crop in Acres	Quantity Harvested in Kgs	Income Gained in Rs
Jasmine (Kakada)	01.00	1727	433000
Marigold	01.00	6147	282000
Sambangi	06.00	32250	3516382
Arali	02.50	23348	2750000
Kozhi Kondai (Celosia)	01.00	3650	168000
Chrysanthemum	00.50	2110	232000
Total	12.00	69232	7381382
Net Revenue from 12 Acres of Land per Annum		Rs. 23,75,000	
Net Revenue per acre of Land per Annum		Rs. 197917.00	

From the above data, one could understand that the flower cultivation seems to be a profitable venture. However, it is a labor consuming activity, the contract laborers from other states are much interested in extracting the flowers in the early hours so as to reach the same to the flower market of Coimbatore and Madurai on time so that the good price realization is possible.

11.7 ORGANIC CERTIFICATION

It is a certification process for producers of organic food and other organic agricultural and its allied produce. In general, any business directly involved in food production can be certified, including seed suppliers, farmers, food processors, retailers and restaurants. Requirements vary from country to country and generally involve a set of production standards for growing storage processing, and shipping.

11.7.1 The Purposes of Certification

Organic Certification addresses a growing worldwide demand for organic food. It is intended to assure quality and prevent any fraud. For organic producers, certification identifies suppliers of products approved for use in certified operations. For consumers, "certified Organic" serves as a product assurance, similar to "low fat", 100 per cent whole wheat or "no Artificial Preservatives".

Certification is essentially aimed at regulating and facilitating the sale of organic products to consumers. Individual certification bodies have their own service marks, which can act as branding to consumers – a certifier may promote the high consumer recognition value of its logo as a marketing advantage to the producers. Most certification bodies operate organic standards that meet the National Government's Minimum requirements.

11.7.2 The Process of Certification

In order to certify a farm, the farmer is typically required to engage in a number of new activities, in addition to the normal farming operations

- **Study** the organic standards, which cover in specific detail what is and is not allowed for every aspect of farming, including storage, transportation and sale
- **Compliance** – farm facilities and production methods must comply with the standards, which may involve modifying facilities, sourcing and changing suppliers etc.

- **Documentation** – extensive paper work is required, detailing farm history and current set-up and usually including the results of soil and water tests
- **Planning** – a written annual production plan must be submitted, detailing everything from seed tot sale; seed sources, field and crop locations, fertilization and pest control activities, harvest methods, storage locations etc.
- **Inspection** – Annual on – Farm inspections are required. With a physical tour, examination of records and an oral interview
- **Fee** – A fee is to be paid by the grower to the certification body for annual surveillance and for facilitating a mark which is acceptable in the market as symbol of quality
- **Record Keeping** – Written, day-to-day farming and marketing records, covering all activities must be available for inspection at any time. In addition, short notice or surprise inspections can be made, and specific tests may be requested. For the first time farm certification, the soil must meet basic requirements of being free from use of prohibited substances for a number of years. A conventional farm must adhere to organic standards for this period, often, three years. This is known as being in transition. Transitional crops are not considered as fully organic. A farm already growing without chemicals may be certified without any delay.

11.7.3 Certification and Product Labelling Requirements

Labelling a Word "Organic" on a food product is a valuable one and fetches higher price per unit while making the qualitative features of the product on the label. Certification is intended to protect consumers from misuse of the term, and make buying decisions. However, the organic labelling made possible by certification itself usually requires explanation. In many countries organic legislation defines three levels of organics. Products made entirely with certified organic ingredients and methods can be labelled "100% Organic". Products with 95% organic ingredients can use the word "Organic". Both may also display organic seal. A third category, containing a minimum of 70 % of organic ingredients, can be labelled "made with organic ingredients". In addition, products may also display the logo of the certification body that approved them (Table 11.3). Products made with less than 70% organic ingredients cannot advertise this information to consumers and can only mention this fact in the product's Ingredient statement.

During the year 2006, India's Organic certification process under National Program on Organic Production has been granted equivalence with European Union. It has also been recognized for conformity assessment by USDA's National Organic Program.

Table 11.3 Details of Organic Logo Practiced in Different Regions / Countries

Sl. NO	Name of the Country / Region	Year of Implementation	Organic Logo for Certification
01	European Union	1992	
02	United States of America	2002	
03	Multinational Corporation	--	
04	Canada	2002	
05	Japan	2001	
06	Australia	2004	
07	India	2006	

In China, the China Green Food Development Centre awards two standards. A and AA: while the former standard does permit some use of synthetic agricultural chemicals, the latter is more stringent.

In Japan, the Japanese Agricultural Standard (JAS) was fully implemented as law in April 2001. This was revised in November 2005 and all JAS certifiers were required to be re-accredited by the Ministry of Agriculture, Japan.

In US, the National Organic Program (NOP) was enacted as federal legislation in October 2002. It restricts the use of the term "Organic" to certified organic producers. Certification is handled by State, Non- Profit and Private agencies that have been approved by the US Department of Agriculture. In India, the following Certifying firms are there for the Organic Certification. The details are presented in Table 11.4.

Table 11.4 Organic Produce Certifying Firms in India

Sl. No	Name of the Firm	Name of the State	Email
01	International Resources for Fairer Trade Control Union Certifications	Mumbai, Maharashtra	Ph: 2267255396,97
02	IMO Control Private Limited	Bangalore, Karnataka	imoind@vsnl.com
03	SKAL International	Bangalore, Karnataka	skalindia@eth.net
04	Lacon Quality Certification Private Limited	Thiruvalla, Kerala	info@laconindia.com
05	Rajasthan Organic Certification Agency	Jaipur, Rajasthan	rocajpr.cb@gmail.com
06	Onecert Asia Agri Certification Private Limited	Jaipur, Rajasthan	info@onecertasia.in
07	Indian Organic Certification Agency	Ernakulam, Kerala	biocosmetics@ indocert.org
08	Natural Organic Certification Association	Pune,	NA

Sl. No	Name of the Firm	Name of the State	Email
09	Bureau Veritos Certification India Private Limited	Mumbai, Maharashtra	NA
10	Ecocert SA	Aurangabad, Maharashtra	NA
11	Vedic Organic Certification Agency	Hyderabad, Andhra Pradesh	92904 50666
12	Food Cert India Private Limited	Hyderabad, Andhra Pradesh	foodcert@foodcert.in
13	Uttaranchal State Organic Certification Agency	Dehradun, Uttaranchal	uss_opca@rediffmail.com
14	SGS India Private Limited	Gurgaon, Haryana	Sudarshan_sharma@sgs.com
15	Aditi Organic Certification Private Limited	Bangalore, Karnataka	aditiorganic@gmail.com
16	APOF Organic Certification Agency	Bangalore, Karnataka	aocabangalore@yahoo.co.in

(NA: Not Available)

11.8 PARTICIPATORY GUARANTEE SYSTEM (PGS)

Participatory Guarantee Systems (PGS), as defined by IFOAM, are "locally focused quality assurance systems. They certify producers based on active participation of stakeholders and are built on a foundation of trust, social networks and knowledge exchange." They represent an alternative to third party certification, especially adapted to local markets and short supply chains. They can also complement third party certification with a private label that brings additional guarantees and transparency. PGS enable the direct participation of producers, consumers and other stakeholders in:

The choice and definition of the standards

The development and implementation of certification procedures

The certification decisions

Participatory Guarantee Systems are also referred to as "Participatory Certification".

The International Federation of Organic Agriculture Movements (IFOAM) and the organic movement remain a leader in the concept of PGS at the international level. IFOAM is running a program to recognize PGS in the organic sector. PGS is a tool that can be adopted not only for organic agriculture but is useful in various sectors.

The organic movement has been a pioneer in the implementation and definition of Participatory Guarantee Systems (PGS). Organic certification started in various parts of the world in the 70s and 80s based on associative systems that were very close to what is now called PGS. Some of these associations are still doing participatory certification today, such as for example Nature and Progrès in France. Even though third party certification (following ISO 65 requirements) has become the dominant form of certification in the food sector, as well as many other sectors, alternative certification systems have never ceased to exist.

In 2004, IFOAM and MAELA (The Latin American Agroecology Movement), jointly organized the first International Workshop on Alternative Certification that took place in Torres, Brazil. It is at that workshop that the concept of Participatory Guarantee Systems was adopted. At this event, an international working group on PGS was established, which later became an official Task Force under the umbrella of IFOAM. The Task Force worked on further defining PGS, and established the key elements and key features of PGS in a document entitled "Shared Visions – Shared Ideals".

Since then IFOAM has continuously supported the development of PGS and are now advocating for their recognition by governments as valid local certification systems in cases where the organic sector is legally regulated. In parallel, other sectors have been looking into the concept to certify various products or processes.

PGS are shaped by the very farmers and consumers that they serve and while the details of methodology and process vary, there is a consistency of core principles which can be adapted and specific to local conditions (communities, geographies, politics and markets). Participatory Guarantee Systems have transparent, systemised decision-making processes and aim to share the responsibility for the organic guarantee and to verify that farmers are consistently maintaining the standards. Trust is created through open information and peer reviews. PGS involve less administration and lower costs than export focused third party certification.

Participatory Guarantee Systems support and encourage producer groups to work together and to improve their farming practices through the sharing of knowledge and experiences. Active participation on the part of the stakeholders' results in greater empowerment but also greater responsibility – there is a high priority placed on knowledge and capacity building of producers and consumers, ideal for the rural communities of the Pacific.

Participatory Guarantee Systems can be used as a tool for improving local socio-economic and ecological conditions by encouraging small-scale production and product processing. In local markets they help smallholders to have their products recognized as organic and can open unique opportunities such as small holder links with the tourism and hospitality industries.

11.9 INSTITUTIONAL SUPPORT FOR PGS INITIATIVES

- Facilitate the development of local markets for organically produced foodstuffs, building on Participatory Guarantee Systems as a credible and affordable method of organic guarantee.
- Encourage diverse production by avoiding a strong focus on cash crop mono-cultures, fostering and safeguarding local agro-biodiversity.
- Bolster peer-support approaches between farmers, by organizing themselves within a Participatory Guarantee System, to achieve food security and food sovereignty, as well as a fair price.
- Encourage public awareness of local agriculture and facilitate access to urban markets for local/regional farmers.
- Revitalize the inclusion of social justice norms and practices as an essential part of organic production systems.

The above actions will ensure the trust among the stakeholders besides fulfilling the goal of better price, quality produce and better satisfaction to the users and the producers.

11.10 THE NATIONAL ACCREDITATION BODY FOR CERTIFICATION

The National Steering Committee would also function as the National Accreditation Body. The members of the National Accreditation Body shall comprise of representatives from Ministry of Agriculture, Ministry of Commerce and Industry, Agricultural and Processed Food Products Export Development Authority (APEDA), Coffee Board,

Spices Board and Tea Board. The Chairman of the Body shall be the Chairman of the National Steering Committee. The Work of the National Accreditation Body will include the following

- Drawing up of Procedures for evaluation and Accreditation of Certification Programs
- Formulating Procedures for evaluation of the agencies implementing the programs
- Accreditation of inspection and certification agencies

11.10.1 Evaluation Committee

Eligible inspection and certification agencies implementing certification programs will be evaluated by an evaluation committee. The Evaluation Committee will be appointed by the National Accreditation Body. The Members of the Evaluation Committee will comprise of members drawn from the APEDA, Coffee Board, Spices Board, Tea Board, Ministry of Agriculture and Export Inspection Council of India / Export Inspection Agencies. APEDA, on behalf of all the National Accreditation Body, will receive and screen the applications from the certification agencies, will coordinate and arrange evaluation visits etc. to ascertain the credentials of Certification Programs of the applicants. The Evaluation Committee will submit its recommendations to the National Accreditation Body for Considering the Accreditation.

11.11 SUMMARY

This paper highlights the concepts of organic farming, the principles associated with organic farming, the need of certification, characteristics of organic farming, the agencies involved in certification of organic foods at national and global level. Besides it had also highlighted the certification process, procedures associated with certification, participatory certification, history of participatory certification and the need of participatory certification, inspection, evaluation and accreditation of the certification related to organic farming. Currently, the success of organic agriculture depends on the certification procedures and the extent of certification initiatives taken up by the national and international organizations.

References

1. IGNOU Reading Materials related to Organic Certification Programs, 2015 (IGNOU: New Delhi)
2. Pasumai Vikatan, (2018) "Profitable Floriculture in Tamil Nadu" (Pasumai Vikatan: Chennai), pp. 39-43
3. Yadav, A.K. (2012). "Training Manual on Certification and Inspection Systems in Organic Farming in India," (Ministry of Agriculture: Department of Agriculture and Cooperation), Utter Pradesh.

Web References:

- http://www.tnocd.net/downloads/Form%201A2.doc
- http://www.tnocd.net/downloads.html

CHAPTER XII

Economic Benefits from Precision Moringa Orchards

Dr. C. Sekhar, N. Venkatesan and Dr. J. Pushpa

Objectives

- To introduce the social and economic benefits of Moringa cultivation to the students and other stakeholders
- To highlight the steps involved in Post harvest Processing of Moringa produce
- To infuse the ideas to the readers on the costs and returns associated with Moringa Cultivation

Organization of the Chapter

- Introduction
- Moringa Farming in India
- Moringa Varieties for Cultivation in Tamil Nadu
- Methodological Approaches to Socio Economic Study
- Social Aspects of Moringa Cultivation
- Factors Influencing Moringa Plantations
- Economic Benefits of Moringa Cultivation
- Post Harvest Processing of Moringa
- Composition of Nutrients in Moringa Produce
- Economics of Moringa Plantations
- Summary

12.1 INTRODUCTION

Now a days, commercial demand exists for Moringa based products produced from India particularly from Tamil Nadu. Because of the increasing demand in lieu of medicinal and aphrodisiac importance and good for general health, it is frequently consumed in different value added forms by the consumers. This has induced the farmers to grow Moringa as a separate crop or in combination with some other agricultural crops under organic condition or the combination of both organic and inorganic to fetch higher return per unit area of land following a precision approach. In this context, one has to assess whether, the farmers economic independence is happening or not is in question. For that a small attempt made to document the socio economic aspects of Moringa orchards. It might be useful to the students and the teachers.

Moringa (*Moringa oleifera* Lam.) belongs to the family 'Moringaceae' is a fast growing multipurpose medicinal tree extensively grown in tropics and subtropics of India and Africa. It is also widely distributed in India, Egypt, Philippines, Sri Lanka, Thailand, Malaysia, Burma, Pakistan, Singapore, West Indies, Cuba, Jamaica and Nigeria. In eastern and southern regions of India, Moringa is widely used as vegetable and grown commercially for its edible pods and leaves. *Moringa oleifera* is an important food commodity which has had enormous attention as the 'natural nutrition of the tropics'. Almost all the parts of this plant: root, bark, gum, leaf, fruit (pods), flowers, seed and seed oil have been used for various ailments in the indigenous medicine of South Asia, including the treatment of inflammation and infectious diseases along with cardiovascular, gastrointestinal, haematological and hepatorenal disorders. Its popularity is increasing steadily because of its nutritional, medicinal value and for its sweetness in curry and slurry preparation along with red gram dhal. From that one could understand the importance of Moringa. *(D'Souza and Kulkarni, 1993; Anwar and Bhanger, 2003; Anwar et al., 2005).*

12.2 MORINGA IN INDIA

India is the prime producer of Moringa (Drumstick) with an annual production of 2.2 to 2.4 million tonnes of tender fruits from an area of 38,000 ha leading to the productivity of around 63 tonnes per ha. Among the different states, Andhra Pradesh leads in both area and production (15,665 ha) followed by Karnataka (10,280 ha) and Tamil Nadu (13250 ha). In other states, it occupies an area of 4,613 ha. Tamil Nadu is the pioneering state as it has varied genotypes from diversified geographical areas, as well as introductions from Sri Lanka.

12.3 MORINGA VARIETIES FOR CULTIVATION IN TAMIL NADU

12.3.1 The Perennial Eco-types of Moringa

The cultivation of Moringa in India occurs mainly in the southern states of Tamil Nadu, Karnataka, Kerala, and Andhra Pradesh. Principally perennial types have been known for cultivation for a very long time. However, perennial types are beset with many production constraints, such as a relatively long pre-fruit bearing period, non availability of planting materials (stem cuttings), requirement of a greater number of rainy days in regions where water is scarce, and vulnerability to pests and diseases. Important varieties of Moringa are Moolanur Moringa, Valayapatti Moringa, Chavakacheri Moringa, Chemmurungai, Jaffna Type, Kattu Moringa, Kodikkal Moringa, Palmurungai, Punamurungai, Palamedu Moringa.

12.3.1.1 *Moolanur Moringa*

Farmers in and around Moolanur, Karur, Dharapuram areas of Tamil Nadu, predominantly cultivate perennial moringa which has a pod length of 45-50 cm, pod weight of 120 g and a yield of 200 kg/tree. Trees are maintained upto 15 years without pruning.

12.3.1.2 *Valayapatti Moringa*

Another perennial type cultivated in and around Usilampatti, Andipatti areas is valayapatti moringa. The pods are around 65 cm long, and weigh about 120 g. Yield is reported to be 1000-1200 pods per tree.

12.3.1.3 *Chavakacheri Moringa*

An ecotype of Jaffna Moringa, it bears pods as long as 90-120 cm. It is also cultivated in Tamil Nadu.

12.3.1.4 *Chemmurungai*

Another ecotype of Jaffna murungai, it flowers and fruits throughout the year, yielding a heavy crop. The tips of the pods are red. The tree is medium sized and bearing long pods.

12.3.1.5 *Jaffna type*

A Yazphanam type moringa introduced from Sri Lanka. Its fruits are 60-90 cm long with soft flesh and good taste. This type yields around 40 pods from the second year of planting which increases upto 600 pods/tree from the third year onwards. It is cultivated commercially in Tirunelveli and Tuticorin districts of Tamil Nadu.

12.3.1.6 *Kattumurungai*

A wild form of moringa, *M. concanensis* found in the forest of Tamil Nadu.

12.3.1.7 *Kodikkal murungai*

It is cultivated predominantly in the betel vine gardens of Tiruchirapalli district of Tamil Nadu. The pods are shorter (20-25 cm long) and are thick fleshed. The pods and leaves are very tasty. Trees are short statured with smaller leaves.

12.3.1.8 *Palmurungai*

It is preferred for its thick pulp and tasty pods.

12.3.1.9 *Punamurungai*

It is grown in home gardens of Tirunelveli and Kanyakumari districts.

12.3.1.10 *Palamedu Moringa*

The pods are 60 cm long, with a pod weight of 95-100 g and an yield of 100 pods/tree.

12.3.2 Annual Moringa Varieties

The Horticultural College and Research Institute of Tamil Nadu Agricultural University, Periyakulam, had an assemblage of 85 Moringa accessions. The germplasm collection block contains perennial and annual moringa accessions with heavy fruit-bearing, cluster bearing, drought tolerance, dwarfing stature and pest and disease resistance. Scientists at the Horticultural College and Research Institute, Tamil Nadu Agricultural University, Periyakulam have succeeded in developing seed-propagated Moringa types, which has revolutionized the Moringa cultivation in the country. By judicious breeding programs, including introduction of elite mother plants, evaluation, selection and hybridization, The Horticultural College and Research Institute of Tamil Nadu Agricultural University has released two improved annual Moringa varieties (PKM-1,

PKM-2) within a span of 10 years, for commercial cultivation. These varieties have developed up well in many traditional and non-traditional areas (Sadashathi, 1995).

12.3.2.1 *PKM - 1 Annual Moringa*

It is a pure line selection developed by continuous selfing for six generations, collected from Eppothumvendran village of Thoothukkudi region. In each generation, only long pods and desirable characters are selected and advanced. It was released in the year 1989. The fruits are fleshy and tasty. It comes to flowering within 5-6 months of sowing and comes to harvest in 7-8 months. The peak harvest is during March-August. The plants grow to a height of 4-6 m in a year and produce 6-12 primary branches. The pinnate leaves are about 40 cm long with small leaflets which are dark green on the upper side and pale green on the lower side. Though the flowers are in clusters of 25-150/cluster, only one pod develops usually and rarely 2-4 develop per cluster. The pods are 75 cm long with a girth of 6.3 cm and weigh 150 g with 70% flesh. The average yield is 220 fruits/tree. The estimated yield is around 52 t/ha. The pods attain edible maturity 65 days after flowering. The leaves can be used as a nutritive green. Ratoon crops can be maintained for 3-4 years. After every year the trees have to be cut back to 1 m from ground level. This variety can be grown as inter crop in coconut orchards during the early period. Chilli, onion and groundnut can be grown as intercrop in Moringa fields.

12.3.2.2 *PKM - 2 Annual Moringa*

It is a hybrid derivative of a cross between MP 31 (Eppothum vendran local) and MP 28 (Arasaradi local). It can be propagated through seeds. The tree comes to bearing six months after planting and is suitable for growing in different cropping systems. The pods are 126 cm long, with a girth of 8.3 cm and individual fruit weight of 280 g with 70% flesh. The pods are less seeded and delicious. The average number of pods per tree is 220/tree/year. It yields 98 tonnes/ha. Ratoon crop can also be taken up for three years. It is suitable for growing in Tamil Nadu and is adapted to most soil types varying from sandy loam to clay loam with good drainage.

In Mulanur block, Moringa is established as an intercrop on field scale and their allies were cropped with vegetables and Sorghum. This system evolved as Moringa offered some protection to alley crops from drying winds during summer and Moringa provided some additional income. With the migration of people from south to north India, and elsewhere in the world, the demand for Moringa picked up.

Farmers found that growing Moringa crops during summer season was remunerative. Thus Moringa gained a foothold as a summer vegetable. Its unique flavour and aroma became very popular. For South Indian any meal without Moringa and pulses is considered incomplete. The demand for the Moringa pod also increased due to increased urban settlements and migration of people to urban colonies.

Any Production activity needs clarity on economics. On realizing the economic benefits that accrue from the production process, either the entrepreneur or the producer of any activity will give his consent in establishing the production unit. In Horticulture, Moringa production is one of the principal activity that is gaining importance across the world and many farmers have resorted their willingness to establish Moringa orchards in their farm lands for maximizing their profits. Many of the small and marginal farmers are unaware on the economics but they continue their activity of producing the crops. Whatever the output that comes to them, they get satisfied. But the enterprising farmers are showing much interest in learning the economics out of Moringa and hence a study has been initiated in the western and southern zone of Tamil Nadu to quantify the exact income generation possibilities out of Moringa.

12.4 METHODOLOGICAL APPROACH TO SOCIO ECONOMIC STUDY

Among the 31 Districts of Tamil Nadu, only 6 districts have the notable area under Moringa cultivation. Among the 6 Districts, the traditional Moringa growers were belonged to Tiruppur and Thoothukkudi Districts. In these two districts, area expansion under Moringa is dismal while other districts take a lead in cultivation of Moringa and hence a study has been initiated to assess the reasons for the same and hence the ***Tiruppur and Thoothukkudi Districts were purposively chosen in the First Stage.***

Table 12.1 Area under Moringa in Thoothukkudi and Tiruppur Districts (2015-16)

Sl. No	Name of the Taluk	Name of the District	Area under Moringa in Ha	% to Total Area under Crops in the District
01	Sathankulam	Thoothukkudi	**1062.86**	**72.55**
02	Tiruchendur	Thoothukkudi	367.81	25.11
03	Others	Thoothukkudi	34.33	2.34
Total Area under Moringa in Thoothukkudi District			**1465.00**	**100.00**

Sl. No	Name of the Taluk	Name of the District	Area under Moringa in Ha	% to Total Area under Crops in the District
03	Dharapuram	Tiruppur	956.25	80.24
04	Kangeyam	Tiruppur	182.61	15.32
Total Area under Moringa in Tiruppur District			**1191.74**	**95.56**

Source: G-Return (2015-16)

Table 12.1 revealed that the total area under Moringa is distributed only in two taluks to the -tune of 98 per cent in Thoothukkudi District. Among the two taluks, Sathankulam found to have 73 per cent of the total area under Moringa. In respect of Tiruppur District, around 80 per cent of the total area under Moringa is vested with Dharapuram Taluk and ***hence the Sathankulam Taluk of Thoothukkudi District and Dharapuram Taluk of Tiruppur District were chosen purposely in the Second Stage.***

In the Third Stage, the villages where the highest area under Moringa available were enlisted after getting the list of farmers from the Department of Horticulture and Plantation Crops of Government of Tamil Nadu and in each village 10 sample farmers were randomly selected following a ***Three Stage Random Sampling Technique for the Study.***

Table 12.2 Details of Samples Selected for the Study

Sl. No	Name of the Villages Selected in Sathankulam Taluk	No. of Samples	Name of Villages Selected in Dharapuram Taluk	No. of Samples
01	*Sasthavinallur*	10	*Kilangundal*	10
02	*Naduvakkurichi*	10	*Komarapalayam*	10
03	*Mudhalur*	10	*Mulayampoondi*	10
04	*Arasoor*	10	*Mulanur*	10
05	*Thatchamozhi*	10	*Puduppai*	10
Total Samples in Thoothukkudi District		**50**	**Total Samples in Tiruppur District**	**50**

Table 12.2 revealed that the samples selected in respect of Moringa are accounted for 50 samples from Thoothukkudi District and another 50 samples were drawn from Tiruppur District. Put together, One hundred samples were selected for the study from these two districts. From the sample farmers, the details of land use, crop management for its production and productivity, marketing of Moringa produce, Cost of production of Moringa and the constraints involved are assessed using the structured and pretested questionnaire exclusively designed for the purpose and the same were analyzed and documented in this study by adopting a descriptive statistics like mean and percentage analysis. In this study, Net Present Worth (NPV), Benefit Cost Ratio (BCR) and Internal Rate of Return (IRR) were estimated to assess the worthiness of investment in moringa plantations.

Net Present Value (NPV)

It was assessed using the following formula

$$NPV = \sum_{t=0}^{n} \frac{B_t - C_t}{(1+r)^t}$$

Where,

B_t = Benefit of Moringa plantation realized during the year 't'

C_t = Costs incurred in Moringa plantations during the year 't'

r = Discount Rate.

t = Time between establishment and harvest.

n = Total number of years.

When NPV is greater than zero, investing in moringa plantation is worthy. If NPV equals zero, the practice is neither worthy nor unworthy. If it is less than zero, the plantation will be unworthy. Among the different systems, one which is capable of generating the highest NPV is chosen.

Benefit Cost Ratio (BCR)

$$\text{Benefit Cost Ratio (BCR)} = \frac{\sum_{t=1}^{n} B_t / (1+r)^t}{\sum_{t=1}^{n} C_t / (1+r)^t}$$

It is the ratio between the sum of discounted benefits and the sum of discounted costs. The same can be expressed as

Where,

B_t = Benefit of Moringa plantation realized during the year 't'

C_t = Costs incurred in Moringa plantations during the year 't'

r = Discount Rate.

t = Time between establishment and harvest.

n = Total number of years.

When Benefit Cost ratio is greater than one the system is beneficial. If it equals one, the system is neither worthy nor unworthy. If BCR is less than one, the system will be unworthy.

Internal Rate of Return

$$IRR = LDR + DBTDR \left| \frac{NPV_L}{ABNPV_S} \right|$$

Where,

LDR = Lower Discount Rate

DBTDR = Difference Between Two Discount Rates

NPV_L = NPV at the Lower Discount Rate

$ABNPV_S$ = Absolute Difference Between the NPVs at Two Different Discount Rates

When IRR is greater than the cut off rate (Bank Interest Rate) the plantation is worthy. If IRR is less than the cut off rate, the system will be unworthy. Among the practices, the one capable of generating the highest IRR is favoured. The cut off rate is the critical minimum rate of return desired, below which the practices may be improved to yield more or the cost of maintenance reduced.

12.5 SOCIAL ASPECTS OF MORINGA CULTIVATION

Under the title of social aspects of farm households, the size of holding available with the Moringa farms, age, educational status, size of family, farming experience, annual income level, asset position, livestock particulars, and irrigation sources available among the heads of the farm households. These will give general idea on taking appropriate decisions on choosing the right crop during the right season and sale of farm output in the right market at the right price. In this context, let us discuss the detailed social aspects of the farm households.

12.5.1 Size of Holding Available with the Sample Farms

The size of holding is the one to be addressed first to learn the status of farm households and the details are presented in Table 12.3.

Table 12.3 Size of Holding Available with the Farm Households

SI. No.	Size of Land Holdings (Ha)*	Number of Sample Farm Households in		Average Size of Holding (Ha)	
		Thoothukkudi District	Tiruppur District	Thoothukkudi District	Tiruppur District
01	Marginal (≤ 1.00)	6 (12.00)	6 (12.00)	0.68	0.51
02	Small (1.01-2.00)	15 (30.00)	16 (32.00)	1.44	1.50
03	Semi-Medium (2.01 - 4.00)	17 (34.00)	6 (12.00)	2.74	2.86
04	Medium (4.01 - 10.00)	9 (18.00)	20 (40.00)	5.43	5.24

SI. No.	Size of Land Holdings (Ha)*	Number of Sample Farm Households in		Average Size of Holding (Ha)	
		Thoothukkudi District	Tiruppur District	Thoothukkudi District	Tiruppur District
05	Large (Above 10.00)	3 (6.00)	2 (4.00)	12.50	10.52
	Total	50 (100.00)	50 (100.00)	2.90	3.35

(Figures in parenthesis indicate percentage to total). (*The Land Use Classification is based on the norms enunciated in the Horticultural Statistics at a Glance – 2015, Ministry of Agriculture, Government of India).

The Ministry of Agriculture through its official publication of Horticultural Statistics at a Glance has classified the land holding into Five different classification of land holding and designated the Marginal holding which possessed less than or equal to one ha. The land holding with the range of 1.01 to 2.00 ha was designated as Small farms and the Semi – Medium category of land was classified with the range of 2.01 to 4 ha and the Medium class of land was described if it is in possession of 4.01 to 10 ha and the Large category of farms were identified if it is cherished with higher than 10 ha. This classification was practiced in the study for analyzing the general characteristics of the sample farms.

Table 12.3 revealed that the semi-medium category of farm holding is predominant followed by small category of holdings which are respectively classified as 2.01 to 4 ha and 1.01 to 2 ha in Thoothukkudi District. But, the medium category of land is predominant with 20 holdings and their average size of holding was found to be 5.24 ha in respect of Tiruppur District followed by small category of land prevalent in 16 holdings of Tiruppur District. In general, Tiruppur district found have higher size of holding. On an average, the size of land holding is found to be 3.35 ha in respect of Tiruppur District followed by 2.90 ha in Thoothukkudi District. From this one could understand that Tiruppur District is in advantageous position for raising different crops like Perennial Moringa in the larger size of holding when compared to Thoothukkudi District. To effectively utilize the lands, the resources like labor and water are found to be important and hence the availability of family labor in the farm households were analyzed and the results are presented in Table 12.4.

12.5.2 Size of the Family among the Sample Farms

The number of persons available per household is an important factor in assisting the farm related events like cultivation of Moringa and realization of proper revenue. In this context, the details of size of family is analyzed and the results are presented in Table 12.4.

Table 12.4 Size of the Family among the Sample Farms

Sl. No.	Family Size (Numbers)	No. of Sample Farms in	
		Thoothukkudi District	Tiruppur District
01	Up to 3	07 (14.00)	05 (10.00)
02	04 to 06	42 (84.00)	43 (86.00)
03	Above 6	01 (2.00)	02 (4.00)
	Total	50 (100.00)	50 (100.00)

Source: Field Survey; (Figures in Parentheses indicate Percentage to Total)

Table 12.4 revealed that the study area is blessed with 4 to 6 persons per household and whom are accounted to be around 85 per cent to the total followed by only 3 persons per household which is accounted for 14 per cent in Thoothukkudi District and 10 per cent in respect of Tiruppur District. In a nutshell, the study area comprised of only 4 to 6 persons and the elders in the family alone are available for agricultural and allied activities like establishment and maintenance of Moringa plantations. However, bringing an activation or innovation in the process of cultivation of Moringa using latest technologies depend on the age factor also. In this respect, the sample farmers in the age group are analyzed and the details are presented in Table 12.5.

12.5.3 Age of the Sample Farmers

Considering the importance of age in taking up different farming practices, the age of the sample farmers is analyzed and the results are presented in Table 12.5. The age classification is arrived based on the norms followed by Durairasu during 2012.

Table 12.5 Age of the Sample Farmers

SI. No.	Age (years)	No. of Sample Farmers in	
		Thoothukkudi District	Tiruppur District
01	Up to 40	14 (28.00)	06 (12.00)
02	41 to 50	09 (18.00)	11 (22.00)
03	51to 60	19 (38.00)	13 (26.00)
04	Above 60	08 (16.00)	20 (40.00)
	Total	50 (100.00)	50 (100.00)

Source: Field Survey; (Figures in Parenthesis indicate Percentage to Total).

Table 12.5 revealed that the farmers in the age group of 51 to 60 years are accounted for 38 per cent in Thoothukkudi District followed by 40 years of age which is accounted for 28 per cent. It implied that around 66 per cent of the sample farmers are actively involved in cultivation of Moringa in their farm lands belonged to the age group of 40 to 60 years. Whereas, the Tiruppur District farmers whom belonged to the age group of above 60 years were involved in the establishment of Moringa plantations in their holdings. It implied that the youth from Tiruppur District is actively involved in other employment activities while the elders in the age group of around 60 years involved in the agricultural activities. In both of the districts, the Moringa cultivating farmers are accounted for 66 per cent to the total. Besides age as one of the factor, equally the educational status of the farmer is also important in deciding the varietal adoption and technology usage. These details are analyzed and the results are presented in Table 12.6.

12.5.4 Educational Status of the Sample Farmers

The details of education, the farmers were in possession are categorized as Illiterate, Primary, Secondary, Higher Secondary and Collegiate Education and the farmers whom are fit into different classification is analyzed and the details are presented in Table 12.6.

Table 12.6 Educational Status of the Sample Farmers

Sl. No.	Educational Status	No. of Sample Farmers in	
		Thoothukkudi District	Tiruppur District
01	Illiterate	02 (4.00)	03 (6.00)
02	Primary (1st to 5th)	09 (18.00)	13 (26.00)
03	Secondary (6th to 10th)	28 (56.00)	31 (62.00)
04	Higher Secondary	03 (6.00)	01 (2.00)
05	Collegiate Education	08(16.00)	02 (4.00)
	Total	50 (100.00)	50 (100.00)

Source: Field Survey (Figures in Parenthesis indicate Percentage to Total)

Table 12.6 revealed that the study area is blessed with very little percentage of illiterates whom are accounted for only around 5 per cent. The remaining 95 per cent are literates in the study area. Among the literate farmers, 56 per cent of the farmers of Thoothukkudi District and 62 per cent of the farmers of Tiruppur District were in possession of Secondary education. Primary education found to be in the second place with 18 per cent of the farmers and 26 per cent of the farmers of Thoothukkudi District and Tiruppur District respectively.

The Farmers with Collegiate Education is found to be in favor of Thoothukkudi District which is accounted for 16 per cent. Whereas, it was only 4 per cent in the case of Tiruppur District. It was mainly due to the employment opportunities available to the resident population of Tiruppur District as it has innumerous textile and spinning units which could offer different levels of employment to the needy. But this was not available appreciably in Thoothukkudi District and hence the offspring of the farmers have undergone higher studies when compared to Tiruppur District. In a nutshell, in terms of education, the Thoothukkudi District farmers had higher educational status when compared to the farmers of Tiruppur District and hence the Thoothukkudi farmers must be more innovative. The details of innovation or adoption of improved practices were discussed under the head of cropping pattern.

12.5.5 Farming Experience of the Sample Farmers

The farm experience gained by the farm household also plays a vital role in deciding the adoption of technologies or innovative practices and hence the experience details are analyzed and the results are presented in Table 12.7.

Table 12.7 Farming Experience of the Sample Farmers

Sl. No.	Experience (years)	No. of Sample Farmers in	
		Thoothukkudi District	Tiruppur District
01	Up to 20	10 (20.00)	07 (14.00)
02	21 to 30	12 (24.00)	11 (22.00)
03	31 to 40	14 (28.00)	12 (24.00)
04	41 to 50	10 (14.00)	11 (22.00)
05	Above 50	04 (8.00)	09 (18.00)
	Total	50 (100.00)	50 (100.00)

Source: Field Survey; (Figures in Parenthesis indicate Percentage to Total)

Table 12.7 revealed that the farmers with 31 to 40 years of experience in agriculture and allied activities was found to be the maximum both in Thoothukkudi and Tiruppur Districts followed by 21 to 30 years of experience could be found among the 24 per cent of the households of Thoothukkudi District and 22 per cent of the households of Tiruppur District. The farming experience is almost distributed among all the households of both Thoothukkudi and Tiruppur Districts revealed that the incumbents are much interested in farm related activities.

12.5.6 Farming Experiences of the Sample Farmers in Moringa Cultivation

The above experience is applicable to establishment and realization of different crops and their output. But it is important to furnish the experiences applicable to establishment of Moringa plantations in their farm holdings and hence these details are analyzed and the results are presented in Table 12.8.

Table 12.8 Farming Experiences of the Sample Farmers in Moringa Cultivation

SI. No.	Experience (years)	No. of Sample Farmers in	
		Thoothukkudi District	Tiruppur District
01	Up to 5 years	20 (40.00)	08 (16.00)
02	06 to 10 years	24 (48.00)	18 (36.00)
03	Above 10 years	06 (12.00)	24 (48.00)
	Total	50 (100.00)	50 (100.00)

Source: Field Survey; (Figures in Parenthesis indicate Percentage to Total)

Table 12.8 revealed that the farm households whom are actively involved in establishment and output realization activities of Moringa was arrived at 6 to 10 years of experience. This could be accounted for 48 per cent in Thoothukkudi District followed by 36 per cent in Tiruppur District. The above 10 years of experience in Moringa cultivation was visible with Tiruppur District farm households revealed that the Tiruppur farmers are pioneer in establishing Moringa plantations in Tamil Nadu. After seeing the economic incentives associated with Moringa plantations, the practice of cultivating Moringa in Thoothukkudi District and other places could have become spread. This was evidenced through the presence of higher number of Moringa cultivators with less experience in Thoothukkudi District.

12.5.7 Asset Position of the Farm Households

The assets available with the farm households in the study area are categorized into movable and immovable assets and the results are furnished in Table 12.9.

Table 12.9 Asset Position of the Farm Households

Sl. No.	Details of Assets	No. of Sample Farmers in Possession of Assets in	
		Thoothukkudi District	Tiruppur District
		I. IMMOVABLE ASSETS	
	Land (Nos.)		
01.	(a) Owned	37 (74.00)	46 (92.00)
	(b) Leased in	13 (26.00)	04 (8.00)
02.	Electric Motor (Nos.)	50 (100.00)	48 (96.00)
03.	Cattle Shed (Nos.)	09 (18.00)	34 (68.00)
04.	Pump House (Nos.)	50 (100.00)	48 (96.00)
05.	Sprayers (Nos.)	49 (98.00)	50 (100.00)
		II. MOVABLE ASSETS	
06.	Tractor (Nos.)	03 (6.00)	02 (4.00)

Source: Field Survey; (Figures in Parenthesis indicate Percentage to Total Sample Farmers)

Table 12.9 revealed that the owned land is on the higher side with the sample farmers of Tiruppur District and the leased-in lands are predominant in respect of Thoothukkudi District. It might be due to the successful establishment and income generation from Moringa plantations motivated the farmers with small holding to get the land on lease from absentee landlords and planted the Moringa plantations in a big way. The other immovable assets are Pump house, Electric Motor for irrigation purposes, are visible almost in all the farmers in the study area both Thoothukkudi and Tiruppur Districts.

Among the immovable assets, the cattle shed is predominantly available with the sample households belonged to Tiruppur District. It might be due to active involvement in Dairying activities of the farmers because of sustainable income generation. The sprayers and its availability among all the farm households is also visible. From that one could infer that the farmers do not have the practice of custom hiring from the Department of Agricultural Engineering or any other agency and they feel pride in owing the sprayers to have timely pest control in the plantations and other agricultural crops established in the farm environment.

Among the movable assets, tractor is found to be the principal asset which is rare among the farmers of both Thoothukkudi and Tiruppur Districts. Only very few farmers, particularly 6 per cent of the sample farmers of Thoothukkudi District and 4 per cent of the sample farmers of Tiruppur District had their own tractors in the farm. Other farmers might have had the practice of using the custom hiring services of Department of Agricultural Engineering and from private sector operators on hourly basis.

12.5.8 Livestock Particulars of the Sample Farmers

Livestock plays an important role in farming activities especially in Moringa plantations because the matured leaves of Moringa were fed to the animals and the young and mature leaves were used for other purposes. Hence, the details of number of animals present in the farm and their details were analyzed and the results are presented in Table 12.10.

Table 12.10 Livestock Particulars of the Sample Farmers

SI. No.	Livestock Particulars	No. of Sample Farmers in Possession of Livestock		Total Number of Birds / Animals
		Thoothukkudi District	Tiruppur District	
01.	Number of Farms having Buffaloes	03 (6.00)	07 (14.00)	10.00
1.1	Number of Buffaloes per Farm	02	03	05.00
02.	Number of Farms having Dairy Animals	09 (18.00)	29 (58.00)	38.00
2.1	Number of Cows per Farm	02	02	04.00
03.	Number of Farms having Sheep and Goats	09 (18.00)	22 (44.00)	31.00
3.1	Number of Sheep and Goats per Farm	05	18	23.00
04.	Number of Farms Raising Poultry	07 (14.00)	20 (40.00)	27.00
4.1	Number of Birds per Farm	06	22	28

Source: Field Survey; (Figures in Parenthesis indicate Percentage to Total)

Table 12.10 revealed that the sample farmers are blessed with Dairy animals, Sheep and Goats predominantly followed by Poultry birds. The Tiruppur District is having higher population of Dairy animals, Sheep and Goats and Poultry which are respectively accounted for 58 per cent, 44 per cent and 40 per cent. Whereas, these population were found to be very less in respect of Thoothukkudi District. It might be due to non-availability of grazing lands with Thoothukkudi District. But, Tiruppur District has higher per cent of grazing lands and the farm lands are blessed with live fence motivated the farmers to raise more animals. Higher number of farm animals resulted in liquid cash generation when emergency situation occurs.

The number of poultry firm in Tiruppur District is found to be on the higher side due to the influence of contract farming arrangements in vogue with M/S Suguna and Sri Venkateshwara hatcheries and the farmers. This arrangement has promoted poultry in a big way in Tiruppur resulting in 1910 birds per farm. During the lean season of agriculture and other activities, it provides additional income to the farmer.

12.5.9 Annual Income of the Sample Farm Households

The distribution of annual income received from the different enterprise components viz., agriculture, dairying, poultry, goat and sheep rearing and non-farm employment were categorized and the details are analyzed and the results are presented in Table 12.11.

Table 12.11 Annual Income Level of the Sample Farm Households

SI. No.	Particulars	Annual Income (Rs per Farm)	
		Thoothukkudi District	Tiruppur District
01	Income from Agriculture	365000 (78.73)	464000 (63.22)
02	Income from Livestock	60600 (13.08)	162000 (22.07)
03	Income from Sheep and Goats	25000 (05.39)	90000 (12.26)
04	Income - Non - Farm Activities	13000 (02.80)	18000 (02.45)
	Total Income	463600 (100.00)	734000(100.00)

Source: Field Survey; (Figures in Parenthesis indicate Percentage to Total)

Table 12.11 revealed that the income generation possibilities in the sample farm were found to be from cultivation of Moringa; rearing of Milch animals; rearing of Sheep and

Goats and from non-farm activities. When comparing the annual income generated by a farm household, the income generation was on the higher side in respect of Tiruppur District which is accounted for 58.33 per cent higher than the counterpart from Thoothukkudi District. The income generation from agriculture component was found to be more in both the districts which are respectively accounted for 79 per cent and 63 per cent to the total annual income generated by a farm household. The second important component of income basket is found to be Livestock capable of generating to the tune of 22 per cent in Tiruppur District. On an average, the annual income generated by a farm household who is cultivating Moringa as his preferred crop could generate 4.63 lakhs which is comparatively lesser than the farm at Tiruppur District who is capable of generating annual income to the tune of 7.34 lakhs.

12.5.10 Irrigation Sources Available to the Sample Farmers

The soundness of an Agricultural Economy of a farm depends on the availability of irrigation potentials within the farm. In this respect, the availability of water sources is classified analyzed and the results are presented in Table 12.12.

Table 12.12 Irrigation Sources Available to the Sample Farmers

SI. No.	Irrigation Sources	No. of Sample Farmers in		Total Samples
		Thoothukkudi District	Tiruppur District	
01.	Open Well			
1.1	Dysfunctional	28 (56.00)	24 (48.00)	52 (52.00)
1.2	Functional	14 (28.00)	22 (44.00)	36 (36.00)
02.	Bore Well			
2.1	Functional	35 (70.00)	32 (64.00)	67 (67.00)
2.2	Dysfunctional	00 (0.00)	01 (2.00)	01 (1.00)

Source: Field Survey; (Figures in Parenthesis indicate Percentage to Total)

Table 12.12 revealed that the irrigation support for farm operations were being carried out through open wells and the bore wells. The open wells are largely available in Thoothukkudi District. Out of this, 14 wells alone found to be functional. In respect of Tiruppur District, 22 open wells are functional. It is mainly due to the flow of Amaravathy River in these tracts and hence the open wells are largely functional and the agriculture and allied services become more functional and productive. Because of the sustained flow of water either from open well or from bore well, the success of agriculture is appreciable which is reflected in annual income generation in the farm household.

12.5.11 Cropping Pattern Prevalent among the Sample Farms

The study has identified that the study environment is blessed with sufficient water source for raising different agricultural and horticultural crops besides livestock and small ruminants. The details of crops raised by the farmers are analyzed and the results are presented in Table 12.13.

Table 12.13 Cropping Pattern of the Samples Farmers (in ha)

SI. No.	Crops	Area under Crops with Sample Farms of	
		Thoothukkudi District	**Tiruppur District**
01.	Moringa	95.50 (65.84)	59.11 (43.61)
02.	Sorghum	03.30 (2.28)	23.28 (17.18)
03.	Coconut	06.29 (4.34)	11.74 (8.66)
04.	Cumbu	00.00 (0.00)	10.93 (8.07)
05.	Gloriosa	00.00 (0.00)	07.39 (5.45)
06.	Groundnut	02.29 (1.58)	05.06 (3.73)
07.	Maize	05.38 (3.71)	04.05 (2.99)
08.	Horse Gram	00.00 (0.00)	03.85 (2.84)
09.	Sugarcane	02.29 (1.58)	02.43 (1.79)
10.	Chilies	04.15 (2.86)	02.23 (1.64)
SI. No.	**Crops**	**Area under Crops with Sample Farms of**	
		Thoothukkudi District	**Tiruppur District**
11.	Cotton	05.43 (3.74)	01.82 (1.34)
12.	Black Gram	00.00 (0.00)	01.62 (1.19)
13.	Sesame	00.00 (0.00)	01.21 (0.90)
14.	Brinjal	04.51 (3.11)	00.81 (0.60)
15.	Cashew nut	02.57 (1.71)	00.00 (0.00)
16.	Mango	04.86 (3.35)	00.00 (0.00)
17.	Banana	04.45 (3.07)	00.00 (0.00)

SI. No.	Crops	Area under Crops with Sample Farms of	
		Thoothukkudi District	Tiruppur District
18.	Watermelon	01.21 (0.88)	00.00 (0.00)
19.	Paddy	02.81 (1.94)	00.00 (0.00)
	Total Cropped Area	145.04 (100.00)	167.51 (100.00)
	Net Area Sown	136.44	135.53
	Cropping Intensity	106.30	123.60

Source: Field Survey; (Figures in Parenthesis indicate Percentage to Total Cropped Area)

Table 12.13 revealed that the Gross Cropped Area is higher in Tiruppur District when compared to the Thoothukkudi District. The Cropping Intensity is also found to be 124 per cent in Tiruppur The Cropping Intensity is also found to be 124 per cent in Tiruppur District followed by 106 per cent in Thoothukkudi District outlined that the farmers in Tiruppur District are very active in utilizing the land for different cropping program by utilizing the available water and other resources. Though the Tiruppur District is engaged with Textile and Spinning related works, it is also busy in establishing different crops in their farm lands.

The area under Moringa is accounted for 66 per cent in Thoothukkudi District. But Tiruppur District had only 44 per cent of area under Moringa. It might be due to much importance to other cash crops like Coconut, Gloriosa and other millet crops in the upland environment. The Thoothukkudi District is concentrating on Vegetable crops also in the study area. Whereas, Tiruppur District is not concentrating much on vegetable crops as it is importing from other villages of neighboring districts like Coimbatore. In a nutshell, Tiruppur is giving priority to perennial crops followed by millets. Thoothukkudi District is giving importance not only to perennial crops like Moringa but also millets, vegetables, oilseeds, cotton and fruit crops due to the demand in the Thoothukkudi town and Tirunelveli Town areas.

12.5.12 Area under Moringa in the Sample Farms of Tamil Nadu

Main aim of the study is to focus on the status of Moringa cultivation in Tamil Nadu. When the researcher could analyze the area under Moringa in the sample districts, the area under Moringa is found to decrease over the years and the area under Moringa in

Tiruppur District is found to increase over the years. It might be due to the enterprising nature of the farmers in Tiruppur District when compared to Thoothukkudi District. On conducting the interview with the farmers, the farmers opined that the youngsters and the middle-aged persons were seeking employment elsewhere particularly in Tiruppur District which has the strength of textile and spinning units. It might be one of the reason (Table 12.14). However, the same will be discussed under the head of factors influencing Moringa cultivation.

Table 12.14 Area under Moringa in the Sample Farms

Year	Area under Moringa (in Ha)		Percentage to State Total	
	Thoothukkudi	**Tiruppur**	**Thoothukkudi**	**Tiruppur**
2010-2011	1457	789	15.93	8.62
2011-2012	1524	885	16.18	9.40
2012-2013	1413	890	14.17	8.93
2013-2014	1398	982	12.45	8.74
2014-2015	1384	1074	11.08	8.60
2015-2016	1465	1191	11.23	9.13
Mean Area	**1440.17**	**968.50**	**13.51**	**8.90**

Source: Season and Crop Report of Tamil Nadu (2015-16)

Table 12.14 revealed that higher area under Moringa plantations is available with Thoothukkudi District which is accounted for 13.51 per cent to the total area under Moringa in Tamil Nadu State. Whereas, the Tiruppur District is blessed only with 8.90 per cent of area. However, the continuous increase in area under Moringa in Tiruppur District revealed that the farmers are much interested in establishing Moringa plantations in their farm holdings. In a year or two the farmers of Tiruppur District could achieve the higher area under Moringa to tap the cost incentives associated with Moringa

12.6 FACTORS INFLUENCING MORINGA PLANTATIONS

The increasing area under Moringa necessarily called for analyzing the influencing factors for establishing the Moringa Plantations in their farm holdings. Few of the factors may be positively influencing and some are negatively influencing. In this respect, these are analyzed and the results are presented in Table 12.15

Table 12.15 Factors Influencing Moringa Cultivation

S. No.	Factors in Favor of Moringa	No. of Farmers Reported	
		Thoothukkudi	Tiruppur
1.	Higher Return from Moringa Orchard	16 (32.00)	18 (36.00)
2.	Availability of Land and Water Resources	08 (16.00)	03 (6.00)
3.	Minimal Pest Problems	02 (4.00)	05 (10.00)
4.	Sustainable Income Generation	18 (36.00)	16 (32.00)
5.	Good Market Demand for Moringa Produce	06 (12.00)	08 (16.00)
	Total	50 (100.00)	50 (100.00)

Source: Field Survey (Figures in Parentheses indicate Percentage to Total)

Table 12.15 revealed that the factors are positively influencing to the farmers in the study area. The influencing factors are delineated as follows

- Higher Return from Moringa
- Sustainable Income Generation
- Availability of Land and Water Resources for establishing Moringa
- Good Market Demand for Moringa Products in the National and International markets
- Minimal Pest Problems and able to control the same through organic way

Among these factors, the Thoothukkudi farmers were of the opinion that the Sustainable Income generation possibilities are available with Moringa and hence they prefer the crop maximum followed by higher return from Moringa. These two are the most influencing factors to the farmers of Thoothukkudi District. Whereas, Higher returns from Moringa followed by Sustainable income generation are the factors dearer to the people of Tiruppur District. Though these factors are very closely related to income realization, the farmers are much interested in Moringa plantations owing to heavy market demand that exists in the National and International Markets. The factor of 'good market demand' is being ranked by them to the fourth rank and indirectly supporting the farmers to enter into higher area under Moringa cultivation. How much income over the costs are being generated by the farmers is important in this context and hence the details on economic incentives are analyzed and details are discussed elsewhere.

12.7 ECONOMIC BENEFITS OF MORINGA CULTIVATION

A developing country like India needs not only food on quantitative terms but also the food must be qualitative and should possess the nutrient requirements of the body which is not available to many in the rural and urban settings. Moringa is one of the best alternative which could serve the nutritious food. Hence in the daily diet of an individual, one should take Moringa products daily to lead a healthy and energetic life. Production of the Moringa as vegetables and conversion of the vegetables into value added products depends on the supply of quality output from the farm. Ensuring the regular supply not only promotes healthy life but also promotes economy and hence one should analyze whether the Moringa plantation and its establishment is dearer to the farm community so that they could ensure sustainable supply. In this respect, the economic incentives associated with Moringa cultivation is discussed under the following heads.

- Establishment Cost for Perennial Moringa Plantations
- Details of Maintenance Cost involved in Perennial Moringa Plantations
- Yield and Income Potentials of Perennial Moringa Plantations
- Cost of Production of Annual Moringa
- Economics of Moringa Plantations

12.7.1 Establishment Cost for Perennial Moringa Plantations

The establishment cost or the cost incurred in the zero time period comprise rental value of land, cost of plant material, cost of farm houses, expenditure on farm equipments, fencing etc assumes greater importance and hence the details of establishment cost incurred by the sample farmers are analyzed and the results are presented in Table 12.16

Table 12.16 Establishment Cost for Perennial Moringa Plantations

SI. No	Cost Components	Thoothukkudi		Tiruppur	
		In Rs/Ha	Per Cent	In Rs/Ha.	Per Cent
01.	Ploughing and Preparation of Land	3230.98	05.12	3705.00	06.19
02.	Digging of Pits	3140.49	04.98	3250.00	05.43
03.	Fencing Charges	19450.50	30.83	18466.91	30.86

SI. No	Cost Components	Thoothukkudi		Tiruppur	
		In Rs/Ha	Per Cent	In Rs/Ha.	Per Cent
04.	Formation of Irrigation Channels	950.00	01.51	850.00	01.42
05.	Cost of Plant Material including Transportation	6350.00	10.07	6446.09	10.77
06.	Cost of Manure	2242.00	03.55	2346.21	03.92
07.	Cost of Filling of Pits, Manuring and Planting	1438.56	02.28	1533.83	02.56
08.	Cost of Irrigation including Labour	3235.60	05.13	3483.34	05.82
09.	Plant Protection Charges while Planting of Limb Cuttings	428.90	00.68	386.77	00.65
10.	Rental Value of Land	22530.58	35.72	19302.30	32.25
11.	Land Cess and Taxes	60.00	00.10	60.00	00.10
12	Total Establishment Cost	**63057.61**	100.00	**59830.45**	100.00

Table 12.16 revealed that the establishment cost incurred for Moringa in Tiruppur District has commanded low cost when compared to Thoothukkudi District. Among the different constituent costs under establishment, the rental value of land and the fencing charges have commanded more than 60 per cent to the total establishment cost. The total establishment cost of Moringa in Thoothukkudi District is arrived at Rs 63058 and Rs 59830 in respect of Tiruppur District. The cost of plant material alone accounted for around 10 per cent to the total establishment cost in both of the districts. From that one could infer that the rental value of land, fencing charges and planting material cost are the major items under establishment cost. The cost of irrigation arrangements during the initial period is accounted for around 5 per cent to the total establishment cost.

12.7.2 Annual Maintenance Cost involved in Moringa Plantations

After incurring the initial investment on establishing the Moringa plantations, the farmer is required to make some of the following investments until the tree reaches the bearing stage. Such costs are known as operational and maintenance costs and are mainly incurred for land preparation, fertilizer application, pesticide spraying, irrigation and manuring the plantations with farm yard manure etc. The analysis of the collected data from the samples revealed that the total investment on the Moringa plantations until getting the yield was averaged and are presented as average maintenance cost up to 3 years; average maintenance cost up to 7 years and average maintenance cost up to the 10th year of maintenance and designated as maturing phase, yielding phase, and economically yielding phase respectively. Similar approach was practiced by ***Ritambhara Singh et.al.*** during 2017. The details are analyzed and presented in Table 12.17.

Table 12.17 Average Maintenance Cost of Moringa in Thoothukkudi District

Sl. No	Particulars of Cost	Economically Yielding Phase (10 Years)	Yielding Phase (7 Years Average)	Maturing Phase (3 Years Average)
01	Manures and Manuring	6896.01	6077.02	8806.98
02	Fertilizer Application	7645.38	6910.62	9359.80
03	Plant Protection	1359.04	1228.43	1663.80
04	Weeding	4549.69	4112.44	5569.92
05	Earthing up and Basin Formation	2413.51	2181.56	2954.72
06	Pruning of Tips	1089.63	960.22	1391.58
07	Irrigation	6924.83	6240.74	8521.04
08	Miscellaneous Charges	1901.28	1758.22	2235.11
09	Land Cess and Tax	74.63	65.77	95.31
Total Annual Operational Cost		32854.00 (58.79)	29535.02 (58.66)	40598.26 (59.01)
	Harvesting and Marketing Cost of Moringa			
10	Harvesting Charges	14558.92	13159.74	17823.66
11	Packing and Grading	6098.63	5512.52	7466.21
12	Transportation Charges	856.41	774.10	1048.45

Sl. No	Particulars of Cost	Economically Yielding Phase (10 Years)	Yielding Phase (7 Years Average)	Maturing Phase (3 Years Average)
13	Loading and Unloading	1518.18	1372.27	1858.62
Total Annual Harvesting and Marketing Cost		**23032.14 (41.21)**	**20818.63 (41.34)**	**28196.94 (40.99)**
Total Annual Maintenance Cost		**55886.13 (100)**	**50353.67 (100)**	**68795.20 (100)**

(Figures in Parentheses indicate Percentage to Total)

Table 12.17 revealed that annual maintenance cost during different phases of Moringa plantation is found to be 59 per cent and the total annual harvesting and marketing cost is arrived at 41 per cent to the total maintenance cost. It revealed that the farmers in Thoothukkudi region is capable of applying same input cost and it did not differ between the maturing phase, yielding phase and the economically yielding phase leaving absolute difference. The absolute difference in the maintenance cost is comparatively higher in the initial stages of maintenance and during the economically yielding stages. It might be due to the healthy establishment of the plantation during the initial stages will withstand some environmental distress like drought and disease. During the economically yielding stage, the increase in cost is visible due to the interest of getting the continuous yield increase in respect of Moringa fresh vegetable pods. In this respect, the annual maintenance cost that are incurred by the farmers of Tiruppur District is much important and the details are analyzed and the results are presented in Table 12.18.

Table 12.18 Average Maintenance Cost of Moringa Plantations in Tiruppur District

Sl. No	Particulars of Cost	Economically Yielding Phase (10 Years)	Yielding Phase (7 Years Average)	Mature Phase (3 Years Average)
01	Manures and Manuring	7276.63	6412.44	9293.08
02	Fertilizer Application	8423.93	7614.35	10312.94
03	Plant Protection	1538.42	1390.57	1883.40

Sl. No	Particulars of Cost	Economically Yielding Phase (10 Years)	Yielding Phase (7 Years Average)	Mature Phase (3 Years Average)
04	Weeding	4949.64	4473.96	6059.57
05	Earthing up and Basin Formation	2491.37	2251.94	3050.04
06	Pruning of Tips	796.39	701.81	1017.08
07	Irrigation	7474.10	6755.80	9150.11
08	Miscellaneous Charges	1946.38	1759.32	2382.84
09	Land Cess and Tax	74.63	65.77	95.31
Total Annual Operational Cost		34971.49 (58.90)	31425.96 (58.76)	43244.37 (59.14)
	Harvesting and Marketing Cost of Moringa			
10	Harvesting Charges	14948.19	13511.61	18300.23
11	Packing and Grading	6332.19	5723.64	7752.15
12	Transportation Charges	934.26	844.48	1143.76
13	Loading and Unloading	2188.59	1978.25	2679.36
Total Annual Harvesting and Marketing Cost		24403.23 (41.10)	22057.98 (41.24)	29875.50 (40.86)
Total Annual Maintenance Cost		**59374.73 (100)**	**53483.95 (100)**	**73119.88 (100)**

(Figures in Parentheses indicate Percentage to Total)

Table 12.18 revealed that the Tiruppur District also had the similar cost use pattern in respect of operational and harvesting related cost. Only the absolute difference is visible between the average costs incurred during different phases. The early stages of maintenance alone attracted higher operational cost as visible in Thoothukkudi District. In this circumstance, one has to analyze the yield and income realized by the farmers from Moringa plantations after incurring heavy investment and hence these data are analyzed and the results are presented in Table 5.18. Though the Moringa plantations are capable of yielding up to 30 years of its life, the yield calculation was made only up to 10 years of establishment due to errors in recollection of data and limited to 10 years.

12.7.3 Yield and Income Potentials from Moringa

The drier environment is preferable to Moringa plantations. Fearing to the drought, the Moringa plants will tend to promote their flowering profusely and capable of producing more number of fresh pods to deliver their offspring to the environment. In this respect, if one could maintain optimum needs of water for their survival coupled with fertigation arrangements, it could yield higher and hence these details are analyzed and the results are presented in Table 12.19 in respect of sample farms.

Table 12.19 Yield and Income Realized from Moringa Plantations

Year	Yield of Moringa Pod in Tonnes per Ha		Gross Income Realized per Ha	
	Thoothukkudi District	Tiruppur District	Thoothukkudi District	Tiruppur District
01	07.20	08.10	75600.00	102465.00
02	12.60	14.40	132300.00	182160.00
03	26.55	27.90	278775.00	352935.00
04	28.35	29.25	297675.00	370012.50
05	30.15	31.05	316575.00	392782.50
06	31.05	32.85	326025.00	415552.50
07	33.75	35.10	354375.00	444015.00
08	36.00	36.45	378000.00	461092.50
09	36.45	37.80	382725.00	478170.00
10	37.80	39.60	396900.00	500940.00
Total	279.90	292.50	2938950.00	3700125.00

(Source: Primary Survey)

Table 12.19 revealed the details of yield and income realization over years from the Moringa plantations. The average yield of fresh Moringa pod is arrived at 28 tonnes in respect of Thoothukkudi District. Whereas, Tiruppur District was capable of realizing 29 tonnes per ha. On examining the yield difference between the Thoothukkudi District and the Tiruppur District, the much difference is invisible, but only marginal when comparing the mean yield.

The yield potential of Moringa in respect of Tamil Nadu state is also arrived and the details are presented in the trading section. On an average, Tamil Nadu could be able to produce around 2.81 lakh tonnes of Moringa pods per annum with the average productivity of 21 tonnes per ha. In this circumstance, whether the Moringa plantations are economically viable or not has to be analyzed and hence the details are analyzed and the results are presented elsewhere.

12.7.4 Cost of Production of Annual Moringa

So far, the discussions were made in favor of perennial Moringa. Very few households particularly in Tiruppur and Thoothukkudi Districts had raised Annual Moringa (PKM-1 and PKM-2). Before analyzing the cost of production of annual Moringa, the discussion on how many farm households are in practice with establishing annual Moringa is analyzed and the details are presented in Table 12.20.

Table 12.20 Number of Households Practicing Annual Moringa

Sl. No	Particulars of Annual Moringa	Number of Households in	
		Thoothukkudi District	Tiruppur District
01	Perennial Moringa	47 (94.00)	42 (84.00)
02	Annual Moringa – PKM - 1	02 (04.00)	05 (10.00)
03	Annual Moringa – PKM - 2	01 (02.00)	03 (06.00)
	Total	50 (100.00)	50 (100.00)

(Source: Primary Survey); (Figures in Parentheses indicate Percentage to Total)

Table 12.20 revealed that the perennial Moringa is cultivated by 94 per cent of the farm households in Thoothukkudi District. Whereas, only 84 per cent of the farm households have raised Perennial Moringa in Tiruppur District. Only 6 per cent of the farm households are having the practice of cultivating annual Moringa in Thoothukkudi District. But 16 per cent of the farm households are having the practice of establishing annual Moringa in their farms. Among the annual Moringa varieties, the PKM – 1 variety found to dominate both in Thoothukkudi District and Tiruppur District because the fresh Moringa Pods are in optimum size which facilitates transportation to different consuming centers. Whereas, the PKM – 2 variety pods are fleshy and lengthy which has bit difficulties in transportation. Sometimes, breakage is also visible in PKM – 2 and hence it is less preferred among the farmers. However, the cost of production of annual Moringa is important and hence these details are analyzed and the results are presented in Table 12.21.

Table 12.21 Costs and Return from Annual Moringa Plantations

Sl. No	Particulars of Cost	Number of Units	Cost per Unit in Rs	Total Cost in Rs per Acre	Total Cost in Rs per Ha
I	Operational Cost				
01	Preparatory Cultivation and Tillage	03	700.00	2100.00	5187.00
02	Cost of Farm Yard Manure	12	400.00	4800.00	11856.00
03	Application of Farm Yard Manure	24	25.00	600.00	1482.00
04	Cost of Seed and Planting Charges	00.50 Kg	2400	1200.00	2964.00
05	Weeding Cost	54	25.00	1350.00	3334.50
06	Hiring Cost of Sprayers and its Operation	04	260.00	1040.00	2568.80
07	Cost of Pesticides and Fungicides			2500.00	6175.00
08	Cost of Neem Cake	150 Kgs	23.00	3450.00	8521.50
09	Cost of Urea	275 Kgs	06.00	1650.00	4075.50
10	Cost of Super Phosphate	100 Kgs	07.75	775.00	1914.25
11	Cost of Potash Fertilizer	130 Kgs	17.00	2210.00	5458.70
12	Cost of Harvesting (Men Hours)	68 Hours	60.00	4080.00	10077.60
13	Cost of Harvesting (Women Hours)	50 Hours	25.00	1250.00	3087.50
	Total Operational Cost (I)			27005.00	66702.35 (77.47)
II	Annual Fixed Cost				
14	Rental Value of Land	01 Acre	7720	7720.00	19302.30

Sl. No	Particulars of Cost	Number of Units	Cost per Unit in Rs	Total Cost in Rs per Acre	Total Cost in Rs per Ha
15	Land Tax, Cess, etc	01 Acre	95.00	95.00	100.00
	Total Annual Fixed Cost			7815.00	19402.30 (22.53)
	Total Cost of Cultivation			34820.00	86104.65 (100.00)
III	Gross Income				
	Income from Main Product (Drumstick)	14605 Kgs	12.65	184753.00	456340.53
	Income from Bi-Product (Leaves)	9100 Kgs	05.00	45500.00	112385.00
	Total Income from Moringa			230253.00	568725.53
IV	**Cost of Production per Kg of Main Product**			02.38	02.38
V	Output – Input Ratio			06.62	06.62

(Source: Primary Survey) (Figures in Parentheses indicate Percentage to Total)

Table 12.21 revealed that the total operational cost incurred in respect of establishing annual Moringa is arrived at Rs. 66702/ per ha which is accounted for 77.47 per cent to the total cost of cultivation of annual Moringa. Whereas, the annual fixed cost incurred was arrived at Rs 19402/ which is accounted for 22.53 per cent. The total cost of establishing the annual Moringa per ha was arrived at Rs 86104/. The gross income realized by the farmer per ha was arrived at 5.69 lakhs in which the revenue earned from main product was accounted for 80 per cent and the bi-product accounted for only 20 per cent. The bi-product harvested was green leaves for local sales. Put together, the cost of production of fresh Moringa pod per kg was arrived at Rs 2.38 per kg. The cost of production per ton was arrived at rupees 2380/. However, the sale of fresh Moringa pods per ton was arrived at Rs 11750/ which has indicated that the cultivation of annual Moringa is highly profitable as there exists good market for the produce. Hence, the farmers can take up annual Moringa cultivation in their farm lands in general. But, specifically, as the perennial Moringa is enriched with much of nutritional and pharmaceutical importance, the cultivation of Perennial Moringa in the sample farm households are predominant. In this context, the demanding places and centers need to be assessed. These are presented in the ensuing section.

12.7.5 Factors Governing Profitability of Moringa

The profitability of Moringa is influenced by certain factors. Though certain farmers are not conversant and has clarity on which parameter mostly influences the profitability, the researcher has given highlights on certain factors and then the farmer could say something about the factors and hence those factors have been analyzed and the details are presented in the Table 12.22.

Table 12.22 Factors Governing the Profitability of Moringa

S. No.	Factors Governing the Profitability of Moringa	No. of Farmers Reported	
		Thoothukkudi	Tiruppur
1.	Selection and Adoption of High Yielding Varieties of Moringa	24 (48.00)	14 (28.00)
2.	Sowing of Moringa in Appropriate Season	03 (06.00)	05 (10.00)
3.	Moringa Variety that should have Greater Consumer Acceptance	09 (18.00)	08 (16.00)
4.	Adoption and Management of Ratoon Crop in Moringa	04 (08.00)	11 (22.00)
5.	Adoption of Appropriate Post Harvest Management Practices	10 (20.00)	12 (24.00)
	Total	50 (100.00)	50 (100.00)

Table 12.22 revealed a different response in respect of sample farms drawn from Thoothukkudi and Tiruppur Districts in respect of Profitability aspects of Moringa. In respect of sample farms from Thoothukkudi District, they were of the opinion that the Selection and Adoption of High Yielding varieties of Moringa is one of the Principal factor that decides profitability. But it is not that much reflected in respect of farms from Tiruppur District. It might be due to the expertise developed in this regard in raising both perennial types and high yielding varieties of Moringa. The higher percentage of respondents from Thoothukkudi reported that the high yielding varieties of Moringa could enhance the profitability.

The Moringa variety practiced should have greater consumer acceptance with regard to taste of the vegetable as well as marketability specifically the size of Moringa. Higher the length of the Moringa vegetable or pod is not much preferred by the farmers. The response of the farmers with regard to this factor is almost equal and hence the breeders who are involved in making new varieties for release should take care of this sensitive factor.

The most important factor is "Adoption of Appropriate Post Harvest Management Practices" which is also contributing to the profitability as reported by the sample farmers of Thoothukkudi and Tiruppur Districts which are also almost equal and ranked as second important factor that decides the profitability. The post harvest measures included are harvesting methods, transportation methods, processing methods like drying and value addition. Around 20 to 24 per cent of the sample farms have expressed this as a most important factor and hence immediately after harvest of Moringa pods or Moringa leaves, the post harvest management issue should be properly looked into and one should see that the aspects are being implemented properly or sequentially or not. Hence those who are planning to establish the Moringa orchards, these factors must also be given a fillip to earn higher profitability.

12.8 STEPS INVOLVED IN POST HARVEST PROCESSING OF MORINGA LEAVES

Various steps are involved in processing of Moringa leaves and Pods for manufacturing various products like Moringa leaf Powder, Moringa Leaf Tablets, Moringa Capsules, Moringa Chips etc. The steps involved are arranged orderly and sequentially and the details are presented in Table 12.23.

Table 12.23 Steps involved in Post Harvest Processing of Moringa Leaves

Steps	Sequence of Operations
01	Cut Big Branches of Moringa Tree
02	Strip the Leaves off the Branches before Transporting to the Processing Centre
03	Freshly Harvested Leaves should be Transported to the Processing Centre as Quickly as Possible
04	Wash the Leaflets in Troughs using Clean Potable Water to remove dirt and impure materials

Steps	Sequence of Operations
05	Wash the Leaves again in One per cent saline solution for 3-5 Minutes to remove Microbial load if any
06	Drain Each Trough after Wash
07	Spread the Leaflets on Trays made with food grade mesh and leave to drain for 15 Minutes before taking them to the Dryer
08	Try to Use Solar Drying or Room Drying. Do not expose the leaflets to sun drying. Use electric or gas hot –air dryers. Drying temperatures should range between 50 degree Celsius to 55 degree Celsius
09	Dry the leaves and see that the Moisture level is less than 10 Per cent.
10	Ensure the Chlorophyll Content to the level of 6 to 7 Per cent. Higher the Chlorophyll Content of leaves will have higher acceptability in Export Markets.
11	Mill dry leaves using a stainless steel hammer Mill.
12	Sieve the leaf powder if needed. Recommended particle Sizes are Coarse (1.0 MM to 1.50 MM); Fine (0.50 MM to 1.00 MM) and Very Fine Powder (0.2 MM to 0.50 MM)
13	Dry the Leaf Powder immediately after Milling. The Drying may be performed at 50 degree Celsius for 30 Minutes to reduce Moisture content to the level below of 7.50 per cent.
14	Do Packing and Store the produce free from Bacteria and Fungi as it could attract these two.
15	Maintain Personal Hygiene by wearing Head Caps, Nose Masks, Disposable Gloves etc.
16	Do Labeling and Branding
17	Get the FSSAI Certification from Government of Tamil Nadu
18	Get HALAL certification to highlight the vegetable origin.

While doing labeling the following points must be taken into account. They are

- Name of the Product
- Net Content in Grams or Mg
- Name and Address of the Producer or the Manufacturer

- Country of Origin
- Lot Number or Code
- Instructions for Use
- Date of Manufacturing
- Date of Expiry of the Product
- Nutritional Information (Optional)

The above aspects of post harvest and value addition measures if practiced scrupulously, the product quality could be ensured and it fetches higher consumer acceptance both in the regional, national and international markets.

12.9 COMPOSITION OF NUTRIENTS IN MORINGA PRODUCE

The Moringa leaves or greens of Moringa, the pods of Moringa contain high levels of energy contributing Vitamins, Minerals, Proteins and Carbohydrates. The people who are regularly taking Moringa in their diet were of the opinion of energy gaining as reported by Ponnuswami during the year 2012. The details of energy contributing Vitamins and Minerals are delineated in Table 12.24.

Table 12.24 Composition of Nutrients from Moringa Leaves and Pods

Sl. No	Available Nutrients per 100 Gram of Intake	Units	Moringa Leaves	Moringa Pods
01	Edible Protein	Percent	75.00	83.00
02	Moisture Content	Percent	75.00	86.90
03	Protein	Grams	06.70	02.50
04	Fats	Grams	01.70	00.10
05	Carbohydrate	Grams	13.40	03.70
06	Minerals	Grams	02.30	02.00
07	Fibre	Grams	00.90	04.80
08	Calories	--	92.00	26.00
09	Calcium	Mg	440.00	30.00
10	Magnesium	Mg	24.00	24.00
11	Oxalic Acid	Mg	101.00	101.00

Sl. No	Available Nutrients per 100 Gram of Intake	Units	Moringa Leaves	Moringa Pods
12	Phosphorous	Mg	70.00	110.00
13	Potassium	Mg	259.00	259.00
14	Copper	Mg/Gram	01.10	03.10
15	Iron	Mg	07.00	05.30
16	Sulphur	Mg	137.00	137.00
17	Vitamin - A	IU	11300	184.00
18	Choline	Mg	423.00	423.00
19	Thiamine	Mg	00.06	00.05
20	Riboflavin	Mg	00.05	00.07
21	Nicotinic Acid	Mg	00.80	00.20
22	Vitamin - C	Mg	220.00	120.00

(Source: Ponnusamy, 2012)

Table 12.24 outlined that the Moringa leaves contain high level of energy contributing vitamins and minerals. In that Vitamin – A acts as a shield against eye disease to the human, skin disease, heart ailments, diarrhea and many other diseases. Similarly Vitamin – C is capable of fighting a host of illness including Colds and Flu. The Calcium which is present in the Moringa is able to build Strong bones and teeth and helps prevent Osteoporosis.

Potassium is an essential element for the functioning of Brain and Nerve cells (Ponnuswami, 2012) and the Proteins, the building blocks of all our body cells and hence every consumer must take the produce of Moringa in either as a juice or side dish or as main dish or soup and or in Capsule or tablet forms. Those who are in the urban fringes may not be able to get the Moringa greens and hence they can go for consuming Moringa Tablets and or the Moringa Capsules which are available in Palamudhir Nilayam and in Big Retail Formats.

12.10 ECONOMICS OF MORINGA PLANTATIONS IN TAMIL NADU

To work out the economics of Moringa production, the year wise and operation wise costs incurred over the years were analyzed and the results are presented elsewhere. In a nutshell, the plantations are economically viable or not has to be discussed and hence the data have been analyzed using the economic appraisal tools which are presented in Table 12.25.

Table 12.25 Economics of Moringa Plantations

Sl. No	Tools of Appraisal	Units per Ha	Thoothukkudi District	Tiruppur District
I	**Perennial Moringa**			
01	Gross Income	Rupees	293895.00	370013.00
02	Net Income	Rupees	231703.00	304660.00
03	Net Present Value @ 12%	Rupees	1128756.00	1507024.00
04	Benefit Cost Ratio	Ratio	04.17	05.07
05	Internal Rate of Return	Per Cent	>100	>100
II	**Annual Moringa**			
01	Gross Income	Rupees	568725.53	
02	Net Income	Rupees	482620.88	
03	Cost of Production / Ton	Rupees	2380.000	
04	Output – Input Ratio	Ratio	06.62	
05	Return on Investment		05.60	

Table 12.25 revealed that the perennial Moringa is capable of generating a net income of Rs 2.31 lakhs in Thoothukkudi District. But the same perennial Moringa is able to provide 3.05 lakhs per annum as its net income per ha. This is comparatively more because Tiruppur farm households able to receive an average price of Rs 12.65 per kg of Moringa pods. Whereas, the farm households at Thoothukkudi District is able to receive an average price of Rs 10.50 per kg of Moringa vegetables. It is mainly due to the export competitiveness available in Tiruppur and the demanding firms are checking the quality of the produce and offer best price. The Tiruppur produce naturally commands best price due to its taste because of its cultivation made in calcium rich soils.

On discounting the values of income and investment over years using 12 per cent discount rate, the Thoothukkudi farm households are able to generate the Net Present Value of Rs 11.39 lakhs while the Tiruppur counterparts are able to secure net present value of Rs 15.07 lakhs. The Benefit Cost Ratio was arrived at 4.17 and 5.07 in respect of Thoothukkudi farm households and Tiruppur farm households revealed that the farmers respectively could earn a profit of Rs 4.17 and 5.07 for a rupee of investment in the establishment and maintenance activities of Moringa. The study conducted by Ritambhara Singh during the year 2017 in Gujarat State has arrived the Benefit Cost Ratio to the tune of 10.76 for a rupee of investment and the NPV under normal cost and return was arrived at Rs 25.24 lakhs per ha and the IRR was estimated at 95.53 per cent. In this study, the returns are valued at lower price per unit of the produce. None other enterprises are able to provide such returns and hence the area under horticulture particularly Moringa can best be encouraged by the extension functionaries in their respective districts.

On analyzing the economic worthiness of annual Moringa, the annual Moringa established by the farm households could generate a net income of Rs 4.83 lakhs per annum with the cost of production of Rs 2380 per ton of Moringa vegetables. On an undiscounted term, the Output-Input ratio is arrived at 6.62 revealed that the farmers are able to generate a gross income of Rs 6.62 per rupee of investment. The Return on Investment figure revealed that the farm households whom are capable of establishing the annual Moringa in their farms, could generate 5.60 as their profit per unit of investment.

12.11 SUMMARY AND CONCLUSIONS

Economic analysis of Moringa orchards revealed that the plantations are capable of generating higher return per unit of the produce. This was proved with the data provided by the Benefit Cost Ratio, Net Present Value and Internal Rate of Return. The Benefit Cost Ratio was arrived at 4.17 and 5.07 in respect of Thoothukkudi farm households and Tiruppur farm households revealed that the farmers respectively could earn a profit of Rs 4.17 and 5.07 for a rupee of investment in the establishment and maintenance activities of Moringa. None other enterprises are able to provide such returns and hence the area under horticulture particularly Moringa can best be encouraged by the extension functionaries in their respective districts by infusing technology to the farm households and the area expansion could be achieved which is one of the focal area of National Mission on Horticulture. With regard to Post Harvest Management practices, the steps involved in processing of Moringa leaves to protect the Chlorophyll content and the appearance of the leaf powder is also discussed which are considered to be the principal factors for export.

References

1. Anwar F, Ashraf M, Bhanger MI. (2005). "Interprovenance Variation in the Composition of *Moringa oleifera* Oilseeds from Pakistan", ***Journal of American Oil Chemists Society,*** **82**: pp. 45-51.

2. Anwar F, Bhanger MI. (2003). "Analytical Characterization of *Moringa oleifera* Seed Oil Grown in Temperate Regions of Pakistan", ***J Agric Food Chemistry,*** **51**: pp. 6558–6563.

3. D'souza J, Kulkarni, A.R. (1993). "Comparative Studies on Nutritive Values of Tender Foliage of Seedlings and Mature Plants of *Moringa oleifera* Lam", ***Journal of Ecology, Taxonomy and Botany***, **17**: pp. 479-485.

4. Ponnuswami, V. 2012. "Advances in Production of Moringa", (All India Coordinated Project on Vegetable Crops – Tamil Nadu Agricultural University: Periyakulam). Pp. 73-83.

5. Ritambhara Singh, M.R. Prajapati and Jemish Savani. (2017). "Economics of Production of Drumstick (*Moringa olifera*) in Vadodara District of Gujarat," ***International Journal of Advanced Biological Research,*** 7(2): pp. 322-328.

6. Sadasakthi, A.1995. Studies on combining ability, Heterosis and prediction model for ideo type in annual moringa (Moringa pterigosperma Gaertn.). Ph.D. thesis submitted to TNAU, Coimbatore.

CHAPTER XIII

Marketing of Precision Vegetables in India

Dr. C. Sekhar and Dr. A. Vidhyavathi

Objectives

- To expose the students and the readers on the Marketing Practices prevalent in respect of Vegetables
- To highlight the Price Spread and Marketing Margin to the different intermediaries in the Marketing Process of Vegetables
- To identify the constraints in Marketing of Horticultural Prodce and to suggest strategies for the same.

Organization of the Chapter

- The Concepts on Marketing
- Marketing Practices in Principal Vegetables
- Price Spread in Marketing and Distribution of Vegetables
- Marketing Efficiency in Vegetable Marketing
- Constraints in Marketing of Horticultural Produce
- Summary

INTRODUCTION

Any vegetables once produced, it has to be processed, graded and should be distributed to the demanding ends. Here a separate chapter on marketing of vegetables were earmarked exclusively to highlight the practices prevalent in respect of vegetables produced out of precision farming technologies. Here, a brief presentation on what is marketing, marketing channels, price spread, marketing cost and marketing margin and the marketing institutions for distribution of vegetables are discussed in detail.

Marketing of vegetables is referred to the business activities that are involved in transferring the Vegetables from the primary producer to the ultimate consumer which offered time, place, form and possession utilities to the user.

With respect to marketing of vegetables, the name of the organizations involved in trading the vegetables produced from the Precision Farms, price per unit of the vegetables prevailed during before and after implementation of precision farming practices are also assessed by interviewing the farmers and the same are appropriately documented in this chapter.

13.1 THE CONCEPTS RELATED TO MARKETING

13.1.1 Marketing Channel

The marketing channel is defined as a path traced in the movement of Vegetables from the primary producer to ultimate consumer

13.1.2 Market Functionaries

13.1.2.1 *Wholesalers*

In the current study, wholesalers were considered as those engaged in procuring and selling of Vegetables to those who demand the same for business need. Wholesaler –cum – exporters were considered as those who do the activity of wholesalers as well as do processing of Vegetables for export purpose. The Wholesalers- cum – Retailers does the functions of wholesaling in addition to the processing activities for sale to the ultimate consumers.

13.1.2.2 *Commission Agent*

Commission agent is an intermediary person handling Vegetables in bulk quantities and sold to the wholesale units.

13.1.3 Marketing Cost

In the present study, marketing cost is defined as the actual expenses incurred by the producers and other agencies such as Wholesalers, Wholesalers-cum-Exporters,

Wholesalers - cum – Retailers, and the Retailers for performing their functions in the movement of Vegetables from the producers to the ultimate consumers.

13.1.4 Marketing Margin

In the present study, marketing margin is defined as the difference between the consumers' price and the producers' price of Vegetables for an equivalent quantity. It included the marketing costs and the profit margin to the intermediaries involved in the marketing process of Vegetables.

13.1.5 Price Spread

Price Spread in general, is referred to as the difference between the price paid by the Consumer and the Price received by the Producers per unit of the commodity. Price Spread analysis would estimate the share of different market intermediaries in the consumer rupee and this would often facilitate the understanding of the relative efficiencies otherwise of alternative channels in marketing.

To analyze the Price Spread in the distribution of Vegetables, the following formula is used.

$$Ps = (Pp / Cp) \times 100$$

Where,

Ps = Producer's share in the consumer money expressed in per cent

Cp = Consumer's Price per unit of Vegetables in Rupees

Pp = Producer's Price per unit of Vegetables in Rupees. Similar approach was used by Sekhar *et.al.* (1998).

The price spread analysis involved computation of different marketing cost and marketing margin at each stage and their expression as a percentage to the consumer's rupee. The various costs incurred in the marketing process were considered for each of the identified channels and separate price spread was worked out. The profit margin for each market functionaries in the different marketing channels was computed by subtracting the price paid and the marketing cost incurred by them from the price received by him on the sale of the commodity. Thus the marketing costs and marketing margins were distinguished

13.1.6 Estimation of Marketing Efficiency

Marketing Efficiency is the degree of Market performance. The movement of goods from producer to the ultimate consumers at the lowest possible cost consistent with the provision of services desired by the consumer is termed as marketing efficiency or efficient marketing. To analyze the marketing efficiency, the formula adopted by Shepherd is used

$$ME = (V/I) - 1$$

Where,

ME = Marketing Efficiency Index

V = Value added to the output in the marketing system. Here it is referred to as the consumers' price per kg of Vegetables

I = Input used in the marketing process. Here it is referred to as the Marketing Costs incurred per kg of Vegetables. Similar methodology was used by Sekhar *et.al.* during 1998.

The lower the value of marketing cost higher would be the efficiency

CALKIN'S INDEX

Calkin's index of marketing efficiency was estimated using the following formula:

ME=1+ [Sum of profit or margin/Sum of Marketing Cost]

The lower the value of the index, higher would be the efficiency.

13.2 MARKETING OF VEGETABLES

Marketing of vegetables is referred to the business activities that are involved in transferring the Vegetables from the primary producer to the ultimate consumer which offered time, place, form and possession utilities to the user.

With respect to marketing of vegetables, the name of the organizations involved in trading the vegetables produced from the Precision Farms, price per unit of the vegetables prevailed during before and after implementation of precision farming practices are also assessed by interviewing the farmers and the same are appropriately documented in this chapter.

13.2.1 Marketing of Brinjal in Tamil Nadu

Marketing is the process of transferring the farm produce from producer - farmer to the ultimate consumers. This section deals with marketing channels through which the produce is transferred, price spread and marketing efficiency among the different channels. Price Spread addresses on the producers' share in the Consumers' Money. Whereas, marketing efficiency addresses on the efficiency aspects that which channel is transferring the commodity at the least marketing cost and fulfills different role players by meeting their margin and gets satisfied.

13.2.2 Marketing Channels Prevalent in Distribution of Brinjal in Tamil Nadu

Here the producers are the farmers who cultivate the vegetables under precision farming and they transfer the commodity to the traders and wholesalers of native state and other states.The commodities produced under precision farming are Tomato, Brinjal and Banana. The onion is an important crop. Though it is not cultivated under precision farming in the study area, channel of distribution in respect of Onion is also presented for discussion. These details are analyzed and the results are presented in Table 13.1.

Table 13.1 Marketing Channels Prevalent in Distribution of SelectVegetables and Fruits

Sl.No	**Channel Number**	**Marketing Channel Prevalent in Marketing of Jasmine**
Tomato		
01	I	Producers → Commission Agent → Cart Vendors → Consumers
02	II	Producers →Commission Agent → Wholesalers → Retailers → Consumers
03	III	Producers → Farmers' Market at (Periyakulam and Theni) → Consumers
Brinjal		
01	I	Producers →Primary Wholesalers (Kochi) →Secondary Wholesalers → Retailers → Consumers

Sl.No	Channel Number	Marketing Channel Prevalent in Marketing of Jasmine
02	II	Producers →Commission Agent → Primary Wholesalers (Kochi) →Secondary Wholesalers→ Wholesaler cum Retailers →Retailers → Consumers
Onion		
01	I	Producers →Commission Agent →Primary Wholesalers (Theni and Oddanchathiram) → Secondary Wholesalers (Other Towns) → Wholesaler cum Retailers → Retailers → Consumers
02	II	Producers → Commission Agent → Retailers → Consumers
03	III	Producers → Commission Agent → Cart (Street) Vendors → Consumers
Banana		
01	I	Producers → Preharvest Contractors → Primary Wholesalers (Theni) → Secondary Wholesalers (Bengaluru) → Wholesalers (Chennai) → Retailers → Consumers
02	II	Producers → Commission Agent → Primary Wholesalers (Theni) → Secondary Wholesalers (Chennai) → Retailers → Consumers
03	III	Producers → Preharvest Contractors → Primary Wholesalers (Theni) → Secondary Wholesalers (Chennai) → Wholesaler Cum Retailers (Palamudhir Nilayam) → Consumers
04	IV	Producers → Preharvest Contractors → Primary Wholesalers (Theni) → Wholesalers cum Retailers (Palamudhir Nilayam) → Consumers
05	V	Producers → Preharvest Contractors → Primary Wholesalers (Theni) → Cart (Street) Vendors → Consumers
Guava (Jennifer 2014)		
01	I	Producers → Local Traders → Commission Agent → Wholesalers → Retailerss → Consumers
	II	Producers →→ Retailerss → Consumers
	III	Producers → Commission Agent → Retailerss → Consumers

Sl.No	Channel Number	Marketing Channel Prevalent in Marketing of Jasmine
	IV	Producers → Local Traders → Wholesalers → Retailerss → Consumers
	V	Producers → Commission Agent → Wholesalers → Retailerss → Consumers

Table 13.1 revealed that the vegetables are reaching the consumers of regional and national through some intermediaries in different marketing channels. In respect of Tomato, three important channels could be found out. Brinjal is being distributed through two important marketing channels. Onion is marketed through three marketing channels. Banana is marketed through five important marketing channels. Among these important channels, only the Channel I and III in respect of Tomato, Channel – I in respect of Brinjal, Channel – III in respect of Onion alone are traceable from producer level to the consumer level and the price prevailed among the different functionaries were analyzed and the results are discussed in the section dealing with price spread. The first channel identified in respect of Tomato is Producers → Commission Agent (Theni and Periyakulam) →Cart Vendors →Consumers and the same is designated as Channel - I. Here, the commission agent at Periyakulam of Theni District used to distribute the Tomato to the Cart Vendors on Commission basis. The Commission charges collected are Rs 10 for every hundred rupees worth of commodity sale. The cart vendors distribute the tomato to the ultimate consumers at their door step.

In respect of Brinjal, the traceable channel identified is designated as Channel – I which is delineated as follows.

Producers →Primary Wholesalers (Kochi) →Secondary Wholesalers → Retailers → Consumers

The Brinjal drawn from Precision Farming was not distributed locally as it commands good price in the state of Kerala and Bengaluru and hence the traders visit the farmers' field and they take the produce to their state for distribution to the consumers.

In respect of Onion, the Channel – III alone is traceable and hence the details of channel are furnished as follows.

Producers → Commission Agent → Cart (Street) Vendors → Consumers

The onion produced in the Periyakulam Block is brought to the Periyakulam Commission mundies. The open auction is conducted in the evening and the vendors use to visit the mundies and participate in the auction and take the title of the goods at their desire and distribute the same to the ultimate consumers.

13.3 PRICE SPREAD IN MARKETING AND DISTRIBUTION OF VEGETABLES

Before addressing the price spread, let us discuss the concepts which are very often used in estimation of price spread of vegetables becomes important

Price spread is said to be a composite of various costs incurred and margin of intermediaries in the various processes, such as those of assembling, processing, storage, transport and retailing. How these are varying between the identified marketing channels in respect of marketing of select vegetables is analyzed and the results are presented in Table 5.25 to 5.27.

13.3.1 Price Spread in Tomato Marketing through Channel – I

Tomato is distributed through three known channels. Out of these three, only one channel is traceable and hence the price spread for the Tomato Marketing Channel – I, is worked out and the details are presented in Table 13.2

Table 13.2 Price Spread in Tomato Marketing through Channel –I

Producers →Commission Agent (Periyakulam) →Cart (Street) Vendors → Consumers

Sl. No	Particulars of Cost	Amount (Rs/Kg)	Per Cent
Producer			
01	Producers' Price	08.00	44.44
1.1	Marketing Cost		
1.1.1	Grading Charges	00.60	
1.1.2	Transportation Charges	01.30	
1.1.3	Commission Charges Mundies	00.70	
1.1.5	Miscellaneous Charges	00.40	
1.1	Marketing Cost at Producers' Level	03.00	16.67
1.2	Marketing Margin	00.00	

Sl. No	Particulars of Cost	Amount (Rs/Kg)	Per Cent
02	Cart (Street) Vendors		
2.1	Transport Charges	00.40	
2.2	Grading Charges	00.60	
2.3	Miscellaneous Charges	01.00	
2.1	Marketing Cost to the Cart Vendor	02.00	11.11
2.2	Marketing Margin to the Cart Vendor	05.00	27.78
03	Procurement Price by the Consumers	18.00	100.00

Table 13.2 revealed that the Producers' Price for Tomato was arrived at Rs 8 per kg which is accounted for 44 per cent of the Consumers' Price. The Consumers' Price is arrived at Rs 18 per kg. The total marketing cost incurred in the channel was arrived at Rs 5 per kg and the net marketing margin was arrived at Rs 5 per kg which is accounted for 28 per cent to the consumers' price.

13.3.2 Price Spread in Onion Marketing through Channel – III

Onion is distributed through three known channels. Out of these three, only one channel is traceable and hence the price spread for the Onion Marketing Channel – III, is worked out and the details are presented in Table 13.3.

Table 13.3 Price Spread in Onion Marketing Through Channel –III

Producers →Commission Agent (Periyakulam) →Cart (Street) Vendors → Consumers

Sl. No	Particulars of Cost	Amount (Rs/Kg)	Per Cent
	Producer		
01	Producers' Price	12.00	60.00
1.1	Marketing Cost		
1.1.1	Packing Charges	00.40	
1.1.2	Transportation Charges	01.20	
1.1.3	Commission Charges Payable in Commission Mundies	00.70	
1.1.5	Miscellaneous Charges	00.70	

Sl. No	Particulars of Cost	Amount (Rs/Kg)	Per Cent
	Marketing Cost at Producers' Level	03.00	15.00
1.2	Marketing Margin	00.00	
02	Cart (Street) Vendors		
2.1	Transport Charges	00.30	
2.2	Cleaning and Processing Charges	00.40	
2.3	Miscellaneous Charges	00.30	
2.0	Marketing Cost to the Cart Vendor	01.00	05.00
	Marketing Margin to the Cart Vendor	04.00	20.00
03	Procurement Price by the Consumers	20.00	100.00

Table 13.3 revealed that the Producers' Price for Onion was arrived at Rs 12 per kg which is accounted for 60 per cent of the Consumers' Price. The Consumers' Price is arrived at Rs 20 per kg. The total marketing cost incurred in the channel was arrived Rs 4 per kg and the marketing net margin was arrived at Rs 4 per kg which is accounted for 20 per cent to the consumers' price. During the peak season of demand, the consumers' price of Onion shoots up to even Rs 80 to Rs 100 per kg. During this time, the marketing cost and margin will be varying differently and the price prevailed and elsewhere is found to be a normal period in the blocks of Periyakulam and Theni of Theni District.

13.3.3 Price Spread in Brinjal Marketing through Channel – I

Brinjal is distributed through two known channels. These channels are operated in other states. Out of these two, only one channel is traceable and hence the price spread for the Brinjal Marketing Channel – I, is worked out and the details are presented in Table 13.4.

Table 13.4 Price Spread in Brinjal Marketing through Channel –I

Producers →Primary Wholesalers (Kochi) →Secondary Wholesalers→ Retailers → Consumers

Sl. No	Particulars of Cost	Amount (Rs/Kg)	Per Cent
01	Producer		
01	Producers' Price	08.00	26.67
1.1	Marketing Cost		
1.1.1	Grading and Processing Charges	00.30	
1.1.2	Assembling and Packing Charges	00.65	
1.1.3	Cost of Labor in Loadingand Forwarding Charges	00.60	
1.1.4	Miscellaneous Charges to the Producer	00.45	
1.1	Marketing Cost at Producers' Level	02.00	06.67
1.2	Marketing Margin	00.00	00.00
02	Primary Wholesaler		
2.1	Procurement Price by the Primary Wholesaler	10.00	33.33
2.2	Marketing Cost		
2.2.1	Transportation Charges to the Kochi Market	02.60	
2.2.2	Auctioning Charges	00.40	
2.2.3	Miscellaneous Charges including Rent	00.50	
	Marketing Cost at Primary Wholesaler Level	03.50	11.67
2.3	Marketing Margin to the Primary Wholesaler	04.50	15.00
03	Secondary Wholesaler		
3.1	Purchase Price Incurred by the Secondary Wholesaler	18.00	60.00
3.2	Marketing Cost		
3.2.1	Loading and Unloading Charges	00.50	
3.2.2	Transportation Charges	00.85	
3.2.3	Miscellaneous Cost including Rent	00.65	
3.2	Marketing Cost to the Secondary Wholesaler	02.00	06.67
3.3	Marketing Margin to the Secondary Wholesaler	03.00	10.00

Sl. No	Particulars of Cost	Amount (Rs/Kg)	Per Cent
3.4	Sale Price to the Retailer	23.00	76.67
04	Retailer		
4.1	Procurement Price by the Retailer	23.00	76.67
4.2	Marketing Cost to the Retailer		
4.2.1	Transportation Charges	00.80	
4.2.2	Cost of Labor in Distribution of Vegetables	00.55	
4.2.2	Miscellaneous Charges Including Rent	00.65	
4.2	Marketing Cost to the Retailer	02.00	06.67
4.3	Marketing Margin to the Retailer	05.00	16.67
5	Sale Price to the Consumer and the Consumer's Price	30.00	100.00

Table 13.4 revealed that the Producers' Price for Brinjal was arrived at Rs 8 per kg which is accounted for 27 per cent of the Consumers' Price. The Consumers' Price is arrived at Rs 30 per kg. The total marketing cost incurred in the channel was arrived at Rs 9.50 per kg and the net marketing margin was arrived at Rs 12.50 per kg which is accounted for 42 per cent to the consumers' price. In respect of Brinjal, there is no much price variation. The consumer price varies from Rs 30 to Rs 40. The variation price over the normal period will be only Rs 10 per kg. Hence, there are no much changes in the marketing cost and the net margin gained by the intermediaries in the channel.

To sum up, the Producers' Price for Tomato was arrived at Rs 8 per kg which is accounted for 44 per cent of the Consumers' Price. In respect of Onion, the Producers' price is arrived at Rs 12 per kg which is accounted for 60 per cent of the Consumers' Price. In respect of Brinjal, the Consumers' Price is arrived at Rs 8 per kg which is accounted for 27 per cent of the consumers' price. The total marketing cost incurred in the respective channel in respect of Tomato, Onion and Brinjal was arrived at Rs 5, Rs 4 and Rs 9.50 per kg and the net marketing margin was arrived at Rs 5, Rs 4 and Rs 12.50 per kg which are respectively accounted for 28 per cent, 20 per cent and 42 per cent to the consumers' price. Higher net marketing margin with respect to Brinjal is mainly due to the number of intermediaries involved in facilitating the trade of Brinjal to meet the consumers of different destinations in Kerala.

13.3.4 Price Spread in Guava Marketing through Channel –I

Table 13.5 Price Spread in Guava Marketing through Channel –I

Producers →Local Traders → Commission Agent → Wholesalers→ Retailers → Consumers (Source: Jennifer, 2014)

Sl. No	Particulars of Cost	Amount (Rs/Kg)	Per Cent
01	Producer		
01	Producers' Price	35.00	60.34
1.1	Marketing Cost		
1.1.1	Transportation Charges	00.84	01.45
1.1.2	Loading Charges	01.10	01.88
1.1	Marketing Cost at Producers' Level	01.94	03.33
1.2	Net Price Received	33.06	57.00
02	Local Trader		
2.1	Procurement Price	35.00	60.34
2.2	Marketing Cost		
2.2.1	Transportation Charges	00.40	
2.2.2	Unloading Charges	00.60	
	Marketing Cost Local TraderLevel	01.00	01.72
2.3	Marketing Margin to the Local Trader	04.00	06.90
03	Commission Agent		
3.1	Purchase Price	40.00	68.97
3.2	Marketing Cost		
3.2.1	Packaging Charges	00.40	
3.2.2	Transportation Charges	00.65	
3.2.3	Loading Charges	00.70	
3.2	Marketing Cost to the Commission Agent	01.75	03.02
3.3	Marketing Margin	04.25	07.33
3.4	Sale Price	46.00	79.31
04	Wholesaler		
4.1	Procurement Price by the Wholesaler	46.00	79.31

Sl. No	Particulars of Cost	Amount (Rs/Kg)	Per Cent
4.2	Marketing Cost to the Wholesaler		
4.2.1	Transportation Charges	00.75	01.29
4.3	Marketing Margin to the Wholesaler	04.75	08.19
4.4	Sale Price	51.50	88.79
05	Retailer		
5.1	Purchase Price by the Retailer	51.50	88.79
5.2	Marketing Cost to the Retailer	00.80	01.38
5.3	Marketing Margin to the Retailer	05.70	09.83
6	Sale Price to the Consumer and the Consumer's Price	58.00	100.00

Table 13.5 revealed that the Producers' Price for Guava was arrived at Rs 35 per kg which is accounted for 60 per cent of the Consumers' Price. The Consumers' Price is arrived at Rs 58 per kg. The total marketing cost incurred in the channel was arrived at Rs 6.20 per kg and the net marketing margin was arrived at Rs 16.30 per kg which is accounted for 28 per cent to the consumers' price. In respect of Guava, there is no much price variation. The consumer price varies from Rs 50 to Rs 60. The variation in price over the normal period will be only Rs 10 per kg. Hence, there are no much changes in the marketing cost and the net margin gained by the intermediaries in the channel. Whatever the variation in price, the intermediaries are keen on generating their share from the consumers' rupee or by lowering the purchase price from the producers.

In respect of organizations, the producer price in respect of Guava is almost fixed and there are little variation in the consumers' price

13.4 MARKETING EFFICIENCY IN VEGETABLES MARKETING

Marketing efficiency is defined as the ratio of output (Consumers' Price) to the input (Marketing Cost). In this study, the consumer price is varied from channel to channel and hence the same were worked out using the Kalkins Index and the results are presented in respect of Tomato (Channel – I), Onion (Channel – III) and Brinjal (Channel – I) in Table 13.6.

Table 13.6 Marketing Efficiency in Distribution of Vegetables

Marketing Channel	Value of Output (Consumers' Price)	Input (Marketing Cost)	Marketing Efficiency Ratio (O/I)
Shepherd Method			
Tomato (C-I)	18.00	05.00	03.60
Onion (C-III)	20.00	04.00	05.00
Brinjal (C-I)	30.00	09.50	03.16
Kalkins Index			
Marketing Channel	**Marketing Margin**	**Marketing Cost**	**Marketing Efficiency**
Tomato (C-I)	05.00	05.00	02.00
Onion (C-III)	04.00	04.00	02.00
Brinjal (C-I)	12.50	09.50	02.32

Table 13.6 advocated that an increase in the marketing efficiency ratio would represent improved efficiency and vice versa. A reduction in the cost for the same level of satisfaction or an increase in satisfaction at a given cost would result in an improvement of efficiency. When comparing the data delineated in the Table 5.28 one could understand that the Onion Marketing Channel – III found to be the efficient one. Because, the Channel – III fulfilled the requirements of functionaries in the marketing channel and keep them satisfied besides meeting the higher percentage share to the producer from the consumers' price.

13.5 MARKETING INSTITUTIONS FOR HORTICULTURAL PRODUCE

India amended APMC act and allowed cooperatives and private entrepreneurs to set up special markets for fruits, vegetables and flowers. India has opened up the economy during 1990, a number of innovative liberalized markets for Fruits and Vegetables in the WTO regime to eliminate middlemen and facilitate direct contact between producer and consumers was made possible. Economic reforms led India to opened up post globalization, a number of new private retail liberalized markets for Fruits and Vegetables came into existence. It is necessary to study and understand these markets and their functions, marketing strategies, their performance and success. These details are presented in Table 13.7.

Table 13.7 Details of Marketing Institutions for Horticultural Produce in India

Sl. No	Details of Marketing Institutions	Status / Key Functions / Mandate / Strategies
1	Rythu Bazaar	Started in Andhra Pradesh To avoid the exploitation of both farmers and the consumers by the Middlemen Right now, 96 Rythu Bazaars are functioning
2	Farmers' Market or Uzhavar Sandhai	Started in Tamil Nadu by Dr. M. Karunanithi, the Then Chief Minister of Tamil Nadu, Chennai To avoid the exploitation of both farmers and the consumers by the Middlemen Right now, 179 Farmers Markets are functioning in Tamil Nadu. Out of which only 103 are found to be effective in their operations. Modern Weighing Machines are provided Currently, the failure of Monitoring Mechanisms resulted in entry of different traders in the name of farmers are visible. Example: Coimbatore Market, The prices of vegetables in the Farmers' Market are almost tallying the prices of Retail Market in many vegetables.
3	Apni Mandi	Started in Punjab. It does away with the middlemen Apni Mandis are similar to the Saturday Markets of UK and USA
4	Hardaspar Vegetable Market	It is a model market for direct marketing of vegetables in Pune City It is an ideal market for vegetables Market is provided with Modern Weighing Machines as practiced in Tamil Nadu

Sl. No	Details of Marketing Institutions	Status / Key Functions / Mandate / Strategies
5	Shetkari Bazaar	It was established in Maharashtra State for Fruits and Vegetables It will eliminate middlemen, links producers and consumers directly, reduce the price spread, Enhance producers' share. Enhancement of Farmers' Income
6	Krushak Bazaar	Established in Odisha during 2001 Purpose is to empower farmer to compete effectively in the open market to get a remunerative price and ensure products at affordable price to the consumers
7	Mother Dairy Booths	Basically handling Milk in Delhi But it was asked to handle retailing of vegetables Mother Dairy management has opened retail outlets in New Delhi for providing vegetables to consumers at reasonable price.
8	Cooperative Marketing Society	Few Successful Cooperative Marketing Societies for fruits and vegetables are there. Example: Maha-Grape growers cooperative federation, Maharashtra Cooperative Marketing of Pomegranate, Banana in Jalgaon District Vegetable Cooperatives in Thane District Milk Cooperatives in Maharashtra and Tamil Nadu Nilgiri Cooperative Marketing Society in Tamil Nadu for Hill Vegetables

Sl. No	Details of Marketing Institutions	Status / Key Functions / Mandate / Strategies
9	Contract Marketing	Essentially is an agreement between Farmer – Producer and the agribusiness firms to produce certain Pre-agreed quantity and quality vegetables at a particular price and time. This is an important initiative for reducing transaction costs by establishing farmer processor linkages. **Example**: Marigold Farmers; Tomato Growing in Punjab and Rajasthan
10	Safal Market	NDDB started a fruits and vegetable unit of SAFAL at Delhi was one of the first Fruits and Vegetables retail chain NDDP has setup an alternate system of wholesale Markets in Bangalore as a pilot project It is a move to introduce a transparent and efficient platform for sale and purchase through producer organization.
11	Forward and Future Markets	Price stabilization and risk management Regulated through forward contracts act, 1952 Narrows the marketing, storage and processing margins
12	Commodity Exchanges	National multi-commodity exchange of India Limited (2002) National Commodity and Derivatives Exchange of India 2003 at Mumbai
13	E-Trading	Trading based on buying and selling of Electronically Warehouse receipts and providing latest price information The services can be accessed by Internet, Telephone
14	Food Retail Supermarkets	Started in 1990 and Early 2000 Higher Standards Lower Prices

Among the 14 models of vegetable marketing in different states, basically the concept of farmers' market is capable of providing 15 per cent lesser price over the retail market to the consumers and at the same time 20 per cent higher price for the vegetables over the wholesale price in the previous day as practiced in Tamil Nadu. However, because of political vindictiveness and poor monitoring mechanisms, the traders have been intruded and almost the prices are on par with the retail in certain commodities. What is needed in this situation is, tightening the monitoring mechanism, periodical checking of traders and the farmers for their original identity, price prevailed on the previous day in wholesale market etc. It seems that few of the personnel in the market are favoring the traders and hence the same should be brought to halt. Besides these, the Private markets are also actively enrolled and they do large scale business in retailing and the details are presented in Table 13.8.

Table 13.8 Growth and Role of India's New Liberalized Private Marketing Institutions

Year	Name of the Retail Store	Ownership	Format	Number of Outlets	Location
1971	The Nilgiris	Local+Actis, UK	Supermarkets, Convenient Store	Around 500 Numbers	Major Cities in South India
1986	Trinethra	Aditya Birla Grup	Supermarkets, Convenient Store	170	South India
1994	Margin Free	Local + Consumer Protection and Grievances Society	Discount Stores, Super Markets	350	Kerala, TN, Karnataka
1996	Spencers	RPG Group	Hypermarket, Supermarkets	Around 140	Major Cities in South India

Year	Name of the Retail Store	Ownership	Format	Number of Outlets	Location
1997	Subhiksha	Subhi Trading Private Limited	Discount Stores	Around 1200	TN, AP, Pondicherry, Delhi
1999	Food World	Subsidary of Dairy Farm International	Supermarket	Around 60	Bangalore, Hyderabad
2001	Trumart	Pyramid Retail	Supermarket, Convenience Stores	Around 100	Maharashtra, Gujarat, Bangalore, Chennai, Hyderabad
2002	Food Bazaar	Future Group	Hypermarkets, Supermarkets	Around 250	National Metros and Large Cities
2003	Metro Cash and Carry	Foreign Metro AG, Germany	Wholesale Cash and Carry	Around 20	Bangalore, Hyderabad, Mumbai, Kolkatta, Chennai
2004	My Dollar Store	Franchisee of My Dollar Store of USA	Convenience Stores	Around 400	Nationwide
2004	Shoprite	Subsidary of Shoprite, South Africa	Hypermarket	03	Mumbai
2004	Star India Bazaar	Trent Division of TATA Group	Hypermarkets	23	Nationwide

Year	Name of the Retail Store	Ownership	Format	Number of Outlets	Location
2006	Reliance Retail	Reliance India Limited	Hypermarkets, Convenience Stores	Around 2500	Nationwide
2006	Spinach	Wadhwan Retail	Supermarkets	1500	Nationawide
2007	Max Hyper	Joint Venture of SPAR International	Hypermarkets	25	Nationwide
2007	Bhartis	Bharti Enterprises	Hypermarket, Supermarkets	NA	Nationwide
2007	Bhart Walmart	Joint Venture of Walmart, USA	Wholesale, Cash and Carry	NA	Nationwide
2007	Birla More	Birla group	Supermarkets	Around 1000	Nationwide

NA: Not Available

(Source: USDA, Foreign Agricultural Service 2006, updated data from various sources)

Table 13.8 revealed that around 8300 Supermarkets / Hypermarkets in private sector other than the government and cooperative marketing institutions are functioning in India promoted by 18 different private organizations had a joint venture approach in the retail marketing of goods and services. Among the 18 private marketing institutions, the Reliance, Subhiksha, Spinach and More groups had the highest share which is accounted for 75 per cent of the total retail stores in their hands. Besides these private organizations at national level, there were many regional marketing institutions in the name of "Pazhamudhir" and in other names are also promoting the sale of goods and services. In addition, the grocery stores in the local are also equally doing their business and trying to procure the grocery items by forming a group and nominating a representative for bulk procurement and redistribution to the concerned retail outlets. Sometimes, this role is performed on rotation basis.

13.6 MARKETING MODELS AND SUPPLY CHAINS IN FRUITS AND VEGETABLES

We have seen the role of different private sector institutions whom have taken a lead role in marketing of goods and services to the consumers. Both private and Government role are existing in the marketing of goods and services. Let us discuss the supply chain related issues and their differential roles in mobilization of fruits and vegetables for trading from the farmer producers. These details are analyzed and the results are presented in Table 13.9.

Table 13.9 Marketing Models of Fruits and Vegetables under Different Supply Chains

Sl. No	Particulars	Farmers' Market	Reliance	Pazhamudhir
1	Marketing Model	Government acts as a Facilitator	Directly Purchase from Farmer Producers	Directly Purchase from Farmer Producers
2	Farmers' Share in the Consumers' Rupee	Prices are fixed around 20 per cent higher than the wholesale market price	Higher than the Farmers' Market	Moderate Share to the farmers. It depends on the produce bought from farmers
3	Role of Middlemen	Partially and Indirectly existing.	Complete Eliminatin of Middlemen	Complete Elimination of Middlemen
4	Consumer Status	Below Poverty Line, Middle Class Consumers	Higher Income Group, Software Professionals, Metro People	Higher Income Group and Upper Middle Class Category
5	Organization	Government	Private	Private

Sl. No	Particulars	Farmers' Market	Reliance	Pazhamudhir
6	Grading	No Grading	Graded Produce	Graded Produce
7	Pricing Status	15 per cent lower price than the retail price	Lower Price and Higher Price for different Grades	Low Price for Second Grade and High Price for First Grade Items

Among the supply chains listed in Table 13.9, the one practiced by Government is found to be effective when compared to the private supply chains of Reliance and Pazhamudhir. Here, Government is acting as a facilitator in providing all the infrastructure to the farmers who migrate along with their vegetables and fruits to the market yard allotted to them on regular basis between 5 AM to 10 AM. The Marketing Yard (Uzhavar Sandhai) officials provide electronic weighing machine, every day new allotment of shops on rotation basis after verifying the identity of farmers who had registered with the office of the Agricultural Officer / Administrative officer of the Uzhavar sandhai. The farmers Market has attracted lower price when compared to the private outlets manned by Reliance and Pazhamudhir. In private the produce are graded and fixed as first, second and third grades and accordingly the produce were priced. These will be comparatively higher than the price prevailed in the private markets. The Farmers' Share in the Consumers' Rupee is also advantageous in both private as well as government. Otherwise, the promoters cannot survive long in the marketing of fruits and vegetables.

13.7 CONSTRAINTS IN MARKETING OF HORTICULTURAL PRODUCE

Among the production problems in horticultural produce, poor access to inputs, high fuel costs, pests and disease control, high input costs, lack of finance, shortage of skilled labour for processing and value addition, and breakdown of irrigation systems were found to be the major production problems. Among the marketing problems identified in horticultural marketing, excess production of tomato like vegetables during the season and some time deficit supply of tomato like produce called for regulation of cropping in each zone of production and high transportation costs, poor processing infrastructure, absence of price policy during the glut situations, high competition with imports, and lack of marketing skills with the farmers and farmer producer organizations were the major marketing problems. Subsidized fuel price, increase area under cultivation, effective implementation of efficient cropping plan, easy access to finance, building new market places, proper dissemination of market information, establishment of

Agro Processing and Service Centers, declaration of support price for the horticultural produce,establishing sanitary and phytosanitary certification and its related training to the farmer producers and the Farmer Producer Organizations and revision of existing import policy were among the suggestions to minimize the marketing problems.

13.8 SUMMARY

This Chapter has highlighted the Price Spread, Marketing Efficiency, in respect of select fruits and vegetables produced out of Precision Farming Technologies. Marketing of Fruits and vegetables are being practiced by both Government and Private organizations whom are spread in different states of the country. The Private marketing organizations are numbering 18 have their retail stores in different name as Supermarkets or Hypermarkets established in 8300 locations. Among these locations, the Reliance, More, Spinach and Subhiksha alone had shared 75 per cent to the total retail outlets in India. Among the several channels considered, efficient marketing channel is visible in respect of Onion which is capable of generating higher share to the producer from the consumers' rupee. In India, organized retailing for FF&V is gaining a lot of impetus with huge investments by leading Indian corporations. Traditional markets are making way for new formats such as departmental stores, hypermarkets, supermarkets and specialty stores and western-style malls. From the above, one could infer that the Farmers' Market are found to be efficient network to meet the requirements of poor farmers and poor consumers. Whereas, the Reliance, Subhiksha, Palamudhir are fulfilling the requirements of upper middle class and the higher income group in respect of graded fruits and vegetables.

References

1. Sekhar, C. (2016) "Enrepreneurship Development in Agriculture" (MJP Publishers: Chennai)

2. Jennier, F.M. (2014). "Value Chain Analysis of Guava in Dindigul District of Tamil Nadu", (Unpublished M.B.A. Thesis Submitted to the Department of Agricultural Rural Management, Tamil Nadu Agricultural University, Coimbatore).

Web References

- http://crida.in/agrl_martng/ISAM/PDF%20FILES/T-II/Dastagiri.pdf

CHAPTER XIV

Trading of Vegetables from India

Dr. C. Sekhar and Dr. A. Vidhyavathi

Objectives

- To highlight the details of Export of Principal Horticultural Produce from India to the stakeholders of the book
- To spell out the organizations involved in promotion of Horticultural Trade in India

Organization of the Chapter

- The Concepts on Export Trade
- Trading Practices of Vegetables Prevalent in India
- Export of Horticultural Produce from India
- Organizations Involved in Promotion of Trade
- Summary

14.1 THE CONCEPTS ON EXPORT TRADE

What is Trade?

Before modern currencies were invented, bartering was a common way to buy and sell goods, that is, to exchange them directly for the sake of mutual benefits. Some 10,000 years B.C., exchanging cacao leaves for shiny turquoise stones was as common as selling your car and buying a new one today.

Bartering, just like trading later on, was based on two main factors. They are products that were needed by others and negotiation between seller and buyer to reach an understanding and close the deal.

What is International Trade?

International trade is the exchange of goods and services across national borders. In most countries, it represents a significant part of GDP. While international trade has been present throughout much of history, its economic, social, and political importance have increased in recent centuries, mainly because of industrialization, advanced transportation / globalization, multinational corporations and outsourcing of need based resources.

Empirical evidence for the success of trade can be seen in the contrast between countries such as South Korea which has adopted a policy of export oriented industrialization and India, which historically had a more closed policy. South Korea has done much better by economic criteria than India over the past fifty years, though its success also has to do with effective state institutions. In this situation, trade between different countries become a recognized and significant one in terms of food and non food items.

14.2 TRADING PRACTICES OF VEGETABLES PREVALENT IN INDIA

Vegetables and fruits become the most important one which are produced in large scale by India and China to feed the ever increasing population through appropriate trading practices which are acceptable to both ends. Vegetables are vital to the general good health of human beings, providing essential vitamins and minerals, dietary fiber, and phytochemicals, and reducing risk from dangerous diseases and other medical conditions. Vegetables are grown world-wide in almost 200 countries. A world vegetable survey indicated 392 vegetable crops cultivated worldwide. Most of the vegetables are marketed fresh with only a small portion processed.

Vegetable consumption is also rising, reflecting the consumer's increased income, desire for diversity, and awareness of nutritional benefits. At the same time, consumers have a rising concern with product safety issues. Traditional marketing practices are giving way in developing countries to the more modern practices of the developed countries: supermarkets, long-distance shipping, international marketing, various processing and packaging practices, safety and quality standards, constant product supply, and product diversity. These changes have increased pressure on traditional, small, and poor farmers to keep up, with the requirements of good seed, efficient practices, hiring help beyond the family, market awareness, and the ability to provide safe and high quality vegetables.

The system of exchanging the vegetables and fruits between the countries by accepting the one's need and conditions are otherwise termed as export. Trading of Vegetables included both import and export from India. In this section, the export of Vegetables in respect of select vegetables is analyzed and the details are presented in this section.

14.3 EXPORT OF HORTICULTURAL PRODUCE FROM INDIA

According to Wikipedia (2013), Export can be defined as the act of sending a commodity abroad for trade or sale. Export is nothing but the foreign demand for goods produced by the home country.

According to On-Line Dictionary (2013), Export is to send or transport a commodity abroad, especially for trade or sale. Export is defined as to cause the spread of an idea in another part of the world. In this respect, the Export is defined as the goods and services that are produced in India are transmitted to the other country by fulfilling their prescriptions on that commodity.

On the back of record production, India's, second largest producer of fruits and vegetables in the world, horticulture exports has helped the country to earn Rs 14,000 crores in 2011-12, according to the Ministry of Agriculture and Food Processing Industries.

India's horticulture production has crossed all time high of over 240 million tonnes and Sharad Pawar, Agriculture Minister, has annouced 2012 as the "Year of Horticulture" to bring horticulture development in the country to the centre stage.

According to Pawar, India have achieved considerable success on the production of horticultural output, but the supply chain issues still need to be addressed in a big way. The Vegetable Initiative for Urban Clusters is an important step in this direction.

The scheme is being implemented with focus on sustained supply of good quality vegetables to urban centres having a population of one million and above and also lays special emphasis on promoting protected cultivation of vegetables in green houses and shade-net houses for enhancing its productivity.

Formation of Farmer Producer Organizations (FPO) and linkage with financial institutions through Aggregators is an innovative feature of the Scheme, Over 62,000 farmers have already been mobilized into over 3000 Farmer Interest Groups (FIG) and 50 Farmers Producers Organizations (FPOs) are also under this scheme.

Meanwhile, Floriculture is another field helping India in generating quick returns per unit of area for the farmer.With the production level at nearly 7000 million cut flowers, there has been significant improvement in their production. These enjoy good demand in the domestic and international markets.

India is producing almost all types of Vegetables and other horticultural produce. They are from leafy vegetables to gourds, flowers and fruits. Many countries are exporting Vegetables across the globe. It is important to identify the quantity of Vegetables and other horticultural output exported from India to other countries in terms of value. In this context, the status of horticultural output exported from India and the same are presented in Table 14.1.

Table 14.1 Export of Horticulture Produce in India (Quantity in Tonnes; Value – Rs in Lakhs)

Sl.No	Name of the Horticulture Produce	2010-11		2016-17	
		Quantity	Value	Quantity	Value
01	**Floriculture**				
1.1	Floriculture	27776.14	28645.41	22020.33	54670.73
1.2	Fruit And Vegetable Seeds	11182.50 (00.41)	17519.52 (02.52)	11288.62	52275.17
01	**Total**	**38958.64**	**46164.93**	**33308.95**	**106945.90**
02	**Fresh Fruits and Vegetables**				
2.1	Fresh Onion	1163472.58	174155.41	2415739.06	310606.44
2.2	Other Fresh Vegetables	490914.05 (17.96)	89293.61 (12.82)	1016437.72	283202.56
2.3	Walnuts	005244.58	15650.59	2191.19	5527.26
2.4	Fresh Mangoes	059220.77	16292.13	52761.00	44366.00
2.5	Fresh Grapes	099311.83	41206.32	231116.86	206528.94
2.6	Other Fresh Fruits	253850.99	48964.74	409081.25	183592.69

Sl.No	Name of the Horticulture Produce	2010-11		2016-17	
		Quantity	Value	Quantity	Value
02	**Total**	**2072014.80**	**385562.80**	**4127327.08**	**1033823.89**
03	**Processed Fruits and Vegetables**				
3.1	Dried & Preserved Vegetables	110173.91 (04.33)	051697.09 (07.42)	86791.22	108170.99
3.2	Mango Pulp	171929.43	081400.66	130886.07	84601.79
3.3	Other Processed Fruitsand Vegetables	340067.97 (12.44)	131635.53 (18.90)	355315.51	314881.22
03	**Total**	708001.24	262119.41	572992.80	507654.00
	Grand Total (1+2+3)	3393602.88	676915.44	4733628.83	1648423.79
Percentage Change in 2016-17				39.49	143.52

(Figures in Parentheses Indicate Percentage of Vegetables to the Total Produce Export)

Table 14.1 revealed the details of export of horticultural produce from India. Among various produce exported, Floriculture, Fresh Vegetables, Dried and Preserved vegetables and other processed fruits and vegetables are found to be significant. Among these items, the percentage quantity exported between 2010-11 and 2016-17 found to be on the increase in respect of Fresh Onion, Other fresh vegetables, Fresh Grapes and other Fresh Fruits. But, the export of fresh vegetables, dried and preserved vegetables and other processed fruits and vegetables are showing a decrease in the export in terms of quantity between the year 2008-09 and 2010-11. But the same were found to be on the increase during the year 2016-17. It is inferred that these items were found to be qualitative and meeting the requirements of foreign countries for export and hence the quantity of export found to be increased. The overall export of horticulture produce between these period in terms of quantity is found to show an increase which is accounted for around 73 per cent to the total quantity of horticultural produce export. The export value between these periods is found to show an increase which is accounted for 137 per cent to the total value of export. In this circumstance, the export of vegetable seed should be enhanced by increasing the productivity of vegetables by practicing the precision farming techniques in a larger scale so that the quantity produced will be enhanced and the export possibilities could be increased. Since the demand for vegetable seeds is significant in foreign countries, the Tamil Nadu Agricultural University and the seed producing firms should concentrate more on vegetable seeds capable of fetching higher productivity may be placed for export requirements.

14.3.1 Quality Standards for Export of Red Chilli

Among the Dried and Preserved Vegetables, Chillies are one of the important and commonly exported spice all over the world and are believed to be originated in the Tropical America (Peru). Major Chilli growing countries are India, China, Pakistan, Indonesia, Korea, Turkey and Sri Lanka, Nigeria, Ghana, Tunisia, Egypt, Mexico and the US in North and Central America; Yugoslavia, Spain, Romania, Bulgaria, Italy and Hungary in Eurpope; and Argentina, Peru and Brazil in South America. Chilli Crop came to the Asian Continent during the sixteenth century with the identification of new sea routes by the Portuguese and the Spanish explorers. A large percentage of Chilli production has shifted to Asia, as the South Asian Climate suits this vegetable crop. In the recent years, high value varieties of Chilli are grown in Asia only. Chilli is cultivated in tropical and sub tropical climates mostly as a rainfed crop in India.

India is one of the major red Chilli pepper exporting countries in the world. They are marketed as dry whole pods as well as in ground form. There are more than fifty Chilli varieties grown in India. Each of these varieties differs from each other in pungency and color. The Naga jolokia and the birds eye Chillies are among the hottest Chillies in the World and the Bydagi Chillies are known for their Bright Red Color. Red Chilli peppers are mostly consumed as a whole or in powdered or flaked or paste form. Whole or ground (powdered), shall be free from living insects and shall be practically free from mould growth, dead insects, insect fragments and rodent contamination. The details of specifications for Export of Red Chilli is presented in Table 14.2.

Table 14.2 Export Specifications for Red Dry Chillies

Sl. No	Particulars of Chilli	Specifications
1	Varieties of Chilli	Sannam S4 Best, Teja Best, Wrinkle 273, Wonder Hot, Tomato Chilli, Mundu, Yellow Chilli
2	Form of Chilli	Whole Chilli with Stem, Whole Chilli Without Stem, Grounded Chilli Powder, Chilli Oleoresin, Chilli Meal
3	Color	Bright Red
4	Length	As per Grade
5	Pods without Stalk	Eight Per Cent Maximum
6	Moisture	12.50 Per Cent Maximum
7	Foreign Material	Maximum of 2 Per Cent
8	Brocken Chillies	Maximum of 7 Per Cent
9	Loose Seeds	Maximum of 2 Per Cent
10	Damaged and Discolored Pods	Maximum of 6 Per Cent
Export Specifications for Chilli Paste		
1	Appearance	Reddish Orange Colored Paste
2	Odour	Characteristic Pungent Aroma of Fresh Chilli Fruits
3	Added Preservatives	One Per Cent
4	Salt	15 Per Cent
5	TSS	20-22 Per Cent
6	P^h for 10 Per cent Dilution	4.50 – 5.00
7	Acidity	0.60 – 0.69
8	Moisture	65 – 70 Per Cent

(Source: Mohan, P. 2016)

Table 14.2 has highlighted the export specifications for Whole Red Chilli from India. It revealed that the Moisture content should not exceed 12.50 per cent of the produce ready for export and at the same time the broken chillies should not cross 7 per cent of the total consignment. The damaged and discolored pods should be less than 6 per cent of the lot and the loose seeds present in the lot and foreign material should not cross 2 per cent level in the export quality. The whole and ground chillies may vary in color from dark blackish red to orange yellow according to the variety. The pungency for the red peppers and the color value for the paprika are the most important parameters. The extractable color of the Red Chilli is usually expressed in American Spice Trade Association (ASTA) color value.

The European Economic Community Countries constitute a major market for Indian spices. Most EEC countries import spices in accordance with prevailing food laws. The Netherlands have specifications for various quality parameters of spices. Germany has prescribed tolerance levels for pesticide residues. The German specifications are most stringent. The Dutch law also prescribes maximum residual limit for pesticides in spices. The food industry and spice processors in United Kingdom follow the limit for pesticides prescribed by the German Countries. Conformance to the quality requirements of the buying countries, meeting the consumer's expectations and ability to maintain price competitiveness are the key factors which determine our survival in the international markets.

14.3.2 Export of Tomatoes from India

India is exporting vegetables to different countries. Among that, Tomato is one of the principal vegetable in the list of export of vegetables. While the general trend in the export of horticultural produce showed a positive trend in respect of quantity exported between 2010-11 and 2016-17, the export of Tomato exhibited also had showed a positive trend. The details of data in respect of Tomato export are analyzed and the results are presented in Table 14.3.

Table 14.3 Export of Tomatoes from India (Product Code Number: 07020000; Quantity in Tonnes; Value – Rs in Lakhs)

Sl.No	Name of the Country	2010-2011		2016-2017		Percentage Change in Quantity during 2016-2017
		Qty	Value	Qty	Value	
1	United Arab Emirates	33902.90	7468.60	30789.61	8948.20	-9.18

Sl.No	Name of the Country	2010-2011		2016-2017		Percentage Change in Quantity during 2016-2017
		Qty	Value	Qty	Value	
2	Bangladesh	19711.90	2048.20	28927.18	6641.75	46.75
3	Saudi Arabia	2261.50	539.40	69.61	21.29	-96.92
4	Oman	2203.60	532.00	659.23	194.60	-70.08
5	Pakistan	3708.90	340.60	190739.69	36845.47	5042.76
6	Nepal	3266.60	227.30	14512.33	1611.13	344.26
7	Maldives	2447.70	169.80	1161.76	485.69	-52.54
8	Kuwait	125.90	29.90	22.61	7.47	-82.04
9	Qatar	19.60	8.00	158.82	44.67	710.31
10	Others	535.10	14.50	157.65	5.77	-70.54
11	Total	68183.70	116.70	267198.49	54806.04	291.88

Table 14.3 revealed that the export of Tomato to different countries between the period 2010-11 and 2016-17 showed a positive trend in export which is accounted for 292 per cent to the total quantity exported between the periods. The value of Tomato export also showed a positive trend which is accounted for many fold increase when comparing the base year of 2010-11. Pakistan found to be the biggest importer of Tomato from India during 2008-09 which was reduced drastically during the year 2010-11 which is accounted for 96.32 per cent to the total Tomato export. Whereas, the Tomato export between the period 2010-11 and 2016-17 is found to be many fold increase which is accounted for 5043 per cent over the base year of 2010-11. It might be due to the failure of Tomato crop due to unfavorable climate in the country of Pakistan. Similarly, export of Tomato to the Qatar, Nepal and Bangladesh is found to be increased which are respectively accounted for 710 per cent, 344 per cent and 47 per cent increase during the year 2016-17 when comparing the base year of 2010-11.

The countries like Saudi Arabia, Oman, Kuwait, Maldives and other countries have showed a positive signals in importing the Indian Tomato during the Year 2010-11. But lateron, these countries have not imported much of Tomato from India during the year 2016-17 and their import level was found to be negative when comparing the base year of 2010-11. In general the trend in export of Tomato during the period 2008-09 to 2010-11 is totally a different picture and the same has been changed during the year 2016-17 and the overall export of Tomato across the nations had shown a positive trend and hence Indian Tomato is having rich market potential in respect of Pakistan, Qatar and Nepal. What is needed here is the Department of Horticulture and Plantation Crops of respective Governments should promote Tomato Cultivation only for export under Protective Cultivation Technique to impart higher share of revenue to the farmers.

14.3.2.1 *Quality Standards for Export of Tomato*

The Tomato should be Whole, Fresh in appearance, Sound, produce should not be affected by rotting and bruising marks, free from crack, clean, practically it should be free of any visible foreign matters, free of abnormal external moisture, excluding condensation following removal from cold storage, practically fre of pests affecting the general appearance, free of damages caused by low and or high temperature affecting the general appearences of the produce and the acceptance by the consumers, free of any foreign smell. For that purpose, the tomatoes shall be carefully harvested at appropriate maturity, suitable to the variety and geographical location. The tomatoes in this standard are classified into three different grades which are delineated below.

- **Extra Class**
- **Class – I**
- **Class – II**

Extra Class

The Tomatoes in this class shall be of Superior Quality. They shall be the characteristic of the variety as regards shape, appearance and development and it can have firm flesh.

Class – I Tomato

The Tomatoes in this class shall be of good quality with firm flesh of the characteristics of the variety. The fruit shall be uniform in terms of size. They shall be free of visible greenback when they ripe.

Class – II Tomato

This class includes Tomatoes which qualify neither for inclusion in Extra Class nor Class – I category, but satisfy the minimum requirements. Slight defect in shape and development, slight defects in coloring, skin defects or brusing marks, provided the fruit is not seriously affected and shallow scars (Govindhan et.al. 2018).

14.3.3 Export of Onion from India

Onion is an important crop which played significant role in changing the Government too due to the increase in the price level of Onion. It is mainly due to the export policies in vogue with the Government. In this respect, the quantity of onion exported to different countries from India needs an analysis and hence the export data in respect of onion was analyzed and the results are presented in Table 14.4.

Table 14.4 Export of Onion from India (Quantity in Tonnes; Value – Rs in Lakhs)

Sl.No	Name of the Country	2010-2011		2016-2017		Percentage change in Quantity during 2016-2017
		Qty	Value	Qty	Value	
1	Bangladesh	391550.20	57233.40	846869.86	97590.93	116.29
2	Malaysia	273232.70	46343.50	371972.16	49308.58	36.14
3	Sri Lanka	114450.40	17599.20	207480.57	26420.65	81.28
4	United Arab Emirates	126228.80	16627.70	302359.80	39928.57	139.53
5	Indonesia	43061.30	9000.00	81871.82	11430.21	90.13
6	Singapore	20876.30	2920.60	38700.78	5713.72	85.38
7	Nepal	29977.60	2530.40	133530.16	16311.05	345.43
8	Oman	17988.60	2266.20	48934.78	6133.86	172.03
9	Vietnam Social Republic	15281.50	2077.50	42881.01	5065.82	180.61
10	Others	130824.10	17557.00	341138.11	52703.05	160.76
	Total	1163472.60	174155.40	2415739.05	310606.44	107.63

Table 14.4 revealed that the Onion is exported to different countries from India. The importing countries of onion from India are Bangladesh, Malaysia, Sri Lanka, United Arab Emirates, Indonesia, Pakistan, Singapore, Nepal, Oman and the Vietnam

Social Republic and other nations. Among all these countries, the quantity of onion exported to Indonesia, Oman and to the Vietnam Social Republic is found to be on the increasing side and the percentage increase in the export is arrived at 238 per cent, 50 per cent and 61024 per cent respectively during the year 2010-11 over the year 2008-09. Whereas, the picture is entirely different between the period 2010-11 and 2016-17. Almost all the nations were significantly importing Onion from India.

Export of Onion registered a dip which is ranging from 4 per cent to 43 per cent during the period 2008-09 and 2010-11. But the export of Onion had registered a significant increae during the period 2016-17 when comparing to the year 2010-11. The increase was accounted to be 108 per cent. It might be due to the poor output in the Importing countries or the policy guidelines of importing countries. However, the reasons need to be identified and the trade promotion efforts need to be taken to the neighbouring countries to promote onion export and at the same time the Production and Productivity of Onion need to be enhanced by adopting the Precision Farming Technologies by extending the quality inputs for the production of Onion across the Onion producing states and the farmer producers should be motivated through some awards in the national arena. Equally important to recognize the role of Exporters who are playing a vital role in the export process.

14.3.3.1 *Quality Standards for Export of Onion*

The quality standards of onion have been fixed by Agmark although now it is not mandatory to obtain Agmark certificate for onion export. It is necessary to maintain the quality by observing the standards. The grades, big, medium, small and mixed are followed for different types of onions which depend on requirements from importing country. In the survey made by NHRDF, it is observed that Middle East countries demand light red to dark red colour, European countries and Japan demand yellowish /brown colour onions having mild pungency, 3-4 cm sized onions are preferred in Bangladesh, 4-6 cm sized are preferred in Middle East and Far East, while European and Japan prefer 6-7 cm sized onions. Rose onions of 2.5 - 3.5 cm size and multiplier onions of bigger size with attractive red colour bulblets are preferred for export. European countries and Japan prefer yellow big-sized onions. The demand in these countries is from February and up to May when it is very easily possible to cultivate onion in Nasik area of Maharashtra.

The bulbs selected for export should be reasonably uniform in shape, size, colour and pungency. They should be mature, solid, and reasonably firm with tough clinging skin, thoroughly cured and dried outer scales free from dirt and other foreign material. Defective, diseased, damaged bulbs caused by seed stem, tops, roots, moisture, dry sunscald, sun burn, and sprouting, mechanical or other injuries and staining, free from moulds, soft rot and insect attack should not be used for export. The trained labourer are thus required for grading and packing of onions for exports. The packing size for export varies from 8 to 25 kg depending on requirement of onion by importing country. Although jute mesh bags are used for export, if there is no restriction for their use due to environmental pollution problems in some countries, plastic-wooven bags since are reusable and attractive, should be introduced (Apmcnagar.com)

14.3.4 Export of Fruits and Vegetable Seeds from India

India is exporting not only the vegetables but also the vegetable seeds for further development of area under vegetables. In this respect, the quantity of fruits and vegetables seeds exported to different countries is analyzed and the reslts are presented in Table 14.5.

Table 14.5 Export of Fruits and Vegetable Seeds from India (Quantity in Tonnes; Value – Rs in Lakhs)

Sl.No	Name of the Country	2010-2011		2016-17		Percentage change in Quantity during 2016-2017
		Qty	Value	Qty	Value	
1	Pakistan	5813.60	6540.20	1250.18	7675.14	(-) 78.50
2	Bangladesh	2031.80	1453.60	6952.87	10415.87	242.20
3	United States	129.90	1291.60	77.36	7276.43	(-) 40.45
4	Netherland	82.40	1219.40	154.87	5055.62	87.95
5	Nepal	361.20	841.70	37.18	460.36	89.71
6	Malaysia	94.60	736.80	131.16	661.10	38.65
7	Japan	91.80	718.80	221.34	2738.09	141.11
8	Italy	232.80	383.30	96.47	591.83	(-) 58.56
9	Kenya	74.80	355.00	156.94	2073.25	109.81

Sl.No	Name of the Country	2010-2011		2016-17		Percentage change in Quantity during 2016-2017
		Qty	Value	Qty	Value	
10	Singapore	64.90	342.50	105.28	1742.34	62.22
11	Others	2104.80	3636.80	2105.00	13585.10	00.01
	Total	11182.50	17519.60	11288.65	52275.13	00.95

Table 14.5 revealed that the fruits and vegetable seeds were exported to Pakistan, Bangaladesh, United States of America, Netherland, Nepal, Malaysia, Japan, Italy, Kenya and Singapore. The export of fruits and vegetable seeds in terms of quantity has shown an increase between the period 2008-09 and 2010-11. In terms of absolute quantity, Pakistan found to be the potential country in importing the Indian vegetable seeds. But the scene was totally different between the period 2010-11 and 2016-17 and it is found to show a negative trend in the import of vegetable seeds by Pakistan. It might be due to the establishment of their own quality seed production and certification facility or otherwise, it would have shown preference of importing higher amount of fruits and vegetable seeds from some other countries due to the bitter relations with India. The overall percentage quantity of fruits and vegetable seeds exported during 2010-11 and 2016-17 is accounted for only one per cent. It is found to be dismal. It implied that India has to make hard efforts to produce quality seeds of fruits and vegetables for the purpose of export by identifying the best performing varieties and the same has to be produced in large quantities for export.

The percentage of fruits and vegetable seeds exported to Bangladesh is accounted to be 64 per cent during the year 2010-11 over the year 2008-09. But the same was increased to 242 per cent during the year 2016-17 over the base year 2010-11 followed by Japan registered with 141 per cent increase of fruits and vegetable seeds during the year 2016-17 over the base year 2010-11. United States of America, the leader of the world in terms of technologies had also imported the fruits and vegetable seeds from India considerably which is accounted for 43 per cent increase during the year 2010-11 but registered a dip of 40 per cent during the year 2016-17.

Among the countries, Nepal found to import considerable quantity of Fruits and Vegetable Seeds from India during the year 2010-11 which is accounted for 196 per cent increase over the year 2008-09. But the same was found to be reduced during the year 2016-17 and it is accounted for only 90 per cent over the year 2008-09. Italy, United States and Pakistan are the countries had showed a negative trend in importing of fruits and vegetable seeds from India. It might be due to the attainment of self sufficiency in respect of fruits and vegetable seeds. However, the APEDA should examine the reasons for the trade dip with Kenya and other countries and an appropriate strategies have to be taken to bridge the gap.

14.3.5 Export of Dried and Preserved Vegetables from India

Besides export of fresh vegetables to different countries, the dried and preserved vegetables are also exported to different countries from India. The details of those countries who are importing the preserved vegetables are analyzed and the results are presented in Table 14.6.

Table 14.6 Export of Dried and Preserved Vegetables from India (Quantity in Tonnes; Value – Rs in Lakhs)

Sl. No	Name of the Country	2010-2011		2016-2017		Percentage Change in Quantity during 2016-2017
		Qty	Value	Qty	Value	
1	Russia	13,477.00	5,962.20	4,707.61	4,873.53	(-) 65.07
2	Germany	5,900.00	5,588.80	12,376.79	18,281.07	109.78
3	Malaysia	7465.1	4063.4	482.92	604.41	(-) 93.53
4	United States	11,164.40	4,037.40	6,862.52	9,934.29	(-) 38.53
5	France	10693	3,869.70	918	1,665.71	(-) 91.41
6	Spain	10,175.80	2,884.10	3,714.98	4,227.33	(-) 63.49
7	United Kingdom	3,454.50	2,337.10	6,725.83	8,566.46	94.70
8	Belgium	5,088.40	1,987.10	3,640.72	4,070.96	(-) 28.45
9	Poland	2,110.20	1,613.50	4,358.26	4,373.24	106.53
10	Brazil	1,834.10	1,443.50	4,348.36	4,556.96	137.08
11	Others	38811.50	17910.30	38655.29	47017.04	(-) 0.40
	Total	110174.00	51697.10	86,791.28	1,08,171.00	(-) 21.22

Table 14.6 revealed that the dried and preserved vegetables exported to different countries are Russia, Germany, Malaysia, United States of America, and France, Spain, United Kingdom, Belgium, Poland, Brazil and other countries. Among these countries, export of dried and preserved vegetables showed a positive sign and the percentage change in the export during 2010-11 was arrived at 199 per cent in respect of Poland, 3050 per cent in the case of Malaysia, 90 per cent in respect of Germany, 123 per cent in respect of Brazil. All other countries including United States of America and France, the export of dried and preserved vegetables during the year 2010-11 showed a decrease which is accounted for 59 and 56 per cent respectively. But the scene was entirely different between the block year 2010-11 and 2016-17.

The overall export of dried and preserved vegetables showed a negative trend during 2016-17 and the percentage change in the reduction of export is arrived at 21 per cent when compared to the figure of 2010-11. The export of these items generated a positive trend in respect of Brazil, Germany, Poland and United Kingdom and their respective hike in per centage of import of the same is arrived at 137 per cent, 110 per cent, 107 per cent and 95 per cent over the base year 2010-11. The export of dried and preserved vegetables showed a dip among Malaysia, France and Russia. It might be due to the change in the consumption pattern and the change in the taste and preferences of these nations. The Trade Promotion Organization should conduct the consumer survey periodically and the change in the pattern of consumption should be properly communicated to the exporters so as to do the needful. Beyond that the monitoring mechanisms with APEDA need to be activated to assess the performance of the TPOs abroad.

14.4 MAJOR DESTINATIONS FOR FRUITS AND VEGETABLES

During the year 2015-16 India exported fruits and vegetables worth Rs 8391.41 crores. Of which Fruits export alone accounted for Rs 3525 crores and the vegetable export worth Rs. 4867 crores.

The major fruits that enter into the export market are Mangoes, Grapes, Apples and Citrus fruits. Other fruits identified for export are Bananas, Sapota, litchi etc. During 2013-14, the total exports of horticultural produce from India was arrived at 3.69 million tonnes which amounted to Rs 143.60 billion. The Major destinations for Indian Fruits and Vegetables are presented in Table 14.7.

Table 14.7 Major Destinations for Processed Fruits and Vegetables from India

Sl. No	Processed Fruits and Vegetables	Major Destinations
01	Dried and Preserved Vegetables	Sri Lanka, USA, United Arab Emirate, Germany, France, Netherlands
02	Mango Pulp	UAE, Saudi Arabia, Kuwait, Netherlands, Hongkong
03	Pickle and Chutney	UK, USA, UAE, Germany, Canada, Netherlands, Saudi Arabia
04	Other Processed Fruits and Vegetables	USA, Netherlands, UK, UAE, Indonesia, Philippines, Russia

(Source: Sugam Birthare, 2016)

Table 14.7 revealed that the dried and preserved fruits and vegetables are exported to the countries like Sri Lanka, USA, United Arab Emirate, Germany, France, Netherlands. Mango pulp is another commodity exported to mostly the Arab countries like UAE, Saudi Arabia, Kuwait, Netherlands and Hongkong. Besides few of the products are being exported to Malaysia, Bangladesh and other countries like Indonesia, Philippines and Russia. The Indian products have high preference in foreign countries if one could assess their preferences and process the products accordingly means it will fetch higher foreign exchange and hence HACCP related issues need to be given priority. More training and skill development issues should be addressed through the APEDA.

14.5 DETAILS OF ORGANIZATIONS INVOLVED IN PROMOTION OF TRADE

The Details of organizations involved in trade promotion activities and its headquarters are delineated in Table 14.8. There were 12 organizations involved either directly or indirectly in promotion of trade and relevant to the export and import of farm produce. Their role and functions are outlined individually for better understanding and developing rapport with the organizations for entering into the export related business.

Table 14.8 Details of Institutions Facilitating the Trade in India

Sl.No	Name of the Institutions	Year of Establishment	Head Quarters
01	Agricultural and Processed Food Products Export Development Authority	1986	New Delhi
02	Export Inspection Council	1964	New Delhi
03	Indian Institute of Packaging	1966	Mumbai
04	Export Credit Guarantee Corporation	1957	Mumbai
05	India Trade Promotion Organization	1992	New Delhi
06	Export Promotion Councils		
07	Federation of Indian Export Organizations	1965	New Delhi
08	National Center for Trade Information	1995	New Delhi
09	Price Stabilization Fund Trust	2003	New Delhi
10	Indian Council of Arbitration	1965	New Delhi
11	GS1 – India	--	New Delhi

14.5.1 Role of APEDA, New Delhi

The Agricultural and Processed Food Products Export Development Authority (APEDA) was established in 1986 as a Statutory Body under an Act of Parliament. The Authority, with its headquarters at New Delhi, is headed by a Chairman. The Authority has five Regional Offices at Guwahati, Hyderabad, Kolkata, Bangalore & Mumbai and is entrusted with the task of promoting agricultural exports, including the export of processed foods in value added form. APEDA has also been entrusted with monitoring of export of 14 agricultural and processed food product groups listed in the Schedule to the APEDA Act. APEDA has been actively engaged in the development of markets besides up gradation of infrastructure and quality to promote the export of agro products. In its endeavor to promote agro products, APEDA provides financial assistance to the registered exporters under its Schemes for Market Development, Infrastructure Development, Quality Development, Research and Development and Transport Assistance. During the year 2009-10, APEDA has been empowered by an Act of Parliament to enforce the Intellectual Property Right (IPR) of Geographical Indications (GI) for Basmati Rice.

14.5.2 Role of Export Inspection Council, New Delhi

The Export Inspection Council was set up as a Statutory Body on 1st January, 1964 under Section 3 of the Export (Quality Control and Inspection) Act, 1963 to ensure sound development of export trade of India through quality control and inspection and for matters connected therewith. The Council is an advisory body to the Central Government, with its office located at New Delhi and is headed by a Chairman. The Executive Head of the EIC is the Director of Inspection & Quality Control who is responsible for the enforcement of quality control and compulsory pre-shipment inspection of various commodities meant for export and notified by the Government under the Export (Quality Control and Inspection) Act, 1963. The Council is assisted in its functions by the Export Inspection Agencies (EIAs), which are field organizations located at Chennai, Delhi, Kochi, Kolkata and Mumbai and have state-of-art and accredited laboratories with the required logistic support for quality certification activities. These Agencies have a network of thirty six sub-offices and laboratories located at different ports or major industrial centres to back up the pre-shipment inspection and certification activities.

14.5.3 Role of Indian Institute of Packaging, Mumbai

The Indian Institute of Packaging was registered in May, 1966 under the Societies Registration Act, 1860. The Institute, with its office located at Mumbai and branch offices at Delhi, Chennai, Kolkata and Hyderabad, is headed by a Director. The main function of the Institute is to undertake research on raw materials for the packaging industry, organize training programmes on packaging technology, consultancy services on packaging problems and stimulate consciousness of the need for good packaging.

14.5.4 Role of Export Credit Guarantee Corporation of India Limited

The Corporation was established in 1957 as the Export Risk Insurance Corporation of India Ltd. Keeping in view the wider role played by the Corporation, the name was changed to Export Credit Guarantee Corporation of India Ltd. (ECGC). The ECGC is the premier organization in the country, which offers credit risk insurance cover to exporters, banks, etc. The primary objective of the Corporation is to promote the country's exports by covering the risk of export on credit. It provides (a) a range of insurance covers to Indian exporters against the risk of non-realization of export proceeds due to commercial or political causes and (b) different types of guarantees to banks and other financial institutions to enable them to extend credit facilities to exporters on liberal basis.

14.5.5 Role of Indian Trade Promotion Organization

India Trade Promotion Organization has been formed by merging erstwhile Trade Development Authority (TDA) with Trade Fair Authority of India (TFAI) with effect from 1st January, 1992. India Trade Promotion Organization is the premier trade promotion agency of India and provides a broad spectrum of services to trade and industry so as to promote India's exports. These services include organization of trade fairs and exhibitions in India and abroad, Buyer-Seller Meets, Contact Promotion Programmes apart from information dissemination on products and markets.

14.5.6 Role of Export Promotion Council

Presently, there are fourteen Export Promotion Councils under the administrative control of the Department of Commerce. Names and addresses of these Councils are given in Annexure 1.2. These Councils are registered as non-profit organizations under the Companies Act/ Societies Registration Act. The Councils perform both advisory and executive functions. The role and functions of these Councils are guided by the Foreign Trade Policy, 2009-14. These Councils are also the registering authorities for exporters under the Foreign Trade Policy 2009-2014.

14.5.7 Role of Federation of Indian Export Organization

The Federation of Indian Export Organizations set up in 1965, is an Apex body registered under the Societies Registrations Act XXI of 1860, of various export promotion organizations and institutions with its major regional offices at Delhi, Mumbai, Chennai and Kolkata. The main objective of FIEO is to render an integrated package of services to various organizations connected with export promotion. It provides the content, direction and thrust to India's global export effort. It also functions as a primary servicing agency to provide integrated assistance to its members comprising professional exporting firms holding recognition status granted by the Government, consultancy firms and service providers. The Federation organizes seminars and arranges participation in various exhibitions in India and abroad. It also brings out 'FIEO News', for creating awareness amongst its member exporters and importers.

14.5.8 Role of Indian Council of Arbitration, New Delhi

The Indian Council of Arbitration, India's premier Arbitral Institution, is a Society registered under the Societies Registration Act, 1860 operating on no profit basis, with its head office in New Delhi and eight branches with a pan India network.

The organization originally established in 1965 promotes and administers the use of Alternative Dispute Resolution mechanisms in commercial disputes, thereby expediting dispute resolution and encouraging greater domestic and international commerce. The main objectives of the Council are to promote the knowledge and use of arbitration and provide arbitration facilities for amicable and quick settlement of commercial disputes with a view to maintaining the smooth flow of trade, particularly export trade on a sustained and enduring basis.

14.5.9 Role of National Centre for Trade Information

National Centre for Trade Information was set up in 1995 with a view to create an institutional mechanism for collection and dissemination of trade data and improving information services to the business community, especially small and medium enterprises. NCTI is a Government of India recognized Trade Point in India under the Trade Efficiency Program of United Nations Conference on Trade & Development (UNCTAD). NCTI is the Operational Trade Point in India and is also the recognized Focal Point of Trade Analysis and Information System (TRAINS) of UNCTAD Trade Point Development Centre (UNTPDC). NCTI is promoted by India Trade Promotion Organization (ITPO) and National Informatics Centre (NIC).

14.5.10 Role of Price Stabilization Fund Trust

The Price Stabilization Fund (PSF) Scheme was launched by Government of India in April 2003 against the backdrop of decline in international and domestic prices of tea, coffee, rubber, and tobacco causing distress to primary growers. The growers of these commodities were particularly affected due to substantial reduction in unit value realization for these crops, at times falling below their cost of production. The objective of the Scheme is to safeguard the interests of the growers of these commodities and provide financial relief when prices fall below a specified level without resorting to the practice of procurement operations by the Government agencies. The Scheme is being operationalize through the Price Stabilization Fund Trust. As on 31 October 2009, the PSF Corpus Fund consists of Rs.435.19 crore, out of which Rs.432.88 crore has been contributed by Government of India and Rs.2.31 crore by growers by way of entry fee.

A Personal Accident Insurance Scheme is also under implementation by PSFT through National Insurance Corporation Ltd., which provides insurance cover to the growers in the sectors of tea, coffee, rubber, tobacco and spices (chillies, cardamom, ginger, turmeric and pepper) having plantations up to 4 hectares. The Scheme also covers all plantation workers working on these plantations regardless of the size of holdings. The insurance cover is up to Rs. 1.00 lakh per person. The premium of Rs.17/- is shared between the beneficiary and the PSF Trust in the ratio 50:50. The target coverage is 57.17 lakh growers and workers.

14.5.11 Role of GS1-India

GS1 India is a not-for-profit standards body promoted by the Ministry of Commerce (GOI) and Indian Industry to spread awareness and provide guidance on adoption of global standards in Supply Chain Management by Indian Industry for the benefit of consumers, Industry, Govt. etc.

GS1 India is the only organization in India authorized to issue company prefix numbers for use in barcodes, RFID tags etc. for unique, unambiguous and universal identification of products, cartons, containers etc. GS1 standards find wide application in Supply Chains across sectors for unique-yet-universal product, consignment and entity identification, EDI (Electronic Data Interchange), product data synchronization etc. GS1 standards are the dc-facto global standards in identification of consumer products in Retail. GS1 India is an affiliate of GS1 Global Office, twin headquartered at Brussels (Belgium) and Lawrenceville, New Jersey (U.S.A.), which oversees operations of a network of over 100 GS1 organizations across the world.

14.6 SUMMARY

In this chapter, the fruits and vegetables which are exported to different countries are presented both interms of quantity and value realization. The potential destinations for the produce were also discussed. It is learnt that the arab countries are concentrating more on Indian processed fruits and vegetables in a big way and hence the APEDA can give much skill development issues to the exporters and the HACCP requirements be fulfilled based on the norms prescribed by the developed world to tap the market further.

References

1. Government of India (2016) "Agricultural Statistics at a Glance" (Ministry of Agriculture, Cooperation and Farmers' Welfare: New Delhi)
2. Govindhan, K., M.Rajasekar, R.Kalaiselvi and S.V. Kottiswaran (2018). "Quality Standards for Export of Horticultural Crops Grown under Protected Cultivation" In: S.V. Kottiswaran, Balaji Kannan, M. Rajasekar and M.Shanthini; A Guide on Protected Cultivation (Tamil Nadu Agricultural University: Coimbatore) pp. 314-324
3. Mohan Paramkusam (2016) "Indian Red Chilli Standards for Export" (http://www.biotecharticles.com/) pp. 01-06.

Web References

- https://www.actahort.org/books/921/921_18.htm
- (http://www.biotecharticles.com/)
- http://apmcnagar.com/quality-onion-exporter-onion-export-export-onion.php

CHAPTER XV

Investments in Horticulture

Dr. C. Sekhar and Dr. A. Vidhyavathi

Objectives

- To delineate the budgetary outlay for agriculture and horticulture under five year plans
- To highlight the share of funding for horticulture under twelfth five year plan and the annual plans
- To highlight the investment progress made by National Horticulture Mission for the horticultural development programs and
- To delineate the policy options for horticulture

Organization of the Chapter

- Introduction
- Plan outlay and Expenditure under Five year Plans
- Trend in Plan Expenditure under Horticulture Sector
- Approved Components of National Horticulture Mission
- Details of Expenditure under NHM
- Policy Options
- Summary

15.1 INTRODUCTION

At the time of Independence in 1947, Indian agriculture was stagnant in every respect. The immediate task of the Indian government in the post-Independence period was, therefore, to initiate growth process in agriculture on modern lines. Th e process of planned development in the fi eld of agriculture began with the launching of successive Five Year Plans. Notable development has occurred over the years, more specifi cally in the past one year. The Government has identifi ed horticultural crops as a means of diversification for making agriculture more profi table by efficient soil management through 'Soil Health Card' Scheme, efficient land and water use through "More Crop per Drop of Water" and creating skilled employment for rural masses. As we move away from an economy of scarcity to an economy of plenty, there is a call for knowing the present status as well as future scope of horticulture sector. In this context, the investment made through five year plans for the development of Horticulture sector becomes important and hence the details of outlay and the investment made is outlined in the following sections.

15.2 PLAN OUTLAY AND EXPENDITURE UNDER FIVE YEAR PLANS

Though India has emerged as a major producer of horticultural crops and the share of horticulture in the economy has been increasing, there is still a lot of scope in harnessing the potential of this sector. The Department of Agriculture, Cooperation and Farmers Welfare (DAC&FW) is the nodal department for overviewing horticultural development in the country. It implements the programmes through the horticulture departments in all the states and provides the leadership to coordinate actives for the promotion of horticulure. Since efficient implementation of programmes and polices call for detailed information in time, one of the key areas to be looked into is developing a robust system of collection of horticultural data related to budgetary outlay and actual expenditure to know the extent of investment in Agriculture and Horticulture sector. This is analyzed and the details are presented in Table 15.1.

Table 15.1 Plan Outlay and Expenditure under Five Year Plans (Rupees in Crores)

Sl. No	Plan Details	Plan Outlay for Agriculture and Allied Activities	Plan Outlay for Horticulture
01	IX Plan – 1997 - 2002	37546	1453 (03.87)
02	X Plan – 2002 - 2007	58933	5025 (08.53)
03	XI Plan – 2007 - 2012	136381	15800 (11.60)
04	XII Plan – 2012 - 2017	363273	16840 (04.64)
Percentage Change in the Outlay in a decade		166.37	06.58

Table 15.1 revealed that the outlay for horticulture sector during the Ninth five year plan and the 12th Five year Plan is dismal and the tenth and eleventh five year plans have considered horticulture appreciably and its share in the agricultural sector outlay is arrived at 8.53 per cent and 11.60 per cent respectively. In twelfth five year plan, the agricultural sector outlay is is an increase which is accounted to be 166 per cent over the eleventh five year plan. This is not much appreciable for the development of Agriculture. However, the allocation of horticulture under the agriculture is only 6.58 per cent change over the eleventh five year plan. But the Government raised the slogan that "Doubling the Farm Production and Trebling the Farmers' Income" through appropriate investments in Agriculture and Allied Sectors. But this is not visibly true based on the statistics on budgetary outlay and expenditure based on the different schemes.

Though there were appreciable schemes are available under the head National Horticultural Mission, the area expansion program has to be strengthened and the rejuvenation of plantations has to be strengthened by means of budgetary allocations and monitoring through effective supervision by mandataing the achievable targets in the ensuing years. So that the horticulture sectoral growth could be visualized appreciably. In this circumstance, the trend in the financial expenditure under different schemes are analyzed and delineated in Table 14.2.

15.3 TREND IN THE PLAN EXPENDITURE UNDER HORTICULTURE SECTOR

The trend in plan expenditure under horticulture sector is presented in respect of different schemes (Figure – 14.1). Though there were yearwise data are available under the head of plan expenditure, the author has chosen the expenditure made during the year 2010-11 and 2014-15 and the percentage change in the expenditure is analyzed and the details are presented in Table 14.2.

Table 15.2 Trend in Plan Expenditure under Horticulture Sector

Sl. No	Name of the Scheme	Actual Expenditure During 2010-11	Actual Expenditure During 2014-15	Percentage Increase in the Expenditure
01	National Horticulture Board	15183.00	20644.00	35.97
02	Coconut Development Board	5339.52	12300.00	130.34
03	Horticulture Mission for North East and Himalayan States	39998.00	40202.40	00.51
04	National Mission on Bamboo Technology and Trade Development	8966.29	8008.00	10.69
05	National Horticulture Mission	97086.00	114281.00	17.71
06	National Mission on Micro Irrigation	99725.00	96356.00	(-) 03.38
07	Central Institute of Horticulture	210.00	408.00	94.29
08	Grant for Replanting and Rejuvenation of Coconut Gardens	3400.00	00.00	(-) 100
09	Grant to Coconut Development Board for Palm Insurance Scheme	085	00.00	(-) 100
	Total Horticulture	269992.81	292199.40	08.22

The Table 15.2 revealed that in the horticulture sector, the following schemes are being implemented over the years. They are presented as follows.

- National Horticulture Board (NHB)
- Coconut Development Board (CDB)
- Horticulture Mission for North East and Himalayan States
- National Mission on Bamboo Technology and Trade Development
- National Horticulture Mission (NHM)
- National Mission on Micro Irrigation (NMMI)
- Central Institute of Horticulture
- Grant for Replanting and Rejuvenation of Coconut Gardens
- Grant to Coconut Development Board for Palm Insurance Scheme

The overall expenditure between the block years in respect horticulture sector showed only an increase of 8.22 per cent. Between the above two block years (2010-11 and 2014-15), the Coconut Development Board has attracted an increased allocation of fund and the expenditure is increased to the tune of 130 per cent during the year 2014-15 over the year 2010-11. This was found to be remarkable in respect of coconut development. Besides, the Government of India has also allocated funds during 2010-11 for the schemes like replanting and rejuvenation of coconut gardens and coconut palm insurance scheme. Whereas the same schemes were dropped during the year 2014-15 for no reasons. Coconut palm insurance should be the continuous process and hence the insurance should be continued and hence a relook in the revised estimate may be given for extending the scheme.

Nationa horticultural Board has received the funding and the expenditure made was accounted to be 36 per cent higher than the expenditure made during the year 2010-11 followed by National Horticultural Mission has attracted an increase of expenditure during the year 2014-15 which is accounted for 17.71 per cent over the year 2010-11. The National Mission on Bamboo Technology and Trade Development has also had an increased allocation and expenditure during the year 2014-15 over the year 2010-11. From that one could understand that the increase in the financial allocation is found to be dismal when comparing the allocations for the other schemes. Horticulture is the only sector capable of meeting the nutrition and food security greatly to the human being and hence higher investment towards meeting the requirements of development of Fruits and Vegetables are warranted.

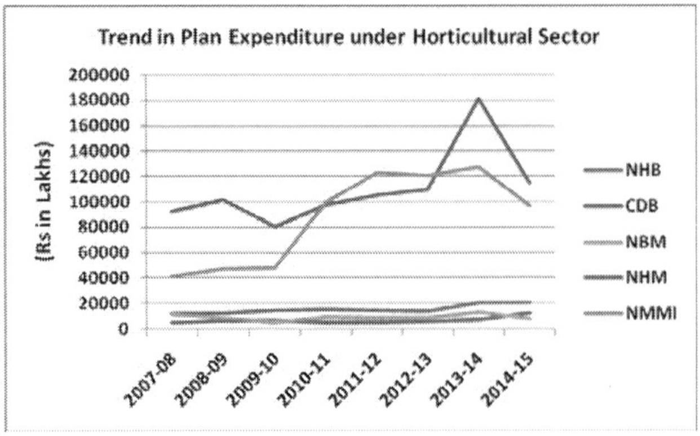

Figure 15.1 Trend in Plan Expenditure under Horticulture Sector

In 2005, The Government of India had announced the National Horticulture Mission, providing for grants and subsidies to farmers for setting up everything from nurseries to plantation developmental activities and even processing units for horticulture produce.

In the past eight years of implementation of National Horticulture Mission, an amount of Rs. 8000 crores (Varma 2013) has been spent for various developmental activities in horticulture. More importantly, the flower production was given a thrust and hence hybrid varieties and hi-tech field operations got a new lease of life in the flower gardens. Before discussing the details of components approved under National Horticulture Mission and their financial and physical progress, let us discuss about the plan wise share of Horticulture in Agricultural outlay. The details are analyzed and the results are presented in Table 15.3.

Table 15.3 Planwise Share of Horticulture in Agriculture Sector (Rs. In Crores)

Year	Agriculture and Allied Activities	Horticulture Allocation	Horticulture as per Demand Book (Grant No.1)
Annual Plan (2012-13)	17692	1954.25	2212.00 (12.50)
Annual Plan (2013-14)	18781	2997.88	2556.00 (13.61)
Annual Plan (2014-15)	11531	3880.83	2263.00 (19.63)

Year	Agriculture and Allied Activities	Horticulture Allocation	Horticulture as per Demand Book (Grant No.1)
Annual Plan (2015-16)	11657	3864.66	2000.00 (17.15)
Annual Plan (2016-17)	19394	4142.20	1620.00 (08.35)
Annual Plan (2017-18)	NA	NA	2329.13
Total Allocation through Annual Plans	79055	16839.82	
XII Plan (2012-2017)	363273	16839.82	10651.00 (02.93)
Percentage Change in the Annual Plan during 2015 over 2013	(-) 78.94	(-) 04.57	

(NA: Not Available; Figures in Parentheses indicate Percentage to Annual Plan Outlay of Agriculture and Allied Activities)

Table 15.3 revealed that the 12th Five year plan allocation for the agriculture and allied activities was arrived at 363273 crores. In that the Agricultural and allied activities has received a considerable allocation only during the two annual plans particularly during 2016-17 and 2013-14 respectively. Whereas, for all the five years, agriculture and allied activities have not received not even one-fourth of the outlay earmarked for agricultural development. This is purely a visionless approach. Similarly the allocation for the horticulture in agriculture is also found to be reduced over a period of time eventhough it is lucrative in terms of production and export. Hence, a vision based agricultural development allocation need to be made in the 13th five year plan at least to bring an all round development of agriculture sector. Many sectors are dependent on agricultural development particularly the agribusiness sectors, textile sectors and coir related sectors. The development of one will be a linkage of another sector that will automatically elevated in the track of development.

15.4 DETAILS OF APPROVED COMPONENTS OF NHM

The Government of India through the National Horticulture Mission has approved some components for the Horticulture development. The components are numbering around eleven items under NHM which are analyzed and the results are delineated in Table 15.4.

Table 15.4 Details of Progress under Different Approved Components of National Horticulture Mission during the Year 2012-2013

Sl.No	Components	Unit	Target		Achievement		% of Expen diture
			Physical	Finan cial	Phy sical	Fin ancial	
01	Nursery	No	05635	7521	0159	1474	01.37
02	Area Expansion	Ha	109493	29198	124362	23804	22.04
03	Rejuvenation	Ha	61734	7469	50667	4802	04.45
04	Protected Cultivation	Ha	19298	27819	20428	20307	18.81
05	Water Resources	No	03916	10959	9224	13153	12.18
06	Organic Farming		70443	5003	13884	0419	00.39
06	Integrated Pest Management (IPM)	Ha	71858	2182	62922	0510	00.47
07	IPM Infrastructure	No	00056	1314	0019	0340	00.32
08	Post Harvest Management (PHM)	No	29568	30060	26754	16524	15.30
09	Markets	No	00664	3875	0147	0791	00.73
10	Bee Keeping	No	123936	1286	79791	0612	00.57
11	Human Resource Development for 2013-14*	No	236629	0038	29914	10.00	02.51
12	Others	--	--	--	--	25245	23.37
	Total	--	--	--	--	107991	100.00

(Source: NHM Update 2012-13; * Data related to the year 2013-14 up to September 2013)

Among the 11 components under NHM, the target fixed for the year 2013-14 is found to be on the higher side in respect of Human Resource Development which is accounted for 2.51 per cent to the total expenditure for the year 2013-14. In the year

2012-13, the human resource component found to be absent and hence the current year component is discussed in this section. In respect of 2012-13, among the 11 components, the number of colonies under Bee Keeping is found to dominate followed by the component, area expansion which is respectively accounted for 0.57 per cent and 22 per cent to the total expenditure. Protected cultivation and the post-harvest management are the next important components for the year 2012-13 under NHM which is respectively accounted for 18.81 per cent and 15.30 per cent to the total expenditure. In respect of achievement, the area expansion program and protected cultivation components were exceeded over the target because of the demand for these two components. Though organic farming has been introduced as one of the component under NHM for the year 2012-13, and the target has been fixed to the tune of 70443 ha, the same could be achieved only to the level of 19.71 per cent to the target area. The reason might be due to the poor takers for organic farming and the resource support other than the investment. Hence the same could be allotted for some other important components like area expansion and protected cultivation which are the important components preferred by many farmers. The take-off in respect of human resource development component is also very poor. It might be due to the delay in decision making on the lines of identifying the training needs and the institutions in which capacity developmental initiatives are to be taken up. The same should be speeded up to tap the fullest benefits of human resource development in horticulture to benefit the farming community.

On examining the components delineated in Table 15.4, both the physical and financial target fixed under the head post-harvest management is found to be very low as the country's need is on the higher side. Recent quote in the dailies revealed that due to perishable nature of the Vegetables, fruits and vegetables produced in India, India is losing 13300 crores of rupees per annum. In India, the wastage is arrived to the tune of 18 per cent to the total out turn of perishables. The main reasons for the wastage is due to the insufficient cold storage facilities and the paucity of refrigerated / insulated air cooled logistic containers for transportation of farm produce from the production site to the consuming ends. Presently, India is blessed with only 6300 cold storage units capable of accommodating only 3.01 crore tonnes worth of food grains. But the need for the cold storage units is 6.01 crore tonnes. For that purpose, an amount of Rs 55000 crore is additionally needed (Daily Thanthi, 2013) and hence the National Horticulture Mission should look into this and can revise the plan document and the need based actions may be incorporated for implementation in the forthcoming financial years.

15.5 POST-HARVEST MANAGEMENT PROGRAM UNDER NHM

Based on the need of Post-Harvest Management Infrastructure, an analysis was made exclusively to highlight what are the infrastructure the National Horticulture Mission is mainly concentrating and what is not included is analyzed and the details are presented in Table 15.5.

Table 15.5 Physical Progress Made under Post Harvest Management Component (2012-13)

Sl.No	Name of the Sub Component under PHM	Number of Units in Tamil Nadu	Number of Units in India	Percentage to Total
01	Pack House	14 (00.54)	2577	09.65
02	Cold Storage Units	01(00.75)	133	00.50
03	Refrigerated Vans	00	015	00.06
04	Mobile Processing Units	00	382	01.43
05	Ripening Chambers	00	034	00.13
06	Low Cost Storage Structures	00	2489	09.32
07	CA / MA Storage	00	004	00.02
08	Evaporative Low Energy Cool Chambers	00	463	01.73
09	Zero Energy Cool Chambers	00	20480	76.67
10	Preservation Units	00	132	00.48
11	Precooling Units attached with Cold Storage	00	002	00.01
	Total	15	26711	100.00

(Figures in Parentheses indicate Percentage to Respective Total)

Table 15.5 revealed that under the head of Post-harvest management component, the zero energy cool chambers was given top most priority and the same attracted around 77 per cent of total infrastructure under the PHM. It is followed by the low cost storage structures and the pack houses respectively accounted for 10 and nine per cent to the total units under post-harvest management erected for the year 2012-13. Among the PHM infrastructure, Tamil Nadu could command only 14 pack houses and only one cold storage unit for the year 2012-13. It might be mainly due to the concept of developed state in the eyes of Government of India. However, the respective Commissioner and Secretary to Government of Tamil Nadu should take all the need based efforts to improve and enhance the quality of PHM infrastructure in Tamil Nadu.

15.6 BUDGET OUTLAY AND EXPENDITURE MADE

It is most important to analyze how much budget outlay fixed for the year 202-13 and how much is released to execute the tasks on time and how effectively the officials of Department of Horticulture and Plantation Crops has utilized the budget outlay and release so as to realize the fruits of Vision components under National Horticulture Mission. These details are analyzed and the results are presented in Table 15.6.

Table 15.6 Details of State wise Budget Outlay, Release and Expenditure Made under National Horticulture Mission during 2016 -2017 (Rs in Crores)

Sl. No	Name of the State	Allocation (GOI)	Amount Released	Expenditure Made	% of Expenditure
01	Andhra Pradesh	64.0	64.00	95.60	11.60
02	Bihar	26.30	18.00	19.50	2.37
03	Chhattisgarh	69.20	61.50	60.00	7.28
04	Goa	2.50	1.60	2.30	0.28
05	Gujarat	71.40	65.50	70.00	8.50
06	Haryana	61.80	59.60	61.50	7.46
07	Jharkhand	40.20	30.10	21.40	2.60
08	Karnataka	73.00	69.50	67.20	8.16
09	Kerala	46.40	15.00	25.40	3.08
10	Madhya Pradesh	45.40	35.60	36.90	4.48
11	Maharashtra	89.00	50.40	49.80	6.04
12	Odisha	49.90	43.70	51.70	6.28
13	Punjab	39.90	30.00	32.00	3.88

Sl. No	Name of the State	Allocation (GOI)	Amount Released	Expenditure Made	% of Expenditure
14	Rajasthan	50.90	43.10	37.20	4.52
15	Tamil Nadu	55.00	47.10	63.30	7.68
	Telangana	32.90	21.00	21.50	2.61
16	Uttar Pradesh	37.30	32.00	30.60	3.71
17	West Bengal	24.90	8.00	29.70	3.60
18	Union Territories	6.80	1.00	1.40	0.17
	Total States	**886.80**	**696.70**	**777.00**	94.31
	Others	61.50	52.80	46.90	5.69
	Grand Total	948.30	749.50	823.90	100.00

(Source: National Horticulture Mission , 2017)

Table 15.6 revealed that the budget outlay for the year 2012-13 under NHM was found to be 1709 crores of rupees. Whereas, the same was only Rs 948 crores during the year 2016-17. In that the amount released by the Government of India for the year to different states is accrued only to the level of 1089 crores which is accounted for only 63.72 per cent to the total outlay fixed for the year 2012-13. But for the year 2016-17, the released outlay was found to be only 750 crores which is very less when compared to the year 2012-13. However, the expenditure made for the year 2012-13 by different states of India found to be 67.43 per cent to the total outlay. It is found to be marginally high. The success of any government program lies only in the hands of concerned secretariat where they have to release the grant for the respective year well in advance so as to plan the programs by the implementing agency at state level. Around 33 per cent of the budgeted outlay was not at all received by the state indicates poor planning and implementation with the coordinating agency, the National Horticulture Mission. The responsibility of equitable distribution based on the resource need and the development rests with the hands of Commissioner, National Horticulture Mission and hence utmost and timely release of fund should be planned and released for implementation at state level.

In respect of Tamil Nadu State, only 70 per cent of the budget outlay was released which is accounted for only 4.48 per cent to the total budget outlay for the year 2012-13. But during the year 2016-17, 85 per cent of the outlay was released and the Department of Horticulture and plantation crops of Tamil Nadu is able to utilize the fund in excess which is accounted for 115 per cent to the total outlay for the year anticipating that it will be released in the next year budget or in the revised budget for the year 2016-17. However, the allocation is very poor for the resource rich state of India. To tap the resource richness of the state, the Government of India through National Horticulture Mission should allocate more funds to increase the area, production and productivity of horticultural crops so as to feed the people with quality resources.

15.7 POLICY EXPECTATIONS IN 2019

According to Economic Times (2017), the Gujarat poll verdict indicating farm distress had put the ruling BJP on a weak wicket in rural areas, the government's next budget is likely to be more farmers' centric focussing on getting them remunerative prices through market reforms and supporting them through robust credit facilities and by increasing the crop insurance footprint. Since agrarian crisis will be one of the key issues in the run up to the assembly polls in Karnataka, Madhya Pradesh, Chhattisgarh and Rajasthan, next year and the general election in 2019, the upcoming budget will provide the government an opportunity to announce its intent.

"India had a good crop year in 2016-17, but farmers had to face problem in getting remunerative prices of their produce. This is something which we need to address and the budget is likely to focus on this aspect", said a senior official who has been part of a team to extend agriculture ministry's input for budget to the finance ministry.

The Centre has already taken several measures to address the issue but will now attempt to strengthen it by taking actions like devising more systematic procurement method and expanding the network of e-NAM (e-National Agriculture Market) from 585 in 2018 to over 1,000 by 2020. The idea will be to bring more and more regulated market on e-NAM platform so that the farmers can get better price of their produce without being exploited by commission agents and cartel of traders.

"The NITI Aayog and the Prime Minister's Economic Advisory Council too are working on it. Though agriculture is the subject where the states have to be on board for success of many schemes, the government's focus in the budget is likely to be on allocating more fund for Centrally-Sponsored Schemes", as per the statement given by the officials. The funds for expanding e-NAM's footprints, cold chain, irrigation infrastructure, micro-irrigation and crop insurance scheme are expected to be increased.

15.8 SUMMARY

Regarding investment in horticulture, the national horticulture mission is responsible for implementation of different components. Though State Government is sponsoring for pay and allowances, the centrally sponsored scheme and its implementation is entirely depends on the investment. Over the years, the NHM share for horticulture development is dwindling for different states. Tamil Nadu is one of the state who has received very poor allocation. With that, the expenditure made for the year 2016-17 is found to be to the tune of 115 per cent in anticipation of revised sanctions in the revised budget. Potential exists in horticulture for increasing the production and productivity of fruits and vegetables under precision farming to feed the millions.

References

1. Economic Times (2017) "Eye on 2019: Government's budget to be more farmers' centric with additional allocation on board", 25th December, 2017.
2. Government of India (2015). Horticultural Statistics at a Glance (Ministry of Agriculture and Farmers' Welfare: New Delhi) pp. 102-111.
3. Government of India (2014). Various Issues of National Horticultural Mission Update (Ministry of Agriculture and Farmers' Welfare: New Delhi)
4. //economictimes.indiatimes.com/articleshow/62237975.cms?utm_source=contentofinterest&utm_medium=text&utm_campaign=cppst

CHAPTER XVI

Strategies

Dr. C. Sekhar and Dr. A.Vidhyavathi

The Project Work on production, marketing and trade aspects of Vegetables with special reference to Brinjal, Tomato and Chilli in Tamil Nadu is an important topic today as most of the farmers have become knowledgeable and started practicing precision farming techniques in their farm holdings with the technical support of Technical Departments like Department of Horticulture and Plantation Crops of Government of Tamil Nadu leading to the success story of the same technique in Dharmapuri District in the yester years made the Government to promote the precision practice in all the districts with appropriate support. The aspects of direct marketing of precision farm produce have not been much explored at national and regional level and hence a special effort has been taken to assess the production potentials, marketing and distribution of select vegetables realized under precision farming techniques at micro level and the export of vegetables at macro level and the details are analyzed in tune with the objectives of this study. The main objectives are to document the experiences of the practitioners of Precision Farming Technologies besides documenting the area, production, productivity and export of select vegetables in India and to document the supporting infrastructure available in the State of Tamil Nadu besides working out the economic aspects of Precision Farming practices were discussed under the various heads.

- Introduction to Precision Farming Technologies
- Area Production and Productivity of Vegetables and Fruits
- Precision Farming Technologies
- Methodology for Assessing the Economics of Precision Farming

- Economics of Precision Farming
- Economic Aspects of Precision Agriculture
- Precision Aquacultural Practices in India
- Precision Silviculture Practices in India
- Precision Approaches to Carbon Neutrality
- Organic Farming and its Certification
- Farmer Producer Organizations in India
- Marketing Practices of Vegetables in India
- Trading Practices of Vegetables in India
- Details of Organizations Involved in Promotion of Trade of Vegetables
- Status of Investment Made in Horticulture to Promote Vegetable Cultivation in India

The data gathered for different periods with respect to the area production and productivity of Vegetables, Export of Vegetables, assembling and trading centers for Vegetables and the importing institutions and states involved in vegetable export were gathered from the secondary sources of Government of India and Government of Tamil Nadu and the data were analyzed using Descriptive Statistics like Mean and the current change in the data in respect of area, production and productivity of Vegetables over the base period was also documented to assess the technological and extension impact due to the Government Intervention through its planned programs in horticulture. The results of this analysis will be better utilized by the policy makers and the planners in horticulture and agriculture for further development of these sectors.

This chapater on introduction to precision farming has addressed the method of classification of vegetables, cropping system prevalent in respect of vegetables, precision farming and the components involved in precision farming of vegetables, History of precision farming, Tamil Nadu Precision Farming Project and its implementation with key components are discussed. Besides the need of documentation of precision farming data and its importance is also highlighted.

In the chapter on production and productivity, the area production productivity of Fruits and vegetable crops have been discussed in respect of India and the same details on Statewise. The selected crops for discussion was Brinjal, Onion and Tomato. In all the crop varieties discussed, the regional varieties which are contributing in terms of higher productivity is not reflected. But at regional level, the regional varieties in respect of Tomato and Brinjal played a dominant role as these varieties are inseparable from the part of farm families as it closely related to taste and preferences among the common public.

The Chapter on Precision Technologies, has discussed the prescribed precision technologies for different vegetable crops like Brinjal, Gourds, Tomato, Cauliflower, Cabbage etc. In that, the fertigation schedule found to be the most important one that should be given in a phased manner in a defined time schedule to realize higher production and productivity from the vegetable crops and hence one should follow the technologies scrupulously.

The Chapater on Methodologies of assessing the Economics of Precision Farming has highlighted the essential tools and techniques which are used to appraise the Technique regarding its economic feasibility. On the lines, the important one is Partial Budgeting Technique which is always prescribed and practiced in the places of using new technology and its impact over the conventional technology which is capable of giving th additional return generation from that technology and hence the partial budgeting technique is the most ideal one in this case and another one is pre and post evaluation of adopting the technology by using the concept of Time Value of Money, one can also appraise for impact assessment.

The success of precision farming technology is not only on the prescription and adoption of technology but also based on the exposure visits and training attended by the farmers / practitioners of precision technology. The perception of sample farmers on Precision Farming Technology and its usage level was ranked as Excellent followed by very few farmers alone have reported that the technology was contributing good to the farmers. Hence the Department of Horticulture and Plantation Crops and Tamil Nadu Agricultural University should have joint venture approach in supporting the farmers in terms of provision of value addition technologies and building the entrepreneurial initiatives.

The Chapter on Precision Agriculture focused the success of System of Rice Intensification and its results concluded that power operated Cono-weeder usage by using either battery or diesel may be activated and supplied to the farmers for efficient operation of Cono-weeder so as to increase the efficiency of the laborers. Skill oriented training should be given to the agricultural laborers regarding SRI practices. This will definitely help the farmers to adopt this technique in a larger way to increase the productivity and net income at farm level and also this technique will be a better solution to meet the food demand. Overall, SRI is a precision technology in which the recommended practices were not implemented by the fellow farmers due to hesitation and reluctance which called for periodic training to them in adopting the technology in full spirit.

The Chapter on economics of aquaculture comprising composite fish culture and the brackish water prawn culture was taken up on case study basis and the details were compared and delineated that the brackish water prawn culture is investment oriented and commands higher profit per unit of investment. However the enterprises chosen are capable of contributing to the practitioners on the higher side revealing that the enterprises are most viable and hence one can take up. Little conflicts are visible in running the enterprise. These are manageable by the policy makers in Tamil Nadu.

The Chapter on Precision Silviculture, analyses the locality factors prevailing in a site and through judicious choice of species and site specific silvilculture technologies will help in generating high utilizable biomass within the shortened rotation period without affecting the environment on a sustainable basis. Hence adequate planning keeping in mind all resources have be to taken up for a successful Precision Silviculture operation.

The Chapter on Precision approaches for carbon neutrality stressed the words of Climate scientists. It revealed that we have years, not decades to stabilize CO_2 and other greenhouse gases. So, they developed an efficient strategy for reducing the CO_2 level in the atmosphere. Carbon sequestration is ought to be a promising solution for reducing the carbon dioxide in the atmosphere by capture and storage of carbon in trees as an effective and efficient strategy. Carbon neutral mechanism help to reduce greenhouse gas emissions where ever and when ever possible, using offset units to compensate for the remainder and achieving net zero emissions. Besides, the steps and strategies to achieve carbon neutrality is prescribed for follow up to reduce the CO_2 level in the atmosphere.

The Farmer Producer Organization (FPO) is one type of PO where the members are farmers. Small Farmers' Agribusiness Consortium (SFAC) is providing support for promotion of FPOs. PO is a generic name for an organization of producers of any produce, e.g.agricultural, non-farm products, artisan products, etc. The advantages of the companies for small and Marginal farmers, and the linkages are also discussed in this chapter.

The chapter on organic farming and its certification highlights the concepts of organic farming, the principles associated with organic farming, the need of certification, characteristics of organic farming, the agencies involved in certification of organic foods at national and global level. Besides it had also highlighted the certification process, procedures associated with certification, participatory certification, history of participatory certification and the need of participatory certification, inspection, evaluation and accreditation of the certification related to organic farming. Currently, the success of organic agriculture depends on the certification procedures and the extent of certification initiatives taken up by the national and international organizations.

The chapter has addressed the impact of precision farming. One of the much known impact is socio economic changes among the farm households. The details of socio economic changes occurred among the sample farmers are Improvements were made in the existing house, purchase of new consumer durables particularly the two wheelers, increase in the frequency of intake of non vegetarian dishes, better linkage developed with the Department of Horticulture and Plantation Crops and enhancement in the consultation, Enhancement in the awareness related to participation with the technical departments and TNAU and Increased the Investment in the Farm and in Savings

The perception of sample farmers on Precision Farming Technology and its usage level was ranked as Excellent followed by very few farmers alone have reported that the technology was contributing good to the farmers. If one could examine the individual parameters, adoption of recommended crop variety scored the excellent and very good scale and only very few reported that the recommended variety is only good. Next parameter is timely supply of inputs which is also ranked as excellent according to majority of the sample farmers. Because of adoption of particular variety, the farmers felt that the yield performance was found to be Excellent when compared to the traditional varieities in practice.

The perception on the technologies, training and the exposure visits to the sample farmers were rated as an Excellent one which is followed by the ranking as Very Good which has been rated by only few respondents. Hence the effort taken up by the Department of Hortiuclture and Plantation Crops is found to be an excellent one according to the respondents and hence the support facility towards these activities should further be extended in a refined manner to the farming community.

Because of the implementation of Technology on Precision farming, the main entrepreneurial initiatives made in the sample farmers were production and distribution of Vegetable nurseries and establishment of horticulture nurseries among the sample farmers. If the farmers were trained in other non farm technologies and value addition issues, the entrepreneurial initiatives could further be enhanced among the farmers. Hence the Department of Horticulture and Plantation Crops and Tamil Nadu Agricultural University should have joint venture approach in supporting the farmers in terms of provision of value addition technologies and building the entrepreneurial initiatives

The constraints faced are Cost of Precision Input is comparatively higher according to cent per cent of the sample farmers. Followed by absence of Know how for maintenance of Drip Systems among the sample farmers were accounted for 85 per cent. The quantum of labor use is comparatively higher than the traditional practice was reported by 76 per cent of the sample farmers. Any new technology which is productivity friendly should ignore labor intensiveness. But the precision farming technology found to attract higher labor for different operations which inturn promote employment opportunities in the rural pockets. However, availability of labor is another constraint. The drip accessories need an intensive care system according to 42 per cent of the respondents

The vegetables are reaching the consumers of regional and national through some intermediaries in different marketing channels. In respect of Tomato, three important channels could be found out. Brinjal is being distributed through two important marketing channels. Onion is marketed through three marketing channels. Banana is marketed through five important marketing channels. Among these important channels, only the Channel I and III in respect of Tomato, Channel – I in respect of Brinjal, Channel – III in respect of Onion alone are traceable from producer level to the consumer level and the price prevailed among the different functionaries were analyzed and the results are discussed in the section dealing with price spread.

The Producers' Price for Tomato was arrived at Rs 8 per kg which is accounted for 44 per cent of the Consumers' Price. The Consumers' Price is arrived at Rs 18 per kg. The total marketing cost incurred in the channel was arrived at Rs 5 per kg and the net marketing margin was arrived at Rs 5 per kg which is accounted for 28 per cent to the consumers' price.

The Producers' Price for Onion was arrived at Rs 12 per kg which is accounted for 60 per cent of the Consumers' Price. The Consumers' Price is arrived at Rs 20 per kg. The total marketing cost incurred in the channel was arrived Rs 4 per kg and the marketing net margin was arrived at Rs 4 per kg which is accounted for 20 per cent to the consumers' price.

The Producers' Price for Tomato was arrived at Rs 8 per kg which is accounted for 44 per cent of the Consumers' Price. In respect of Onion, the Producers' price is arrived at Rs 12 per kg which is accounted for 60 per cent of the Consumers' Price. In respect of Brinjal, the Consumers' Price is arrived at Rs 8 per kg which is accounted for 27 per cent of the consumers' price. The total marketing cost incurred in the respective channel in respect of Tomato, Onion and Brinjal was arrived at Rs 5, Rs 4 and Rs 9.50 per kg and the net marketing margin was arrived at Rs 5, Rs 4 and Rs 12.50 per kg which are respectively accounted for 28 per cent, 20 per cent and 42 per cent to the consumers' price. Higher net marketing margin with respect to Brinjal is mainly due to the number of intermediaries involved in facilitating the trade of Brinjal to meet the consumers of different destinations in Kerala.

The Onion Marketing Channel – III found to be the efficient one. Because, the Channel – III fulfilled the requirements of functionaries in the marketing channel and keep them satisfied besides meeting the higher percentage share to the producer from the consumers' price.

The overall export of horticulture produce between these period in terms of quantity is found to show a decrease which is accounted for around 19 per cent to the total quantity of horticultural produce export. The export value between these periods is found to show only a marginal increase which is accounted for only around three per cent to the total value of export. In this circumstance, the export of vegetable seed and other fresh vegetable export should be enhanced by increasing the productivity of vegetables by practicing the precision farming techniques in a larger scale so that the quantity produced will be enhanced and the export possibilities could be increased

Recent quote in the dailies revealed that due to perishable nature of the Vegetables, fruits and vegetables produced in India is losing 13300 crores of rupees per annum. In India, the wastage is arrived to the tune of 18 per cent to the total out turn of perishable. The main reasons for the wastage is due to the insufficient cold storage facilities and the paucity of refrigerated / insulated air cooled logistic containers for transportation of farm produce from the production site to the consuming ends. Presently, India is blessed with only 6300 cold storage units capable of accommodating only 3.01 crore tonnes worth of food grains. But the need for the cold storage units is 6.01 crore tonnes.

The chapter on investment on horticulture has made a highlight on the extent of investment made for the different components for the successful establishment and development of Orchards and Vegetable gardens in India. The investment pattern revealed the details that they are very inadequate in certain areas particularly the post harvest management and area expansion activities. Considerable focus may be given to the agriculture and allied sectors focusing horticulture and animal husbandry and fisheries would permit not only rural development but also the economy as a whole. In respect of Tamil Nadu State, only 70 per cent of the budget outlay was released which is accounted for only 4.48 per cent to the total budget outlay for the year 2012-13. It is very poor for the resource rich state of India. To tap the resource richness of the state, the Government of India through National Horticulture Mission should allocate more funds to increase the area, production and productivity of horticultural crops so as to feed the people with poor resources.

Recommendations and Policy Implications

Precision farming involves not only efficient management of resources but also application of technologies for improving crop performance and environmental quality. This advanced crop production encompasses a variety of interventions such as micro irrigation, fertigation, protected green house cultivation, micro propagation, genetically modified crops etc., This calls for not only creating awareness among the farmer producers but also involves giving training, exposure visits to demonstration sites to them. Precision farming also warrants continued research and development efforts in various fields which needs heavy investment from Government side.

The success of any government program lies only in the hands of concerned secretariat where they have to release the grant for the respective year well in advance so as to plan the programs by the implementing agency at state level. Around 33 per cent of the budgeted outlay was not at all received by the state indicates poor planning and implementation with the coordinating agency, the National Horticulture Mission. The responsibility of equitable distribution based on the resource need and the development rests with the hands of Commissioner, National Horticulture Mission and hence utmost and timely release of fund should be planned and released for implementation at state level.

Government of India and Tamil Nadu Government are allotting and investing money on micro irrigation, various schemes through National Horticultural Board. NHB is implementing many schemes namely Commercial horticultural scheme, Cold storage scheme, Technology Development and Transfer scheme, Market information service scheme for horticultural crops, Horticulture promotion services. Apart from these schemes, Recognition of horticultural nurseries, and accreditation and rating of nurseries of cut flower plants grown under protected cover are the new schemes which concentrate more on development of standard quality nurseries for horticultural crops.

Availability of true to type, quality planting material is crucial for success of commercial horticulture. This necessitates having a net-work of Horticulture Nurseries which conform to Model Nursery Standards in terms of Infrastructure, Quality of Seed and Planting Materials and adoption of Nursery Management Practices. State Governments ensure supply of quality planting materials for fruit crops by enactment of Fruit Plant Nursery (Regulation) Act and enforcement of its provisions through licensing of horticulture nurseries. However, all the States have not enacted Horticulture Nursery Acts. As per the information available in report of the Working Group on Horticulture for the XI Five Year Plan, at present only 8 states have adopted Nursery Act and in 9 states some system of registration / monitoring exists for nurseries. Whereas in 13 states there is no nursery act at present. Further, rules framed under existing acts generally do not provide for Technical Specification for Quality of Planting Material and Process of Production, Infrastructure required for production of quality planting material and Good Nursery Management Practices. The Acts in present form also do not deal with planting materials of flowers, ornamental plants and vegetables. It warranted to find a solution to the problem by putting in place a system of Recognition of Horticulture Nurseries. National Horticulture Mission has taken initiative to specify the Nursery Standards by specifying Infrastructure required for setting up of Model Horticulture Nursery.

ICAR has recently come out with technical specification of horticulture planting material and its production procedure through a recently published document titled as "Handbook of Seed and Planting Material Testing Manual for Horticultural Crops". Existing Rules framed under Horticulture Nursery (Regulation) Acts provide for record keeping part for such nurseries. However, each one of them individually does not completely define a Model Horticulture Nursery. Therefore, it is imperative to define a Model Nursery in all the three aspects of Nursery Infrastructure, Production System and Quality Parameters of Planting Material and Good Nursery Management

Practices in a comprehensive manner and put a Nursery Recognition Regime in place. A recognized Model Horticulture Nursery should function as a reliable source of supply of quality planting material for horticulture crops. With a view to ensure availability of Good Quality planting material as outlined above, NHB has started a system of Recognition of Horticulture Nurseries on voluntary basis.

Unlike agricultural crops more than eighty per cent of the farmers who cultivate horticultural crops especially vegetables and fruits, fully depend on private nurseries for seedlings. But there is no quality checking for the seedlings raised by these nurseries. Most of the farmers are not aware about the quality standards or measures to check the quality seedlings. All the State government should come forward to implement Horticulture Nursery Act like New Seed Act. Training programmes should be arranged to producers to create awareness about getting quality seedlings from recognised horticultural nurseries, ways and means to avail subsidy under different horticultural schemes.

A Centrally Sponsored Scheme of MIDH has been launched for the holistic development of horticulture in the country during XII plan. The scheme, which has taken off from 2014-15, integrates the ongoing schemes of National Horticulture Mission, Horticulture Mission for North East and Himalayan States, National Bamboo Mission, National Horticulture Board, Coconut Development Board and Central Institute for Horticulture, Nagaland.

All the centrally sponsored development schemes warrants online application which is developed in English language only. Preparation of project report requires assistance from outside. It is necessary to develop online application in regional languages so that farmers themselves can fill the application without outside help. Farmer producers and agribusiness entrepreneurs find it very difficult to prepare the project report. Trainings and hands on sessions will help them to prepare bankable project report. A separate wing can be formed in agricultural and horticultural departments to provide technical assistance and guidelines for submitting project report to avail credit assistance and subsidy to the farmers turned agripreneurs at some minimal cost so that the Agripreneurship Development can be visualized.

Researchable areas where R & D efforts needed are standardizing post harvest management techniques and protocols, prescribing critical storage conditions for fresh produce, bench marking of technical standards for cold chain infrastructure etc and transfer of technology to producers / farmers and service providers such as gardeners, nurserymen, farm level skilled workers, operators in cold storages, work force carrying out post harvest management including processing of fresh horticulture produce and to the master trainers. Changes in consumption of agricultural and horticultural produce especially processed fruits and vegetables consumption should be promoted. Infrastructural facilities especially transport and packing warrants more attention. Carrying out studies and surveys to identify constraints and develop short and long term strategies for systematic development of precision farming in all agriculture and allied activities and providing technical services including advisory and consultancy services is the need of the hour.

Made in the USA
Monee, IL
04 November 2026

67374674R00268